SILAGE EFFLUENT
Problems and Solutions

Proceedings of a conference and exhibition held at the
Staffordshire College of Agriculture
2 February 1988

Edited by

B A Stark

and

J M Wilkinson

CHALCOMBE PUBLICATIONS

contain yourself!

ADAS SPECIALIST SERVICES WILL HELP YOU CONTAIN YOUR SILAGE EFFLUENT PROBLEMS

- **BUILDING A NEW SILO**
- **REPAIRING EXISTING SILOS**
- **USE OF ABSORBENTS**
- **FEEDING TO LIVESTOCK**
- **SAFE AND EFFECTIVE DISPOSAL**

ADAS specialists can prepare all plans; organise contractors to do the work; survey and specify maintenance work; design waste management systems.

AND FOR OTHER FARM WASTES . . .

ADAS will help you avoid pollution risks with a service tailored to your precise needs.
We can provide a few ideas or full consultancy

THE PROBLEMS COME TO US, NOT TO YOU

Contact your local ADAS office now

A D A S

CONTENTS

PREFACE

The effluent or liquor from silage poses a number of problems. Perhaps the most serious is that its polluting power is often underestimated. Not only is the liquid some twenty times more potent, weight for weight, than cow slurry, but it comes in a rush at a time of the year when water courses may be relatively slow-running. Further, leaky silo floors can result in effluent reaching field drains without the farmer noticing that anything ontoward has occurred.

The water authorities view with concern the increasing amount of silage made by farmers. Yet not only is the amount of silage increasing year by year, but the increased use of fertiliser nitrogen, earlier cutting and the use of later flowering types of ryegrass all contribute to the standing crop being wetter at the time of cutting than the more mature grass which was commonly ensiled in clamps and bunkers ten years ago, or which is harvested today as big bale silage.

Of concern also is the trend towards ensiling grass with little or no field-wilting. Research in Northern Ireland has demonstrated clearly the adverse effects on crop yield of prolonged wilting, and also of separate mowing and harvesting. The implications of this research are that maximum output per hectare is achieved with a system of direct harvesting, in which the crop is cut and picked up in a single operation. Clearly the widespread adoption by farmers of this research could lead to even greater risks of water pollution from silage effluent. A serious pollution incident can set back by decades the development of a vigorous population of fish and other aquatic life in a river.

Against this rather depressing scenario, what solutions are available? There is, as Mr Merriman points out in his contribution, the further use of legislation to increase the awareness of farmers to the risks associated with being careless over silage effluent. There are compromise strategies which can reduce the production of effluent, such as minimal-wilting, big bale silage, and cutting strategies which allow for pauses in the silage-making campaign when the weather is too wet. There are ways of absorbing excess liquid in the silo, and of containing it by adequate sealing of silo floors and walls. Some rethinking of current advice on silo design may be necessary, since the objective at present seems to be to accelerate, rather than deccelerate the loss of effluent from the silo, by providing ample drainage channels within and outside the silo walls.

Ultimately the solution lies, as Mr Gummer stresses, in voluntary action by the individual as a result of putting pollution control high on the agenda for action. Advice and a wide range of products are available to assist farmers in tackling the problem of silage effluent. It is my hope that this book, by bringing together a considerable amount of up-to-date information and advice on the subject, will be a valuable contribution towards reducing the number of pollution incidents caused by silage effluent.

I am most grateful to all the organisations and companies which supported the conference and exhibition. In particular, the help from the staff of the Staffordshire College of Agriculture was invaluable. The information about the products and processes has been kindly supplied by the manufacturers and distributors, and no responsibility can be accepted for errors or inaccuracies.

Mike Wilkinson

February 1988

Chapter 1

SILAGE EFFLUENT: THE REAL PROBLEM

The Rt Hon J Selwyn Gummer MP

Minister of State, Ministry of Agriculture, Fisheries and Food,
Whitehall Place, London SW1A 2EY

Edited transcript of the Minister's speech:

I am gratified at the large number of people who feel that this is a subject worthy of discussion. It is always a problem to talk about something like silage or effluent; people suggest all kinds of merry quips on the subject. I tried very hard to think of what my title should be; almost every title I produced resulted in gales of laughter in the office so I didn't have a title at all except "The Real Problem". Silage effluent *is* a real problem, not least because we have not taken it seriously for so many years.

For decades water was taken as something that came naturally, was very easily produced and very simply used and was something which really needed nobody much to look after it. It is now more and more clear that the provision of drinking water and the use of water for irrigation, leisure, boating and angling is something which needs very careful protection and very careful conservation.

Now I am one of those who believes very strongly that the farming community needs constantly to remind the public that the only reason that they have the luxury of attacking farmers is that they have full stomachs. It is the success of agriculture which gives people the opportunity to be as awkward, difficult and carping as they sometimes are about farmers. It is the success of the industry which has enabled people to be safe in attacking surpluses and complaining about modern methods of intensive farming.

I start from the assumption that farmers have always developed and refined their means of production, ever since the first clearings were made in the forests. Thus the countryside of today is an artificial creation which is the result of agriculture and horticulture's impact upon what was previously a wholly different landscape. Much of what we love in the countryside, indeed some would say almost all of what we love in the countryside, is the product of man's interaction with nature.

So if we start with those two assumptions, first, that the production of food is a necessary and vital part of our life and that we ought to be disposed to be grateful and thankful rather than disdainful because of the success of the industry and, secondly, that the countryside which we seek to conserve and which we love to enjoy is the product of agriculture, then when we come to look at the difficulties of pollution we do so, I think, from the right perspective.

The truth is that with the intensification of agriculture has come an intensification of danger. If you produce more and more from less and less acreage, then it is obviously true that the waste products become more concentrated and the dangers of them polluting become greater. There still are, I'm sorry to say, some people who have taken this problem far to lightly. Even though, as I have said, UK farmers have received more than their fair share of criticism, there are some who deserve a good deal of criticism because they have been unwilling to realize that the countryside depends for its continuity upon the sharing of resources. Water is a vital part of the countryside and a crucial resource for the nation.

There is a growing pressure now for further regulatory controls. More and more people are pressing the Government to ensure that we strengthen the code of good agricultural practice and produce new regulations on effluent containment. Both of those, for example, were recommendations from last year's report of the House of Commons Environment Committee on river and estuary pollution. Now we haven't yet produced our response to that report but even if we were to decide to follow the advice of that committee it is one thing to make the law and quite another thing to make it work.

What we have really got to concentrate upon are two things. The first is how we ensure that the code of good agricultural practice is followed by the majority of farmers. The second is how we bring home the code to that minority of farmers whose inadequate facilities for handling silage effluent bring the whole of the farming community into disrepute.

It is a very important thing for the farming community to create an atmosphere which isolates those who do pollute. There is a very important job for the community itself to do to ensure that we are not let down by that small number.

I think the farmer must recognize three dangers. The first is the danger that public opinion will become exasperated by well publicised incidents and will demand the kind of restrictions which would make normal farming practice extremely difficult. The second is that the effects of pollution will combine with other things to make people less willing to understand the general difficulties of farmers, and the third is the very simple matter that the pleasure, the interests, the livelihoods and the health of other people will be clearly damaged by the incidence of pollution deriving from farming.

My Department, together with the Department of the Environment and the water authorities, are seeking to give publicity to the problems of effluent. Farmers will increasingly need advice on this. In addition we need to know what are the key dangers—what goes wrong. The answer to many of these problems starts at the design stage—using the right materials, getting the right sizes. ADAS can offer a very valuable service here and, considering the penalties and costs a farmer may eventually face if he gets it wrong, I very much hope that more farmers will avail themselves of these services in the future.

Increasingly the problems of effluent are technical problems which need expert technical answers. Even the best equipment needs proper maintenance and management. The problem which very often happens is that the equipment has been well bought but the maintenance and management, the sealing of cracks and the rest, has not been as efficient and effective as it ought to be. And it is not just a question of maintenance, it is also a question of use. Whatever size the effluent collecting tank may be, it has to be emptied before it overflows. If you don't put some effort into the normal operations, then naturally your equipment, however good, will fail to work.

I recognize of course that cost is a major factor and I think here that one has to say something pretty tough because farmers ought to recognize that every other industry has to take necessary action to prevent pollution. Farming and agriculture cannot be excluded from the principle that the polluter pays, that the polluter does have to protect the rest of the community from the consequences of his money-making activity.

The fact is that we do make available special help to farmers. The present Ministry grants of 30% for storage and treatment of wastes and 60% in less favoured areas, are the highest grants that we have ever offered. These grants are at the top end of all our grants, and that is because we believe that pollution control and pollution protection is so important.

The trade exhibits here today illustrate that the right products are available to cope with silage effluent efficiently. So we have the technology, we can offer the advice and there is even help to meet any costs which may arise. You can have an initial ADAS farm visit to look at the effluent problem absolutely free of charge.

I am most grateful that so many should have turned up for this important meeting but in a sense you are a self-selected group. You have chosen to come because you think the issue is important. You have chosen to come because you have had problems, or may have problems and you want to prepare yourselves to deal with them. You have chosen to come because you are in the industry or concerned with water. You are here because you have an interest. Outside there are large numbers who either have no interest or have so little that this is not a problem high on their agenda. I do believe that unless the farming industry can effectively meet the problems of effluent control and of water pollution then the farming industry will be bound, as night follows day, to be restricted by legislative action which will make farming more difficult. It is in our hands.

We either make sure that we solve these problems ourselves, voluntarily within the code and within such restrictions as we have now, with the help of grants and advice or we are bound to find that the rest of the community demands that we be restricted and policed in a way which we would find burdensome and restrictive. That would mean that the small minority which fail at the moment to meet the normal reasonable requirements which society places upon any who operate within its bounds will have vested the rest—the majority who act sensibly and responsibly—with a burden which farmers could well do without. That is why I thought it worthwhile coming here. That's why I am a supporter of today's event, and why I hope that it will be mirrored by others of a similar kind throughout the country. Our great industry *can* ensure that we are, as we must always be, the best conservers of the countryside and that we are the people most concerned to ensure that our water and our natural resources are protected and preserved, not only for our own use, but for the use of our neighbours and for the the urban dwellers for whom the countryside is so essential.

Chapter 2

WATER POLLUTION BY SILAGE EFFLUENT

R P Merriman

Welsh Water Authority,
Hawthorn Rise, Haverfordwest, Dyfed SA61 2BH

SUMMARY

Silage effluent, or liquor, is the most polluting waste produced on farms. It can cause very serious pollution and damage if it enters any water. The number of pollution incidents attributed to silage liquor has risen markedly since 1979. The impact of discharges is often exacerbated because they predominate in the summer months when dilution in receiving waters is generally lower. Silage liquor is extremely corrosive, making it difficult to ensure safe collection and storage in improperly-protected systems. Increasingly, problems are arising because of leaks in silos, in storage facilities and in collection systems. Poor planning in respect of the siting of silos and land drains can result in silage liquor finding its way into land drains, either adjacent to, or even under the base of silos.

The production of silage has increased greatly over the last 15 years, and at an even faster rate since the introduction of milk quotas in 1984, when there was also a move away from wilting prior to harvesting. The risk of pollution arising from silage has therefore risen considerably. Water authorities, faced with increasing pollution problems, have undertaken positive measures to reduce the risk of pollution from farms. Despite these efforts and the common knowledge of the polluting strength of wastes such as silage liquor, pollution from farm wastes remains a very serious problem. Overall the attitude of water authorities appears to be hardening in respect of pollution by silage liquor. In the absence of positive measures to reduce pollution from farm wastes, the water authorities are left with little alternative but to pursue further the use of legislation.

Despite its corrosive nature, silage liquor can be contained relatively easily at new installations, providing they are carefully planned, constructed and maintained. The recently improved consultation arrangements in respect of grant aid are welcomed, giving farmers the opportunity to receive water authority comments on proposals before construction. However, in this difficult economic climate the level of grant aid on pollution prevention measures needs to be increased outside the Less Favoured Areas, to encourage farmers to install correct systems at the outset and to improve existing defective systems.

INTRODUCTION

Silage effluent, or liquor, is the by-product of anaerobic crop preservation to provide winter feed in the form of silage. Virtually all of the silage in England and Wales is made from grass, which is usually cut at an early stage of growth when its nutrient and moisture content is high.

The production of silage liquor depends largely upon the moisture content of the ensiled crop. Hardwick, Lang and Nielsen (1987) state that the volume produced may be very large, from 500 litres per tonne for grass ensiled at 12% dry matter (DM) content to only 50 litres at 25% DM. Above 25% DM, production is practically nil.

The maximum production rate of liquor occurs in the first few days after ensiling, some pollution cases have even arisen when liquor has apparently been produced within hours of ensiling. However, several pollution cases have also been recorded after silos were topped up by subsequent cuts, the increased pressure forcing the liquor from an earlier cut out of the silo.

CHANGES IN SILAGE PRODUCTION

Silage production is less vulnerable to unpredictable weather than hay making, and earlier cutting enables faster regrowth of grass. These advantages, coupled with changes in animal housing have resulted in a large increase in production during the last 10 to 15 years, especially in the main dairying areas such as Wales (Figure 1).

Silage production has increased enormously since the implementation of milk quotas in 1984. Although the increase is partly due to climatic conditions favouring grass growth, it also reflects MAFF advice to farmers to increase silage production, and to feed more silage with fewer concentrates in order to reduce costs. There has also been a move away from wilting advocated by suggestions of increased milk production both per animal and per hectare from direct cut crops compared to those that are wilted (Gordon, 1981).

The potential for pollution arising from silage production has therefore increased considerably.

FIGURE 1   **Silage production in England and Wales 1978 to 1986, relative to production in 1977**

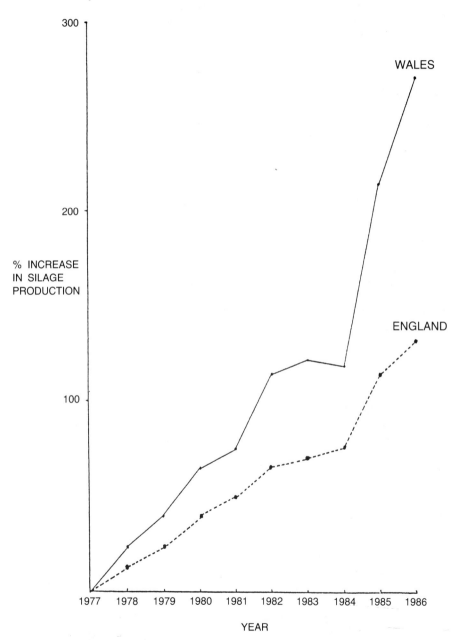

POLLUTION INCIDENTS

There has been a tremendous increase in the number of recorded farm pollution incidents in England and Wales, from 1484 in 1979 to 3510 in 1985, with a very small decrease to 3427 recorded in 1986 (WAA/MAFF, 1987). About 80% of these incidents are attributed to dairy and beef farming. Although the 3427 incidents in 1986 only represent about 17% of all recorded pollution incidents in England and Wales, farm waste discharges are extremely polluting and so the damage caused can be very serious. Farm slurry is up to 80 times more polluting than *untreated* sewage in terms of biochemical oxygen demand. Silage liquor is even stronger and it is the most polluting waste produced on a farm—*up to 200 times stronger than untreated sewage*. The pollution load from a 500 tonne clamp of unwilted silage is equivalent to that of a day's *untreated* sewage from a town of 200,000 people. In England and Wales in 1986 there were about 33 million tonnes of silage produced!

The number of pollution incidents attributed to silage liquor increased markedly between 1979 and 1981 (Table 1). It would appear at first sight that, apart from 1985, the problem has since stabilised despite increased production of silage and a move away from wilting. However, interim figures for 1987 produced by some of the authorities who have experienced the greatest silage liquor problems suggest that the number of incidents recorded in 1987 have considerably exceeded the 1986 figure. Studies presently being undertaken by Welsh Water Authority scientists also suggest that the number of reported farm pollution incidents is an underestimate of the severity of farm pollution.

TABLE 1 **Number of pollution incidents due to silage liquor 1979 to 1986**

Year	Number of pollution incidents due to silage liquor
1979	269
1980	338
1981	577
1982	373
1983	575
1984	573
1985	1006
1986	592

IMPACT

Because silage liquor is so polluting the damage caused can be consider-able. The impact is often exacerbated because the discharges usually occur into watercourses which are relatively unaffected by other forms of pollution and so have a high value with respect to fisheries, potable water supplies, amenity use and farming. In the Welsh Water Authority area in 1986, for example, 7 of the 8 pollution incidents causing closure of potable water supply intakes and over half of the largest fish kills were due to discharges of farm wastes.

Silage liquor causes particularly serious problems. In addition to being the most polluting waste produced on a farm in terms of oxygen demand, it is also acidic and contains high concentrations of nutrients, especially ammonia, which give rise to characteristic growths of "sewage fungus". Ammonia is directly toxic to fish, although smaller fish kills can go unreported as dead fish may be hidden by the very fast growing "sewage fungus". The high ammonia content of silage liquor may also cause serious problems to water authorities when they treat the water.

The impact of silage liquor is compounded further because discharges predominate in the summer months, when dilution provided by receiving watercourses is generally lower.

CAUSES OF POLLUTION

Pollution incidents may arise for several reasons. There may be inadequate provision to store liquor. This is particularly the case on farms where the intention is to wilt and so storage for liquor is not provided. Serious problems then arise if wilting is not undertaken for any reason. The move away from wilting is likely to have serious repercussions due to increases in the volume of silage liquor produced.

Silage liquor is extremely corrosive, making if difficult to ensure safe collection and storage in systems which are not adequately protected. Its corrosive nature has been increased on some farms by the use of home-made acid-based additives.

Increasingly problems are arising because of leaking silos, leaking storage facilities and leaking collection systems, especially those that are below ground and those which may not have been designed to carry a highly corrosive waste.

Presumably this problem has become more prevalent as silos and collection tanks have aged and have been damaged by the aggressive nature of the silage liquor which is now produced in larger quantities. In such situations very serious pollution can arise without the knowledge of the farmer who may, from cursory inspections of the silo face or collection/storage system, be misled into assuming the situation is satisfactory. This is a very important point as during and immediately after silage-making farmers are working long hours and so spend little time checking or maintaining collection systems and ditches. Additionally many silos are only empty for a matter of weeks in the spring. Consequently there is very little time available to attend *properly* to deficiencies in the silo that become apparent with ageing.

A frequent problem is that silage liquor finds its way into land drains either adjacent to or even under the base of silos. This situation usually arises on farms that are close to sources of water for man and livestock. In the wetter western part of the country, where topography is steeper and the land close to farmsteads is often wet, the temptation arises to utilize areas of wet land for silos rather than to "sacrifice" dry land. Cases also arise where silos are constructed literally alongside watercourses and effective containment of the silage liquor proves impossible. This can only be described as poor planning or the wrong choice of site, and it is a situation which should not have arisen if the advice of the local water authority or the Agricultural Development and Advisory Service (ADAS) had been sought.

Problems may also arise due to the use of purpose-made stores to contain silage liquor for subsequent disposal. These stores have advantages in that storage may be provided for the first time and, because the stores are usually sited close to silos, collection systems are short, thus reducing both the chance of leakage and the time required for checking. The collection tanks for silage liquor are, however, generally of limited capacity, often being versions of septic tanks (capacity about 3 cubic metres). They therefore demand frequent emptying at what can be the busiest time of the year for farmers, and sometimes this is ignored. Farmers may be lulled into a false sense of security, forgetting that the tanks may have been installed several years ago and now receive significantly greater quantities of liquor. In addition if the tanks serve an open silage clamp, they are liable to be swamped by rain water.

ACTION BY WATER AUTHORITIES

Pollution arising from farming activities is one of the most serious environmental problems facing several water authorities. Some have undertaken positive initiatives in an attempt to reduce the risk of farm pollution. For example, South West Water has organised a Farm Campaign, and Severn Trent gives prizes to farmers who have carried out work to reduce the likelihood of pollution. Welsh Water appointed an Agricultural Liaison Officer in 1984, whose role is to advise the farming community on measures to prevent pollution.

Welsh Water and other water authorities have worked closely with ADAS and with other farming organisations to make farmers aware of the problems caused by farm waste discharges, together with ways of overcoming these problems. This campaign has included the production of advisory leaflets for widespread circulation, and the insertion of articles and even advertisements in newspapers and the farming press. Television and radio interviews have also been given.

In addition, farming groups have been addressed on the subject of pollution and water authorities have taken part in farm open days or agricultural shows where the emphasis has been on the prevention of pollution.

A Water Authorities Association Farm Waste Working Group, which includes representatives of water authorities, has been formed. The group is presently looking at a number of measures that may combat farm pollution, including the production of a video on the subject.

Despite these efforts and common knowledge of the polluting strength of wastes such as silage liquor, pollution from farm wastes remains a very serious problem. The move away from wilting and the increased production of silage is viewed with real concern by the water authorities. This whole issue needs to be the subject of further consideration by ADAS, including research into methods of reducing the volume and even the polluting strength of silage liquor.

The risk of pollution from silage liquor can, given sufficient investment, be reduced relatively easily, especially at new installations. It is, therefore, very encouraging that water authorities are now being consulted over grant-aided schemes involving the construction of silos, and so they are able to comment on measures to reduce the risk of pollution before work commences. A major problem, however, is that in this difficult economic climate sufficient investment is not being made to remedy deficient waste management systems, especially outside the less favoured areas (LFAs).

Whilst the "polluter pays" principle must still apply, the difficult economic climate needs to be recognised, with grant aid for pollution prevention measures outside LFAs being increased to that which exists within LFAs. Also, a range of pollution prevention measures, such as roofing over silos and yard areas, needs to be given the highest rates of grant. This would provide positive encouragement for farmers to install systems correctly at the outset and to improve defective systems before serious pollution and consequential damage to downstream interests arise.

Prosecution, which consumes resources of both the water authorities and farmers after pollution has arisen, is not necessarily the answer. However, pollution caused by silage liquor is particularly serious, as reflected by farm pollution statistics (WAA/MAFF, 1986, 1987) which show that in 1985 and 1986 silage liquor discharges were responsible for 32.6 percent of serious farm pollution incidents (17.3 percent of total farm incidents). Overall, therefore, it is not surprising to see that the water authorities' attitudes appear to be hardening in respect of silage liquor pollution, resulting in an increased likelihood of prosecution of farmers (Figure 2).

FIGURE 2 **Percentage of farm pollution incidents resulting in conviction (1979 to 1984) or prosecution (1985 and 1986)**

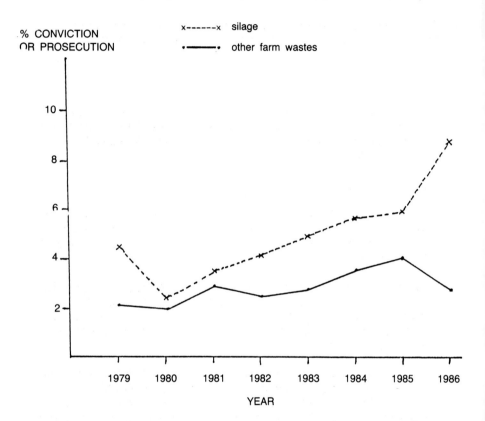

In the absence of positive measures to reduce pollution from farm wastes, water authorities may be left with little alternative but to pursue the increased use of legislation, even though this may prejudice efforts being made to improve relationships with the farming community. Measures could include increasing the level of prosecutions, the use of Regulations under the Control of Pollution Act 1974, and pressing for amendments to legislation, such as altering the status of the Code of Good Agricultural Practice away from being a defence for farmers against prosecution. Water authorities are now actively considering the possible use of regulations to encompass controls and precautions to be taken in relation to a number of aspects of pollution, including silage-making.

Some water authorities are seeking to recover their costs following pollution incidents, irrespective of whether the cases are taken to court. These costs can be considerably greater for the farmer than the fines imposed by courts.

PREVENTING POLLUTION

Despite its very corrosive and polluting nature, silage liquor need not give rise to pollution incidents if the following points are observed:

(i) **Wilting**

Effective wilting can virtually eliminate silage liquor and the associated problems of liquor collection, storage and disposal. The cost and effort of repairing structures, especially expensive silos, damaged by silage liquor is often overlooked.

(ii) **Plan carefully**

Situations will certainly arise when it is impractical to wilt, especially in the wetter western parts of the country. Adequate facilities must therefore be provided to collect, store and dispose of liquor safely on these occasions.

Careful planning can anticipate problems and help to prevent pollution occurring. The cost of proper pollution prevention measures is usually very small when carried out at the stage of constructing silage storage facilities, but it is considerably greater when a system already installed has to be altered.

Farmers should follow guidelines provided by the Ministry of Agriculture, Fisheries and Food (MAFF) and obtain advice from ADAS before starting work. The cost of such advice is small compared to the overall cost of the project. The water authorities' views should also be sought at the planning stage.

(iii) **Silo construction**

A properly constructed silo is absolutely essential. A silo should never be built over land drains or springs. The underlying ground must be firm as pollution cases have arisen where silos have "broken their backs" when constructed on built-up land.

(iv) **Collection systems**

Collection systems must be close to the silo, and, if possible, above ground to minimise the risk of leakage. They should extend around the back and sides of silos to collect leakage through the walls.

(v) **Storage**

The volumes of silage liquor produced are often not large and can be predicted reasonably accurately, which assists management of storage. Purpose-made stores must be sized according to the ability of the farmer to empty them and the time available for this task. Stores with a capacity less than 10 cubic metres may not be worthwhile unless they are fitted with level alarms, which are inexpensive.

Account must be taken of any rainfall that may mix with the liquor and overwhelm stores. An open slurry store can be used providing that it is impermeable.

Storage tanks must be properly protected against attack by acidic liquor, not only to prevent pollution but also to preserve the life of the tanks. All too often the success of a storage system is impaired by it being too small or improperly protected, when the increase in cost by installing a good system may be small.

(vi) **Repairs to systems**

Silos need to be thoroughly checked for leaks and cracks well in advance of silage making, to allow time for *proper* repairs. In particular the joints and bases of side walls must be examined and suspect joints and damaged areas carefully repaired or replaced.

(vii) **Checks after harvesting**

The whole system, together with drains, ditches and streams, needs to be checked *frequently* for any evidence of problems. If problems arise the appropriate water authority must be informed so that downstream interests can be warned and emergency measures agreed. Storage tanks must not be allowed to overflow.

(viii) **Disposal**

Silage liquor, having a high concentration of nutrients, is a very useful fertilizer if diluted to avoid scorching. It needs to be applied well away from watercourses and land drains, at a rate which does not lead to run-off.

CONCLUSIONS

Silage liquor is the most polluting waste produced on a farm. It can cause very serious pollution and damage to downstream interests if it enters any watercourse or body of water.

The potential for pollution can be virtually eliminated if adequate wilting is undertaken. However, the recent shift away from wilting and the massive increase in silage production in some areas strongly suggest that, in the absence of positive actions as outlined above, silage liquor will become an even more serious pollution problem.

REFERENCES

GORDON, F.J. (1981) The effect of wilting of herbage on silage composition and its feeding value for milk production. *Animal Production, 32*, 171-178.

HARDWICK, D.C., LANG, J. and NIELSEN, C. (1987) Agricultural Pollution. *Institute of Water Pollution Control Year Book 1987.*

WATER AUTHORITIES ASSOCIATION AND MINISTRY OF AGRICULTURE, FISHERIES AND FOOD (WAA/MAFF) (1986) *Water Pollution from Farm Wastes 1985—England and Wales.*

WATER AUTHORITIES ASSOCIATION AND MINISTRY OF AGRICULTURE, FISHERIES AND FOOD (WAA/MAFF) (1987) *Water Pollution from Farm Wastes 1986—England and Wales.*

Chapter 3

SILAGE EFFLUENT AND THE SILAGE MAKER

R J Longman

Bagborough Farms Ltd., Bagborough Farm,
Pylle, Shepton Mallet, Somerset

SUMMARY

In June 1984 effluent from silage made on my farm polluted a local watercourse, and in December that year I was prosecuted and fined. With the help of my local Water Authority, I found that there were two main factors which had caused the pollution problem. These were inadequate maintenance of the existing effluent collection system, and a lack of appreciation of the effect on effluent production of changes in my silage-making techniques. I took immediate action to deal with disposal of the effluent being produced, then planned long-term constructional changes and repairs to the silage clamps and effluent collection system. My advice to other farmers is to inspect all constructions concerned with silage and effluent storage, and to be aware of the consequences of changes in silage making methods.

INTRODUCTION

On June 8th 1984 I was presented with a bottle containing a greeny-brown liquid by a very agitated gentleman from the local Water Authority.

Six months later, on December 4th 1984, I was summoned to the local Magistrates' Court and ordered to pay over £3000 in *fines, costs* and *compensation*. If pollution had been a capital offence, the Chairman of the Bench would probably have donned his little black cap!

Very simply, I had been caught polluting a watercourse and punished—and punished very heavily.

THE FARM

My farm consists of 243 hectares (600 acres) of intensive grassland. This provides feed for two herds each of 200 cows, and heifer replacements. I use a simple system of paddock grazing, self-feed and easyfeed silage, with cubicle and kennel housing.

Geographically, the farm is just south of the Mendip Hills and at the head of a valley which slopes gently towards the Somerset Lowlands. There are three important features of this farm setting, namely:

* A high rainfall—approximately 97 cms (38 inches) per annum.

* The soil is a loamy clay on layers of limestone. This allows surface water to filter down some 18 inches, but the water then follows the rock layers to run off into ditches.

* The ditches merge into a stream which is at the head of a river system. In other words, from a pollution point of view the buck stops at me!

My silage making system is fairly standard. I cut with a large mower conditioner, pick up with a full chop harvester and store the silage mainly in covered silage clamps.

THE PROBLEM

Returning to the greeny-brown liquid. What had gone wrong?

With the help of the Water Authority, I traced the silage effluent pollution—it was obviously silage liquor from its smell—back from the sampling point in a ditch until it disappeared underground. Then, by a process of elimination, I found the liquor was seeping into the subsoil from the collection system I had installed some fifteen years earlier.

I found many of the concrete silo floors and collection pipes were cracked, and the joints had been eaten away. Also the concrete ring pits were not big enough and some overflowing had occurred, especially at night.

Quite frankly, I was guilty of:

a) not maintaining my existing collection system, and
b) not being aware of the effect of the changes I had made in silage-making techniques.

The silage clamps had been built in an era when only 5 to 6 tonnes of silage per cow were made. In addition, originally I wilted grass to at least 30 percent dry matter and only took two cuts per year, starting in late May or early June.

Now I make 10 tonnes of silage per cow, with a dry matter content rarely above 20 percent. I also take 3 or 4 cuts a year, starting in early May. In other words, I make 4000 tonnes of good quality but wet silage, which produces a tremendous amount of greeny-brown liquor!

THE SOLUTION

The measures I took to rectify the problem were as follows.

First of all, I did the obvious things like digging holes to act as sumps to catch the liquor, and sucking them out regularly. I then dammed the ditches and sucked them out until the liquor had been cleared from the rock strata. This took several weeks!

Meanwhile the Water Authority constantly monitored the farm and even started aerating the stream lower down with compressed air—hence the large sum for compensation which was added to my fine.

Next, I called in the experts from A.D.A.S. for advice and we drew up plans for the future. Following this, four improvements were made:

1. Eight new fibre glass collection tanks were installed.

2. New *open* collection channels were built around each silage clamp.

3. The clean rainwater was diverted by reguttering the nearby buildings, so that clean water did not mix with effluent and thus overload the tanks.

   I do not believe that the often used maxim "The solution to pollution is dilution" should be applied to silage liquor—rarely can enough water be found to dilute the liquor to satisfy the Water Authority. The answer must therefore be collection and planned disposal.

4. The cracks in the concrete floors were repaired. I grouted out the cracks and joints using a compressed air grouter and filled them with hot pitch. I also repaired the concrete fillets and channels with a mixture of sand, cement and, later, compound. After two seasons these repairs still appear to be quite satisfactory.

ADVICE FOR OTHERS

I would like to end by emphasising two points.

Firstly, *inspect* your silage stores *now*, and make plans for any repairs or additions which are necessary. There is only a very short period in which to do such work, i.e. the last two weeks of April when silage pits are empty and, hopefully, cows are out to grass. It is virtually impossible to repair damaged concrete when there is still some effluent weeping from silage in store.

Finally, *be aware* of any changes you may have made in your silage making techniques, i.e. making earlier cut, wetter silage. Plan the collection system to cope with much more effluent than you ever thought possible.

You will then save yourself the worry of being presented with a bottle of disgusting greeny-brown liquid!

Chapter 4

DEALING WITH SILAGE EFFLUENT: TECHNICAL OPPORTUNITIES

P A Mason

ADAS Farm Buildings Unit, Government Offices,
Coley Park, Reading RG1 6DT

SUMMARY

Approximately 1.5 billion litres of silage effluent are estimated to have been produced from silage made in England in 1987. Silage effluent is highly polluting when discharged to water courses, and in the year to April 1986 there were 592 reported pollution incidents due to faulty design, construction or maintenance of silos.

A major problem is that silage effluent is highly corrosive to many building materials, but risks of pollution should be low if attention is paid to the siting, design and construction of new clamps, and to their correct operation and maintenance. Critical aspects of silo design are the correct specifications for materials and construction joints to resist effluent attack, and the provision of appropriate drainage systems and collection tanks. The risk of leakage of effluent due to structural deterioration can be reduced by following the designer's recommendations for use, particularly on maximum loads, and by inspecting the empty silo facilities each season as part of a maintenance routine.

Hot rolled asphalt has been introduced into the UK recently as an alternative to concrete for the construction of silos, and its durability is currently being monitored.

Silage effluent may be used as a fertiliser on land, although certain precautions must be taken to reduce the risks of pollution or of damage to crops. It may also be fed to livestock as a supplementary liquid feed. Silage absorbents in the clamp retain many of the nutrients in silage effluent, but do not remove the need for leakproof silos. Stored silage effluent can produce highly toxic gases and safety precautions are essential.

INTRODUCTION

In 1987 farmers in England made about 25 million tonnes of silage in conventional silage clamps, producing in the process some 1.5 billion litres of potentially highly polluting effluent[1]. In the year to April 1986 the Water Authorities Association reported 592 instances of significant pollution in England and Wales which involved silage effluent and where faulty design, construction or management of silos were to blame (WAA/MAFF, 1987).

The indications are that silage production will continue at current levels, at least, for the forseeable future while general concern over pollution is likely to increase. Part 2 of the Control of Pollution Act (1974), which has now been law for over 3 years, allows for substantial penalties against polluters.

This paper looks at the nature of silage effluent against such a background—what it is, how much is produced, what makes it so difficult to handle—and describes how it may be collected and disposed of safely. Some practical advice is given on how to design and construct silos and effluent tanks to minimise the risk of pollution.

SILAGE EFFLUENT

Effluent production

Silage effluent is a by-product of the fermentation process whereby grass and other herbage plants are made into silage. Favourable conditions for this fermentation occur when grass is compacted in a silage clamp and stored under airtight conditions; the sugars present in the herbage are converted to acids. It is necessary to achieve the correct level of acidity to bring about stable fermentation, otherwise undesirable clostridia develop, leading to protein breakdown and butyric acid production. The level of acidity required for stable preservation depends largely on the dry matter content of the grass; dry matter contents less than 20 percent require an acidity of pH 4.0 or less, while grass with a dry matter of 30 percent will be stable at about pH 4.5.

The liquid exuded during the fermentation process, silage effluent, is a highly concentrated and corrosive pollutant. The quantity of effluent produced and its chemical composition are influenced by a variety of factors, but the most important of these is the dry matter content of the grass when it is ensiled.

Effluent composition

The difficulty with silage effluent is that it is not only highly polluting when

---

[1] Author's forecast based on the December 1986 Agricultural Census for the United Kingdom

discharged to watercourses, but it is also severely corrosive to many common building materials.

The reason that silage effluent is so highly polluting is that it contains large concentrations of organic compounds such as sugars and other plant material, which provide an ideal food source for the microorganisms which live in streams, rivers and other watercourses. Any effluent leaking into a watercourse is decomposed by these microorganisms and, during the process, part or all of the oxygen dissolved in the water may be used up. Even a small quantity of effluent discharged into a watercourse may remove sufficient oxygen to cause fish and other aquatic life to die over a large area. Regeneration can be a very lengthy process.

The polluting strength of an effluent is described by its Biochemical Oxygen Demand (BOD). Values for the BOD of a number of pollutants, given in Table 1, show how aggressive silage effluent is. In particular, silage effluent is some 200 times more polluting than raw domestic sewage.

TABLE 1 **The biochemical oxygen demand of various wastes and effluents**

Pollutant	BOD (mg/litre)
Untreated domestic sewage	300-400
Vegetable washings	100-300
Parlour and yard washings	1000-2000
Animal slurries	10000-35000
Silage effluent	12000-83000

The corrosivity of silage effluent is a consequence of its acidity. Table 2 lists the main aggressive constituents likely to be present.

TABLE 2 **Aggressive constituents of silage effluent**

Constituent	Concentration
Lactic Acid	4 to 6%
Acetic Acid	1 to 2%
Butyric Acid	normally less than 0.5%
Sulphates	100 ppm
Formic acid)	
Sulphuric acid	Present according to
Formaldehyde	additive used
Ammonia	
Hydrogen sulphide	
Ethanol	

Overall acidity: pH 3.7 to 5.5

Most other farm effluents are far less acidic than silage effluent, with pH values in the range 5.6 to 8.0. They are therefore less corrosive.

The use of silage additives affects the composition of the effluent to some extent. Acidifying additives are often used on herbages with low dry matter and low sugar contents, and it is under these circumstances that effluents with the highest acidity are likely to be produced.

Quantities of effluent produced

Fermentation takes place within the first few days of ensiling and this is when the majority of effluent is produced. Figure 1 shows the pattern of effluent production from a herbage of approximately 18 percent dry matter content.

FIGURE 1   **Pattern of effluent flow from a clamp silo**

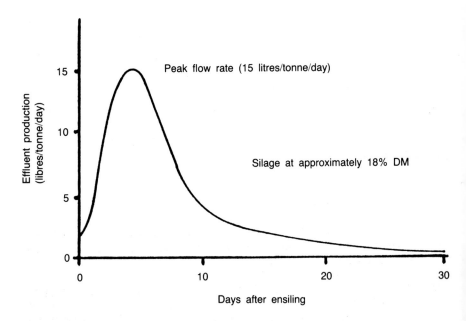

Source: Bastiman (1976)

The total quantity of effluent produced depends principally on the dry matter content of the herbage at ensiling (see Figure 2), but there is evidence that the use of some acid-based additives increases the quantity of effluent produced.

FIGURE 2 **Effect of dry matter content on effluent production**

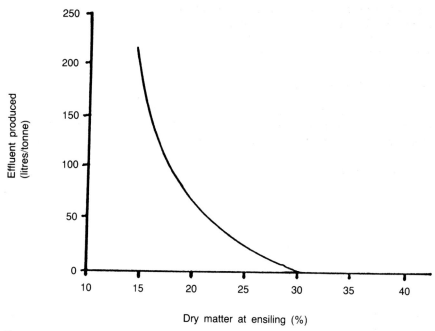

Dry matter at ensiling (%)

Source: Bastiman (1976)

Wilting herbage to raise its dry matter content from say 15 percent to 20 percent has the effect of reducing effluent production from about 200 litres per tonne to 60 litres per tonne.

For quite some time, conventional wisdom has been to field-wilt grass to achieve a target 25 percent dry matter content within 24 hours. This leads not only to a reduction in silage effluent produced in the silo but also to an increased concentration of sugars in the grass, generally faster field work and an increased dry matter content of the silage. However, recent studies on silage losses (Crawshaw, 1987; Mayne and Gordon, 1986) have shown significant reductions in total dry matter conserved and nutritive value from even limited field wilting, and direct cut silage is being made by an increasing number of farmers. This change in silage-making strategy is also leading to greater interest in feeding silage effluent to livestock and in the use of effluent absorbents, as outlined later in this paper.

BUILDING SILOS TO CONTAIN EFFLUENT

By paying proper attention to the siting, design, construction, operation and maintenance of silage clamps, it is possible to reduce the risk of pollution to a very low level.

Siting

Since it can never be guaranteed that a silage clamp will not leak effluent, it should be sited as far as possible from watercourses. Sites with free-draining soils should be avoided. Whatever the site, any underdrains should be blocked off and, if necessary, re-routed. Sites liable to flooding are obviously unsuitable, as are sites where differential settlement is likely to occur, leading to cracking in the floor and a leakage path for the effluent.

Design

The aim when designing silos is to channel all the effluent released by the silage, and effluent-polluted rainwater, to a suitable collection point for safe storage or disposal. To achieve this it is necessary to specify materials which can withstand attack by the highly acidic effluent, and to design features such as floors, drains and gullies which channel the effluent to the collection point quickly. Equally, the aim is to avoid effluent ponding or becoming trapped within the structure, when it will begin to cause damage.

Most leakage occurs through cracks in concrete floor slabs and at the joints between the walls and the floor. In these areas proper design is particularly important.

Critical aspects of silo design are as follows:

a) *Silo drainage*

It is good practice to extend the silo base slab beyond the clamp walls, to form a perimeter drain to catch any effluent seeping through. If this is not provided it is essential to make the walls effluent-tight, but this can be difficult. Timber walls (prefabricated panels, sleepers etc) need lining internally with an impervious sheet. Concrete and masonry walls may need protecting from attack by effluent using special paints and/or renders, whether or not an external effluent drain is provided.

Figure 3 shows a satisfactory arrangement for draining the effluent away from a typical clamp. Bricks laid loosely on edge in a channel, land drainage tiles and slotted PVC pipes all make suitable drainage paths. Positioning the drain near the inside face of the wall has the additional advantage of reducing wall loads, and thus the risk of cracking and effluent leakage.

FIGURE 3   **Example of drainage from a silage clamp**

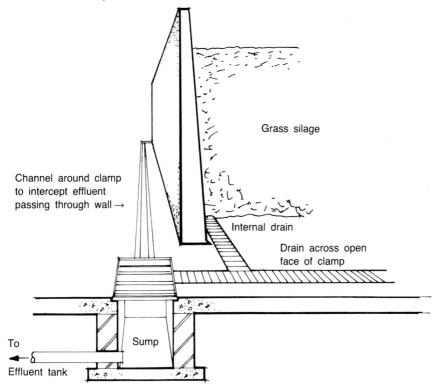

Grass silage

Channel around clamp
to intercept effluent
passing through wall →

Internal drain

Drain across open
face of clamp

To

Sump

Effluent tank

b)   *Effluent tanks*

Effluent tanks must be of a size appropriate to the amount of effluent and effluent-polluted water expected, and to the frequency of emptying. The farm drainage system should be designed so that all other wastes are kept out of the silage effluent system. For a given situation it is possible to calculate fairly accurately the size of effluent tank required, but a general recommendation is to provide 3 cubic metres of storage per 100 tonnes of silage.

Tanks may be constructed of *in situ* concrete or masonry, but very careful design and construction is necessary to prevent effluent leaking from them, or ground water leaking into them. Masonry tanks require rendering internally and externally with cement mortar, then coating internally with an acid-resistant paint. Concrete surfaces need to be similarly painted. All construction joints need to be sealed, as described later for the silo slab.

As an alternative to concrete or masonry, special acid-resistant prefabricated tanks are available. These provide a far better safe-guard against pollution.

28

c) *Concrete specification*

Concrete for the floor slab can be specified in a number of ways, but the main requirement is for low permeability to prevent the highly acidic effluent seeping in and accelerating decay. There is a direct and very sensitive relationship between the water/cement ratio and concrete permeability, as shown in Figure 4.

FIGURE 4. **Effect of water/cement ratio on permeability**

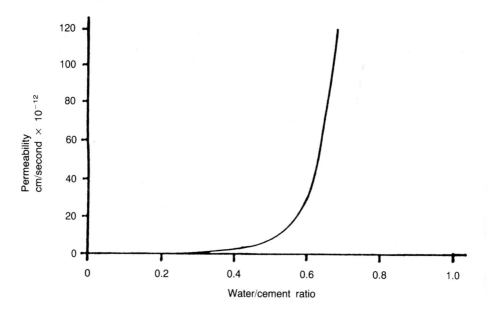

Concrete in silos should have a water/cement ratio not greater than 0.4:1. Figure 4 shows that adding extra water has a very undesirable effect. As a ratio of 0.4:1 gives a rather stiff mix which becomes unworkable relatively quickly, plasticisers, superplasticisers and retardants may be specified to increase workability. These additives are a particular advantage when concrete is laid by inexperienced labour. ADAS is currently testing a number of cement additives, partial cement replacements and surface hardeners, which it is hoped will improve the resistance of concrete to effluent attack. The current conventional specification is a medium workability C25P mix to BS5320 using ordinary Portland Cement.

To minimise the risk of cracking, the floor slab should be between 125 mm and 175 mm thick depending on site conditions and the amount of reinforcement provided.

d) *Contraction joints*

Contraction joints must be specified for concrete silo slabs to prevent random shrinkage cracking. The required frequency of these joints depends on whether, and to what extent, the slab is reinforced. Unreinforced slabs need contraction joints every 5 metres, while slabs with a light square mesh reinforcement (eg A142) only require joints at 15 metres. A reinforced slab is therefore generally preferred.

Contraction joints are formed by pressing a joint former into the wet concrete. This encourages the formation of a single, relatively wide shrinkage crack, which can be sealed, at each joint. The presence of contraction joints relieves the shrinkage stresses which would otherwise result in numerous fine cracks, which cannot be sealed effectively, scattered randomly over the silo. Figure 5 shows a section through a typical contraction joint and Figure 6 indicates the positions of contraction and construction joints in a typical silo.

FIGURE 5   **Section through a typical contraction joint**

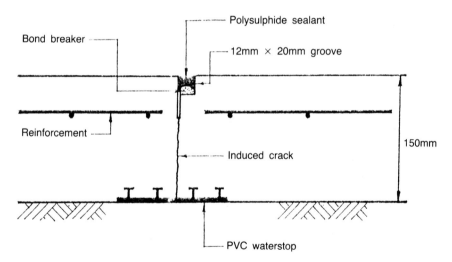

There is a wide choice of sealants on the market for filling these joints. The most reliable to date appear to be the polysulphides and polyurethanes which, if correctly applied, can be expected to give many years of service. Some cheaper sealants, including the commonly used poured bitumens, are not so durable.

e) *Waterstops*

Whilst a carefully formed contraction joint reduces the risk of effluent

FIGURE 6 Layout of joints in a reinforced silo slab

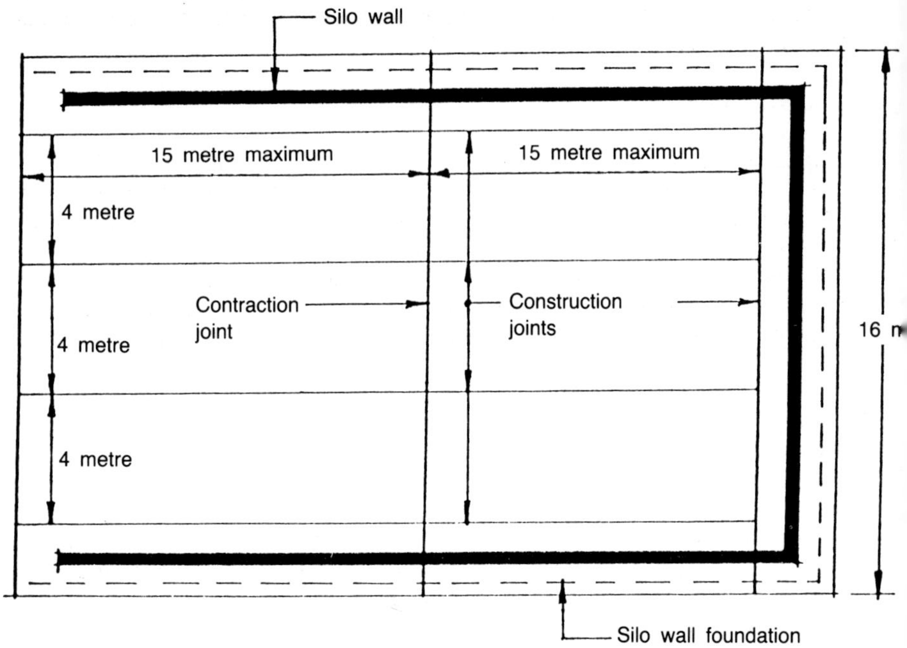

seepage, it is necessary to incorporate a "waterstop", as shown in Figure 5, at the bottom of the slab to form a truly effluent-tight joint. Waterstops should always be used on those sites where the consequences of effluent seepage would be particularly serious.

Construction

To produce a silo which will not leak effluent requires not only proper design, as just described, but careful construction right through from preparation of the site to application of the protective coatings to the walls and floors. Particular points to watch are:

a) *Proper site preparation*

The first step in silo construction is thorough preparation of the site to give adequate support to the floor slab and walls. This entails:

* Removal of topsoil;

* Digging out soft spots and replacing them with suitable hard material;

* Laying a base of selected hardcore, which should be compacted with a vibratory roller and blinded with sand.

The base should be finished to the correct falls across and along the silo to encourage controlled drainage of effluent. A 1,000 gauge polythene sheet should be laid over this base, lapped and taped to form a continuous impermeable membrane.

b) *Concreting*

Concrete slabs must be laid quickly and accurately. Normal practice is to lay slabs in 4 to 5 metre wide bays, forming effective contraction joints between and along bays, as previously described. The C25P concrete mix usually specified for silo slabs must be compacted thoroughly, which requires a mechanical vibrating beam to achieve the desired low permeability. On no account should extra water be added to the mix to increase workability as this greatly increases the permeability of the finished product and allows effluent to seep through.

Particular care needs to be taken over the formation of the contraction joints. Experience has shown that poor formation and poor sealing of these joints has been responsible for many slab failures.

The concrete slab must be cured under a polythene sheet or by using a proprietory spray-on curing compound, to prevent premature drying and excessive shrinkage and cracking. The slab should not be used for silage-making until 28 days after casting.

c) *Application of surface sealants*

Penetration of concrete by effluent can be reduced by applying suitable surface treatments. Such treatments should be applied before the first use of a silo, and strictly in accordance with the manufacturer's instructions. There have been many cases where otherwise suitable treatments have broken down prematurely as a result of being applied to poorly prepared surfaces, or to surfaces which have already been weakened by silage effluent.

For concrete walls the best long-life coatings are epoxy resins and pitch modified epoxy resins. The latter are cheaper and easier to apply. Bituminous paint is very cheap and easy to apply, but it is likely to need recoating within 2 to 3 years. Chlorinated rubber paint has a similarly limited service life.

For concrete floors, any treatment has to contend with the added problem of abrasion. It is also undesirable to lower the floor friction with a smooth coating, otherwise silage slip and other problems can occur. For these reasons penetrative sealants are preferred. Two suitable examples are magnesium silica fluoride in solution and polyurethane resin.

Operational use

It is important to use a silage clamp only as intended by the designer. Typical, and not uncommon, forms of abuse are:

* Overloading by filling deeper than intended, such as doming steeply from the top of the walls;

* Overloading by compacting with a vehicle heavier than the maximum allowed;

* Overloading by not providing effective drainage channels for effluent;

* Mechanical damage from impact by loading vehicles.

Overloading, in addition to being dangerous, can cause excessive cracking in the silo. Effluent quickly penetrates and opens up such cracks.

A most important aspect of pollution control is to minimise, as far as possible, the quantity of effluent and effluent-polluted rainwater. With a roofed silo rainwater should not be a problem, but with an unroofed clamp, particularly during heavy rain, it can increase the volume of effluent which reaches the tank, causing disposal difficulties. One method of overcoming this problem is to install a 'diverter' drainage system. This arrangement directs the relatively slow trickle of effluent to the effluent tank but automatically diverts any stronger flow, such as that associated with heavy rainfall, to a separate storage or soakaway system.

Maintenance

Silos should be inspected each season for signs of deterioration and leakage. Any leakage through concrete or masonry walls is generally evident from staining on the outside. Floor slabs should be examined closely, with particular attention paid to the condition of sealants at contraction and construction joints. As well as being a possible source of pollution, seepage through ineffective seals causes rapid deterioration of the concrete and the need for expensive repairs in the future. Any new cracks which have occurred in the base should be opened up using a concrete grooving machine and then sealed in the same way as for contraction joints. Extensive new cracking is indicative of breakdown of the slab, which needs to be either relaid or overlain with a new base.

Effluent drains should be kept clear to allow rapid drainage. They should be checked for cracks regularly, although internal inspection of effluent tanks should never be attempted without obtaining specialist advice on the precautions necessary against toxic fumes.

Assessment of the condition of a silo and the specification of preventive and/or remedial work should be left to an experienced surveyor or engineer.

SILO CONSTRUCTION USING HOT ROLLED ASPHALT

A number of silos have recently been constructed in the UK using Hot Rolled Asphalt (HRA) instead of concrete as a base material. HRA has been used successfully for a number of years in North America under maize silage and shows great promise so far in the UK. It offers a number of advantages when compared with concrete:
* Requires no formwork
* Inherently resistant to acid attack
* Can be laid in a continuous process and does not shrink. Consequently, fewer construction and shrinkage joints are required and the risk of leakage is reduced
* Cures within 24 hrs. No 28-day wait as with concrete
* Not susceptible to frost damage.

Work involved in site preparation for an HRA base is the same as for concrete, but the base itself must be laid by a specialist contractor using a pavior machine and roller. It is important to note that the specification for HRA for silos differs significantly from that for HRA for other applications such as roadways.

HRA has also been used successfully to resurface badly corroded concrete floors in silos. Provided the existing surface is properly prepared and the new HRA course, typically 40 mm thick, is specified and applied properly, extremely good and durable results can be obtained.

For both new work and refurbishment work HRA costs are competitive with conventional concrete. ADAS is monitoring the performance of a number of HRA silos to assess their long term durability.

EFFLUENT DISPOSAL AND USE

Spreading on land

The usual method of disposal is by spreading on the land. The following precautions need to be taken to reduce the risk of water pollution and crop scorch:

* Dilute effluent at least 1:1 with water before spreading

* Avoid applying in hot dry weather

* Never apply more than 25 cubic metres per hectare

* Never apply close to watercourses

* Never apply to land which forms the catchment area for bore holes, springs or wells

* Avoid applying to dry, cracked soils.

The fertiliser value in each cubic metre of silage effluent is approximately 3 kg N, 1 kg P and 4 kg K.

Silage effluent can also be mixed with slurry in open air slurry stores, for eventual disposal along with the slurry. However, mixing effluent with slurry produces hydrogen sulphide gas, which can be lethal at very low concentrations. Never, therefore, mix effluent and slurry in enclosed tanks, tanks within buildings or tanks below animal slats. It is also dangerous to allow silage effluent and slurry to mix in drains leading to an open store, since this can cause hydrogen sulphide to back-up through the pipes.

Direct feeding to livestock

Fresh silage effluent may be fed directly to pigs and cattle as a supplementary feed. It may be stored for up to 1 year by the addition of formalin at a rate of 3 litres per 1000 litres of effluent.

Trials have shown that effluent fed *ad lib* can replace 0.34 to 0.45 kg of pig finisher ration without adverse effect. With beef cattle a voluntary daily intake of 14 litres of effluent can replace 0.75 kg of concentrate. High potassium and nitrate levels can, however, cause problems and expert advice should be sought before using effluent routinely as a supplementary feed. In general, intakes should be restricted to 5 litres per 100 kg body weight.

The cost of providing extra effluent storage tanks needs to be set against any savings in feed when assessing the economic viability of such a system.

Effluent absorbents

A means of reducing problems with effluent, while at the same time utilising its feeding value, is to line the base of the clamp with a layer of straw bales to soak up the effluent. Although effluent-soaked straw can be expected to have a feeding value approaching that of ammonia-treated straw, the benefit has to be weighed against the loss of silo capacity available for silage. As a general rule, one layer of straw bales, placed on edge, is required to soak up the effluent from silage 1.8 metres deep.

A more satisfactory means of soaking up effluent is to mix a suitable absorbent with the grass as it is ensiled. Chopped barley straw in layers has been shown to be a much more effective absorbent than a bottom lining of baled straw. There is also a range of proprietory products available, some of which are based on beet pulp. These products not only have a high capacity for absorption, but also a considerably higher metabolisable energy (ME) content than that of barley straw, typically about 12 MJ ME per kg dry matter.

WARNING—POISONOUS GASES

Stored silage effluent may produce toxic gases which, if inhaled, may very rapidly cause unconsciousness and death. Special precautions must therefore be taken if it is necessary to enter, or even to peer into, an enclosed effluent tank. Silage effluent mixed with slurry releases hydrogen sulphide and other poisonous gases. Therefore, under no circumstances should silage effluent be added to slurry stores within livestock buildings or in enclosed storage tanks.

CONCLUSIONS

Silage effluent is a highly polluting liquid. Wilting grass to dry matters of 25 percent and greater before ensiling reduces the quantities of effluent produced at the silo to very low, and possibly insignificant amounts. However, trends in silage-making appear to be toward less field wilting which will be accompanied by increased effluent production.

It is perfectly feasible to construct silos which will contain, without leakage, the silage effluent produced by low dry matter herbage, but proper design and careful construction are required. There is little excuse for leaky, polluting silos.

Provided certain precautions are taken, silage effluent can be disposed of safely either by spreading on the land, or by direct feeding to livestock as a supplementary feed.

Silage absorbents, while not removing the necessity for leakproof silos, have the ability to reduce the effluent load and at the same time reduce nutritive losses.

REFERENCES

BASTIMAN, B. (1976) Factors affecting silage effluent production. *Experimental Husbandry*, **31**, 40-46.

WATER AUTHORITIES ASSOCIATION AND MINISTRY OF AGRICULTURE, FISHERIES AND FOOD (WAA/MAFF) (1987) *Water Pollution from Farm Wastes 1986—England and Wales*.

CRAWSHAW, R. (1987) Reducing losses during ensiling. In: *Developments in Silage 1987.* Marlow Bottom, Bucks: Chalcombe Publications, 23-36.

MAYNE, C.S. and GORDON, F.J. (1986) The effect of harvesting systems on nutrient losses during silage making. 1. Field losses. *Grass and Forage Science,* **41,** 17-26.

Chapter 5

THE SILAGE MAKER'S GUIDE TO DEALING WITH SILAGE EFFLUENT

J M Wilkinson

Independent Agricultural Consultant
Chalcombe, Highwoods Drive, Marlow Bottom,
Marlow, Bucks SL7 3PU

SUMMARY

The solution to the problem of silage effluent lies in the adoption of three strategies:

* *Reduce effluent production*
* *Contain effluent if it is produced*
* *Manage effluent effectively*

Reduction of effluent involves harvesting less grass at the first cut, wilting for up to 24 hours, chopping not too short, making silage in big bales, and using an effective absorbent during ensiling.

Containing effluent comprises constructing bunker silos so that concrete is resistant to corrosion and has contraction joints to eliminate random cracking. A wide range of products is available to assist the farmer in making concrete more durable, and for sealing contraction joints or cracks. Hot rolled asphalt looks promising as a superior flooring material to concrete. Complete lining of silo walls can assist in containing effluent. Adequate storage for effluent is essential.

Managing effluent means regular inspection of collection tanks. Careful spreading away from rivers and streams is the most common method of effluent utilisation. The net value of effluent as a fertiliser is estimated to be £0.48 per thousand litres. By contrast, the net value for effluent as a feed is estimated to be £6.33 per thousand litres. Farmers faced with vast amounts of effluent may be well advised to invest in low-cost storage facilities so that effluent can be recycled as feed.

REDUCING SILAGE EFFLUENT

The real problems of pollution by effluent from silos follow the harvesting of huge amounts of first cut grass in wet weather during May. Farmers faced with the daunting job of ensiling 60 to 70 percent of their total annual silage crop in a 10 day period are understandably reluctant to delay harvest in the hope that the weather will improve. Often it doesn't. Delayed harvest after ear emergence means that the energy value of the crop can fall by as much as 0.5 MJ of metabolisable energy (ME) per kilogram of dry matter (DM) per week. Further, the drive to reduce feed costs has led to the use of greater quantities of fertiliser nitrogen in spring to increase silage yields. This has meant that grass crops are leafier, denser, and therefore wetter pre-cutting.

Several options are open to reduce the risk of having to ensile enormous yields of grass in a rush at first-cut harvest time:

* *Graze some fields early in the spring*, before closing them up for silage. This job is best done by sheep. Cattle may be grazed at low stocking rates, but there is a greater chance of damaging the grass sward through poaching (treading). Grazing in April will delay flowering by 7 to 10 days. The grazed fields can then be cut after the ungrazed ones, at a similar herbage quality.

* *Sow grasses with different dates of flowering*. There is a considerable range in flowering date among the ryegrasses, from mid-May for Italian and early perennial ryegrasses to the end of May for the latest perennials. By harvesting the early flowering types before the later ones, harvest may be extended without loss of quality.

* *Be ready to start harvesting early*. There are good reasons for being prepared to start harvesting early in May. In the drier parts of the country the grass has more chance to recover after being cut before the soil dries out. A week's earlier harvest can make the difference between a decent second cut yield and an indifferent yield which is not worth harvesting. In a wet spring an earlier start allows more time to pause between periods of rain.

* *Apply less nitrogen before the first cut and more for the second cut*. The strategy for spring application of fertiliser should be as flexible as possible, bearing in mind differences between years in temperature and in rainfall. The total amount of nitrogen should be split into two doses, the first at T-sum 200, or thereabouts depending on the weather. If the weather is cold and/or wet thereafter, the second dose should be delayed until the weather improves. If this does not occur until the last week in March, the second application should be postponed until after the first cut of silage. Pollution of water by nitrate nitrogen is likely to be reduced if a flexible fertiliser strategy is adopted. Coupled with earlier cutting, the grass crop is likely to be able to make more effective use of fertiliser if it is applied after the first cut. In consequence, the yield of second cut silage will be greater.

In view of the marked reduction in effluent production as crop dry matter is increased up to 30 percent, wilting should be high on the list of strategies to be adopted by the silage maker. It is notable that much of the improvement in output of milk per hectare recorded by Professor Gordon and his team in Northern Ireland can be attributed to an increase in total annual yield of herbage in crops which are harvested direct by flail harvester, compared to wilted crops which are mown and picked up in separate operations (Gordon, 1987). Thus yield decreases of up to 19 percent were recorded in areas of the field which were overlain by the mown swath during extended periods of wilting in poor weather. In addition, yield decreases of up to 8 percent were recorded in wheel tracks.

The results from the Eurowilt series of experiments (Zimmer and Wilkins, 1984) revealed a trade-off between loss of dry matter in effluent from direct cut silage on the one hand, and loss of dry matter in the field from wilted crops on the other. Overall, total losses were similar. A model of losses under conditions of good management and dry weather, derived from the Eurowilt series of 36 trials (Figure 1), shows clearly that the optimum dry matter content for minimum total losses is between 25 and 30 percent dry matter. In dry weather this can be achieved within a 24-hour period. Therefore, the strategy to adopt is to wilt for no more than 24 hours, and to accept that on some days little wilting will occur, whilst on others the crop may reach 30 percent dry matter or more. To maximise the chances of reaching 25 percent dry matter, the crop should be mown when it has reached its driest point in the day. This is normally in the afternoon.

Chopping of grass at harvest releases plant cell contents and speeds up the fermentation in the silo. It also increases the rate at which effluent is produced. Not only are plant juices released more extensively, consolidation of the crop is greater, increasing the pressure on the material at the base of the silo. The farmer worried about coping with the initial rush of effluent would be well advised not to chop the crop as short as possible. A reasonable target average chop length would be as long as the thumb—about 5 cm.

Big bale silage is not chopped at all; in addition, many farmers make big bale silage out of longer, more mature grass than that which is harvested for ensiling in bunker silos. Stem-held water is less easily lost from the plant than water associated with leaf tissue, and stemmy grass compacts to a lesser extent than leafy grass. Further, there is relatively little pressure exerted on big bales during their storage, unless the bales are stacked very high. With good management, the majority of big bales can be prevented from damage during the storage period, so that any effluent is contained completely within the bag or wrapped bale. Even when effluent is discharged from the bale on opening, it is more easily handled because few bales are opened at a time.

Absorbents can reduce substantially both the volume and pattern of effluent

40

FIGURE 1 **Model of dry matter losses in well managed[1] silage systems**

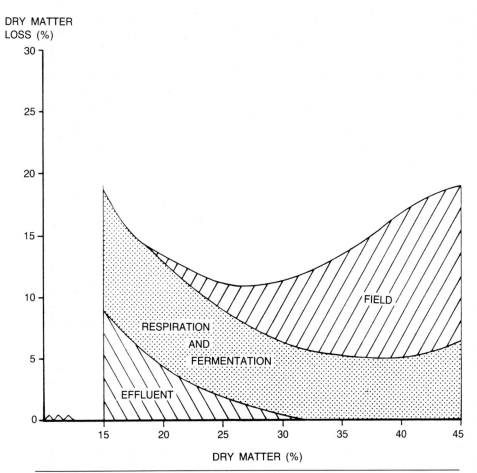

DRY MATTER
LOSS (%)

DRY MATTER (%)

[1] Dry weather, metered chop harvester, effective additive, well-sealed bunker silo.

Source: Wilkinson (1987)

production. But the effect of silo pressure can reduce absorptive capacity. Thus Kennedy (1988) noted that one product only absorbed 1 litre of effluent per kg when added to grass of 19 percent dry matter in 70-tonne silos. Further, the concentration of organic matter in the effluent from silage made with absorbents can be greater than that from untreated silage (Offer and Al-Rwidah, 1988). On the upside, absorbents can increase animal perform-ance (Offer and Al-Rwidah, 1988; Jones and Jones, 1988; Kennedy, 1988). They can also be used effectively in big bale silage.

CONTAINING EFFLUENT

Farmers with older silos are tempted to cram as much silage into them as they possibly can. To achieve this they employ large tractors with weighted wheels or ballasted tyres to increase the pressure on the crop in the silo. But in so doing, greater pressure is placed on the silo walls. Further, the harvesting of wetter grass crops means that the hydraulic pressures exerted on silo walls are increased, compared to those from wilted crops. Drainage channels inside the walls of the silo (Mason, 1988) can relieve the pressure on walls somewhat, but they accelerate the rate of release of effluent. Clearly, it is important not to weaken or to damage the structural strength of the silo wall, so that effluent remains contained in the silo, or is channelled properly to appropriate containers.

It is important to ensure that the floor of the silo does not leak. Dr Mason's contribution highlights the need to control contraction and corrosion of concrete. Hot rolled asphalt offers considerable advantages over concrete, though it involves greater capital outlay.

Complete lining of silo walls not only protects the wall from corrosion by effluent, it also reduces nutrient loss through exposure of silage to air. The lining should extend from the floor, half a metre in from the wall, up the wall and over the shoulder one metre on to the top surface to give a double seal over the shoulder. Old top sheets can be used for the walls, or special silo wall liners may be used to provide more permanent protection for the walls.

Adequate storage is vital for successful containment of effluent. Even with 3 cubic metres of storage per 100 tonnes of silage, the tank will need to be emptied almost daily for the first 10 days after filling the silo.

MANAGING EFFLUENT

Many farmers are unaware that they are causing a pollution problem in their local watercourses until the pollution control officer arrives with the bad news (see page 17). Gradual deterioration of silos, drainage channels and storage tanks can easily pass unnoticed. Regular inspection is therefore an essential part of managing effluent, can save the embarrassment of prosecution, and can reduce repair bills in the long run.

Most of the effluent which is collected in tanks is spread on the land as fertiliser. But if spread in dry weather, effluent can scorch grass, so in dry conditions it is necessary to dilute effluent 1:1 with water before spreading. The choice of land on which to spread is important—away from rivers, and not on porous soils.

The net value of effluent as fertiliser has been estimated at £0.48 per thousand litres (Weddell, Mackie and Sutherland, 1988). By contrast, the same authors estimated the value of effluent as a feed to be £6.33 per thousand litres. With such a large difference in value it is hardly surprising that there is interest in using effluent as a feed for livestock. This involves collecting and preserving large volumes of effluent unless it is given to stock as it is produced. On larger farms there may be merit in storing effluent for use in the winter feeding period, but on smaller farms the best option is to use it as it comes. Weddell *et al* (1988), Davies and Clench (1988) and others have found that effluent is readily consumed by animals. Farmers have noted that animal performance may be improved when effluent is given in addition to the normal diet, which suggests that it may in some situations act as a feed supplement.

FUTURE DEVELOPMENTS

Will the trend towards increased production of silage continue? Or will the water authorities harden their hearts and force farmers to restrict their silage making activities in sensitive areas of the country? A disturbing report (Pearce, 1986) quoted the Department of the Environment as being keen to activate a clause in the Control of Pollution Act, 1974 which gives the Secretary of State power to prohibit or to restrict the carrying on in a particular area of activities which he considers are likely to result in pollution. A prime use of this clause could be to prohibit the construction of silos inside protection zones, and to insist on costly improvements to existing ones.

With surplus land likely to be the central political issue in farming in the coming years, there is a case for reducing the intensity with which grassland is farmed. If at the same time there is a need to reduce costs, then this implies a reduction in stocking rates, lower yields of grass per hectare, less silage made in bunkers and clamps and more silage made in big bales.

New grasses could be developed which combine high digestibility with high dry matter content. With the advent of gene transfer, it should soon be possible to produce grass varieties which typically contain 25 percent dry matter pre-cutting. These grasses would be very popular in the wetter parts of Europe.

It should also be possible to engineer the design of trailers so that the excess water is squeezed out of the grass crop in the field, before it reaches the silo. This development would have the additional benefit that more grass could be harvested per trailer load.

Finally, so long as surplus food production is a feature of European agriculture, the public will continue to urge farmers to put pollution control higher on their priorities for action than hitherto. This, combined with squeezed profits, means that farmers must consider the consequences very seriously before they embark on a programme of expanding silage production, or on further intensification of livestock production.

REFERENCES

DAVIES, O.D., and CLENCH, S.F. (1988) Feeding silage effluent to dairy cows. In: *Silage Effluent, Problems and Solutions* (Ed B.A. Stark and J.M. Wilkinson). Marlow Bottom, Bucks: Chalcombe Publications, p 55.

GORDON, F.J. (1988) Methods of producing grass silage for dairy cows. In: *Nutrition and Lactation in the Dairy Cow* (Ed P.J. Garnsworthy) London: Butterworths (in press)

JONES, R., and JONES, D.I.H. (1988) Effect of absorbents on effluent production and silage quality. In: *Silage Effluent, Problems and Solutions* (Ed B.A. Stark and J.M. Wilkinson). Marlow Bottom, Bucks: Chalcombe Publications, pp 47-48.

KENNEDY, S.J. (1988) "An absorbing experiment". In: *Silage Effluent, Problems and Solutions* (Ed B.A. Stark and J.M. Wilkinson). Marlow Bottom, Bucks: Chalcombe Publications, pp 52-53.

MASON, P.A. (1988) Dealing with silage effluent: technical opportunities. In: *Silage Effluent, Problems and Solutions* (Ed B.A. Stark and J.M. Wilkinson). Marlow Bottom, Bucks: Chalcombe Publications, pp 21-36.

OFFER, N.W. and AL-RWIDAH, M.N. (1988) Some consequences of ensiling grass with absorbent materials. In: *Silage Effluent, Problems and Solutions* (Ed B.A. Stark and J.M. Wilkinson). Marlow Bottom, Bucks: Chalcombe Publications, pp 50-51.

PEARCE, F. (1986) A green unpleasant land. *New Scientist*, **111**(1518): 26-27.

WEDDELL, J.R., MACKIE, C.K. and SUTHERLAND, R.M. (1988) Silage effluent—collection, storage and feeding to cattle. In: *Silage Effluent, Problems and Solutions* (Ed B.A. Stark and J.M. Wilkinson). Marlow Bottom, Bucks: Chalcombe Publications, p 54.

WILKINSON, J.M. (1987) Silage: trends and portents. *Journal of the Royal Agricultural Society of England*, **148**, 158-167.

ZIMMER, E. and WILKINS, R.J. (1984) Efficiency of silage systems: a comparison between unwilted and wilted silage (Eurowilt). *Landbauforschung Volkenrode. Sonderheft*, **69**.

Chapter 6

CURRENT RESEARCH ON SILAGE EFFLUENT

USING ASPHALT IN SILAGE PIT CONSTRUCTION

K Higgins and C A Johnston

Scottish Farm Buildings Investigation Unit,
Craibstone, Bucksburn, Aberdeen AB2 9TR and
West of Scotland Agricultural College,
Auchincruive, Ayr KA6 5HW

Due to its acidic nature, silage effluent corrodes concrete silage pits. The floor and walls of silage pits are often rendered inadequate as containers of effluent, and sometimes the concrete is structurally weakened to the point of being dangerous.

ASPHALT AS A FLOORING MATERIAL

The Scottish Farm Buildings Investigation Unit in Aberdeen and the West of Scotland Agricultural College have monitored asphalt as an alternative flooring material in 4 silage pits (2 self-feed). Asphalt has some distinct advantages over concrete:

* Asphalt floors can be laid in one day and, unlike concrete, do not require a 28-day cure period.
* Asphalt does not have shrinkage or expansion joints. Construction joints are easily and efficiently sealed using hot poured bitumen.
* Asphalt is acid-resistant and durable.
* Asphalt withstands the impact of buckrake tines.

Asphalt can be used either in the construction of new silo floors, or as a remedial measure in the form of 'topping' on an existing defective concrete base.

SPECIFICATION OF ASPHALT FLOORS FOR SILAGE PITS

For a new asphalt floor to be constructed, all black soil or other unsuitable material has to be removed from the solum. If necessary, infill must be brought-in. A granular sub-base of 200 mm is then prepared, which is overlaid with 60 mm of a 20 mm dense bitumen macadam basecourse. Finally 40 mm of hot rolled asphalt is laid to schedule 1B of B.S. 594, with a 40% stone content of 14 mm nominal size.

A concrete foundation is needed under the wall. This concrete is overlaid with 40 mm of hot rolled asphalt to protect it from effluent. A braced, treated timber wall is recommended as the treated timber does not corrode. The bracing avoids the use of steel stanchions for support and the problem of sealing stanchion sockets. The timber bracing is held in position by bolts fixed to the concrete base through the asphalt layer; these are sealed with hot poured bitumen.

Drainage channels to direct the effluent to the storage tank are difficult to form in asphalt, and an alternative design strategy is required. Instead of using drain channels to direct effluent flow, falls in the asphalt are used. 'Sleeping policemen' or speed-ramp ridges of asphalt can be used to retain the effluent within the confines of the pit/effluent tank system.

COSTS

The cost of an asphalt floor to the above specifications is different for each silage pit. It depends on the floor area and distance to the site from the source of asphalt. Typical prices in Scotland are:

Type of floor	Area (m^2)	Unit cost ($£/m^2$)	Total cost ($£$)
Topping	300	6.00	1800
	500	5.60	2800
New	300	10.40	3120
	500	9.35	4675

FURTHER READING

Silage pits—construction and repairs (1986) *Farm Building Progress*, **84**, 7-8.

Surface coatings for concrete silo walls (1987) *Farm Building Progress*, **87**, 19-22.

Silo floor repairs using asphalt (1986) Scottish Farm Buildings Investigation Unit publication.

EFFECT OF ABSORBENTS ON EFFLUENT PRODUCTION AND SILAGE QUALITY

R Jones and D I H Jones

AFRC Institute for Grassland and Animal Production,
Welsh Plant Breeding Station,
Plas Gogerddan, Aberystwyth, Dyfed SY23 3EP

A series of experiments have been conducted from 1985 onwards to determine the efficacy of dried absorbents on effluent production, silage fermentation and nutritive quality assessed by beef production trials.

In the first experiment, carried out in 1985, silages were made from an autumn cut of hybrid ryegrass (Table 1). The silages were made in 50-tonne clamp silos with (a) no-additive (control), (b) formic acid applied in the field at 5 litres/tonne grass and (c) rolled barley applied by hand in the silo at 50 kg/tonne grass.

The three silages were fed *ad libitum* to groups of 12 housed Friesian x Hereford beef steers (400 kg) for a period of 8 weeks. The barley silage was fed unsupplemented and the other silages were supplemented with 1.5 kg rolled barley/head/day—an equivalent amount of barley to that included in the barley-added silage.

The silage with added barley had a lower ammonia level, and higher dry matter (DM) and metabolisable energy (ME) contents than the control silage (Table 1). Effluent production was halved by the addition of barley (Table 2).

RESULTS

TABLE 1 **Chemical composition of grass and silage**

	Dry matter	Sugars	Digestibility of DM	Ammonia	Predicted ME
	(%)	(%)	(%)	(% total N)	(MJ/kg DM)
Grass	16	11	64	—	—
Control silage	16	—	—	11	10.1
Silage with formic acid	16	—	—	5	10.3
Silage with barley added	19	—	—	8	10.6

TABLE 2 **Silage effluent production**

Silage	Effluent production (litres/100 tonnes silage)
Control	5000
Formic acid added	6000
Barley added	2600

Live weight gains were similar for the silages with barley and formic acid (1.00 and 0.96 kg/head/day) but lower for the control silage (0.82 kg/head/day). Total DM intakes were 9.0, 8.6 and 8.8 kg DM/head/day for the added barley, formic acid treated and control silages respectively.

Calculation of feed costs indicated a cost per kg of liveweight gain of £0.60 for the barley silage compared to £0.66 and £0.69 respectively for the control and formic acid treated silages.

FURTHER WORK

In a second experiment, now in progress, dried sugar beet pulp pellets applied at the rate of 50 kg/tonne grass (18% DM) were used as an absorbent. This silage is being compared to a control treatment of grass with a formic acid additive (3 litres/tonne).

Effluent production over 4 weeks was halved with the sugar beet treatment (1500 litres/100 tonne) compared to the control (3100 litres/tonne). The difference in effluent production was, however, much reduced over a 16 week period.

Other experiments have indicated an advantage in the liveweight gain of beef cattle by inclusion of sugar beet pulp in the grass at ensiling compared to feeding an equivalent amount as a supplement to grass silage.

CONCLUSIONS

The experiments conducted to date clearly indicate a potential for reducing effluent production by the use of materials such as rolled barley or dried sugar beet pulp. It has been found, however, that the efficiency of absorption is greatly influenced by a variety of factors. These include the method of application and silo design, particularly in relation to the drainage system and physical characteristics of the crop. The studies reported showed total effluent production from experimental 100 tonne silos which were consider-ably lower than values reported in the literature. Further studies are needed to define more precisely the optimum conditions for absorption.

THE EFFECT OF ABSORBENT ADDITIVES ON SILAGE QUALITY AND ON EFFLUENT PRODUCTION

Diana L Done

MAFF/ADAS Liscombe Experimental Husbandry Farm,
Dulverton, Somerset TA22 9PZ

Concern over pollution of the environment by silage effluent is growing. Early experimental results suggest that incorporation of rolled barley with grass at ensiling could reduce effluent flow by up to fifty per cent. The cost of this technique is relatively high, £5 to £6 per tonne of grass treated, but may be justified if additional benefits can be identified. The two most obvious possibilities are an improvement in fermentation quality, thereby reducing or eliminating the need for a conventional silage additive, and/or an increase in animal performance over and above that which would be expected if silage and barley were fed as separate components of the ration.

PROCEDURE

In 1987, unwilted, first cut herbage (14% DM) was ensiled in 130 tonne bunker silos as follows:

(i)   Untreated (control);
(ii) With rolled barley incorporated at 40 kg/tonne fresh grass;
(iii) With molassed sugar beet shreds incorporated at 40 kg/tonne fresh grass.

No conventional silage additive was used on any of the treatments.

The resulting silages were evaluated through finishing lambs and 6 month old cattle fed either:

(i) The above silages *ad libitum* as the sole feed (lambs and cattle), or
(ii) Control silage *ad libitum* with rolled barley or sugar beet shreds fed separately (lambs only)

RESULTS

Preliminary results indicate a significant reduction in effluent production when barley or sugar beet shreds were incorporated. The incorporation of sugar beet shreds also improved fermentation quality, resulting in significantly higher daily liveweight gains in weaned, suckled calves. Similar responses were shown in lambs, although differences were not statistically significant.

SOME CONSEQUENCES OF ENSILING GRASS WITH ABSORBENT MATERIALS

N W Offer and M N Al-Rwidah

West of Scotland Agricultural College, Auchincruive, Ayr KA6 5HW

PROCEDURE

The relative absorbency of barley (whole or rolled), dried distillery by-products, straw (Viton cubes, chopped or baled), molassed sugar beet feed (shreds or nuts) and shredded paper were tested in 200 litre drum silos. Twelve silages were made in roofed 10 tonne pit silos. *In vivo* digestibility, *ad libitum* intake and liveweight gain measurements were made on the pit silages using Friesian calves of approximately 120 kg liveweight.

RESULTS

TABLE 1 **Silage composition and effluent losses from 10 tonne silos**

Treatment[1]		Volume (litres/ tonne)	Effluent				Silage	
			OM (%)	OM (kg/tonne silage)	DM (%)	NH₃-N (% CP)	ME (MJ/kg DM)	Density (m³/ tonne)
1985 silages[2]								
No absorbent		137	2.9	4.0	15.6	7.0	10.8	1.18
Viton cubes	L	114	4.1	4.7	20.1	8.1	9.9	1.32
Viton cubes	B	130	4.7	6.1	15.5	8.0	10.6	1.43
Chopped straw	L	53	2.5	1.3	16.4	6.1	9.8	1.90
Straw bales	B	146	2.5	3.6	16.8	7.9	10.4	2.20
Beet shreds	L	71	4.2	3.0	16.9	6.9	11.1	1.53
1986 silages[3]								
No absorbent		195	2.9	5.7	15.2	13.7	12.8	0.92
Straw bales	B	220	3.0	6.5	14.7	15.2	—	1.60
Chopped straw	L	113	2.9	3.3	19.4	12.1	9.8	1.59
Chopped straw	L	98	3.6	3.5	20.6	15.3	—	1.70
Beet shreds	L	79	5.5	4.3	20.6	10.7	12.2	1.13
Beet shreds	L	98	4.6	4.5	18.6	11.1	—	1.11

[1] L = absorbent in layers, B = absorbent at bottom

[2] Second cut perennial ryegrass (DM 14.8%) treated with 4.5 litres/tonne Add F; absorbents added at 60 kg/tonne of grass fresh weight.

[3] First cut perennial ryegrass (DM 12.2%) treated with 3.0 litres/tonne Add F; absorbents added at 75 kg/tonne of grass fresh weight.

TABLE 2 **Intake and weight gain of calves fed 1986 silages supplemented with 1 kg/day mineralised rolled barley plus 0.2 kg/day soya bean meal**

Treatment	Mean silage DM intake (kg/head/day)	Liveweight gain (kg/day)
No absorbent	2.27	0.97
Chopped straw	2.14	0.86
Beet shreds	2.62	1.11
SEM	0.05	0.04

CONCLUSIONS

Chopped straw, dried distillers grains and sugar beet shreds proved most absorbent. Viton straw cubes and straw bales may *increase* the loss of organic matter (OM) in effluent. Beet shreds *must* be used in sufficient quantity totally to prevent effluent loss, as the concentration of any effluent produced is increased.

Work with drum silos suggests:

Kg beet shreds/tonne grass = 419—19.1 DM (%).

All the silages were well preserved. Chopped straw inclusion reduced silage metabolisable energy (ME) value and animal performance, whilst sugar beet shreds improved intake and liveweight gain even though the effect on silage ME value was small. At 70 kg per tonne grass, chopped straw and sugar beet shreds reduced the tonnage of grass that could be stored in a given silo by approximately 45% and 20% respectively.

"AN ABSORBING EXPERIMENT"

S J Kennedy

Greenmount College of Agriculture and Horticulture
Antrim BT41 4PU, Northern Ireland

Several factors have contributed to the silage effluent problem that farmers in Northern Ireland face today, viz. ensiling grass without wilting, increasing fertiliser rates, cutting younger grass and using acid-type additives. Absorbent-type products based on sugar beet pulp have recently become available which, if successful, could help to reduce effluent loss. One of these products, an absorbent/feed source/silage additive 'Sweet 'n Dry', was examined in 1987 in a feeding trial involving mature beef cattle.

PROCEDURE

First harvest, direct-cut, double-chop grass was ensiled without an additive or with 30 kg/tonne Sweet 'n Dry (Thompsons Farm Feeds). Evaluation parameters included preservation and nutritive value of silage, in-silo loss (including silage effluent production), silage intake and animal performance.

Each of 4 silos (2 replicates x 2 treatments) was filled from 20 to 23 May 1987 with approximately 70 tonne of grass with a dry matter (DM) content of 190 g DM/kg, water soluble carbohydrate content of 150 g/kg DM, crude protein (CP) content of 180 to 190 g/kg DM and a D-value of 68 to 70. Effluent flow from the silages was monitored throughout a 144-day storage period, after which each silage was fed to 460 kg beef cattle (24 animals per silage), supplemented with 0, 1 and 2 kg/head/day of a concentrate containing 150 g CP/kg throughout a 75-day feeding period.

RESULTS

Preservation and nutritive value of silage

Both silages displayed good preservation characteristics. However, the absorbent-treated silage displayed a significantly higher dry matter content and significantly lower crude protein, ADF and NDF contents than the untreated silage. All other chemical analyses and in-vivo DOMD using sheep were similar for the two silages.

In-silo loss

Although there was a 20% reduction in the volume of effluent produced from the absorbent-treated silage compared with the untreated silage (121 litres/tonne compared with 152 litres/tonne), the effluent produced from the absorbent-treated silage had a slightly higher DM content. At the application rate used in this experiment, 1 kg of Sweet 'n Dry absorbed approximately 1 litre of effluent.

The absorbent-treated silage displayed 8.2% lower total DM losses than the untreated silage.

Intake and animal performance

The intake and animal performance data are summarized in Table 1. Cattle fed the absorbent-treated silage, both without supplementation and with 1 kg/head/day supplement, displayed significantly higher (17 to 20%) daily DM intakes than those fed the untreated silage with similar levels of supplementation. Cattle fed the absorbent-treated silage without supplementation displayed similar daily liveweight gains and similar daily carcass gains to those fed the untreated silage with 1 kg/head/day supplement.

The daily DM intakes, daily liveweight gains and daily carcass gains of cattle fed the silages with 2 kg/head/day supplement did not differ significantly.

TABLE 1 **Intake and animal performance data**

| Additive | Silage | | | | | | SEM | Signi-ficance |
	Untreated			Absorbent-treated				
Supplementation level (kg/head/day)	0	1	2	0	1	2		
Daily dry matter intake (kg/head/day)	6.12^a	6.21^a	6.40^{ab}	7.65^b	7.45^b	7.11^{ab}	0.229	***
Daily live-weight gain (kg/head/day)	0.22^a	0.56^b	0.81^b	0.59^b	0.74^b	0.81^b	0.071	***
Daily carcase gain (kg/head/day)	0.31^a	0.47^{ab}	0.67^{bc}	0.47^{ab}	0.61^{bc}	0.70^c	0.058	***

SILAGE EFFLUENT—COLLECTION, STORAGE AND FEEDING TO CATTLE

J R Weddell, C K Mackie and R M Sutherland

North of Scotland College of Agriculture,
581 King Street, Aberdeen AB9 1UD

Silage effluent is readily consumed by cattle. However, it is best utilised during the winter housing period, necessitating the storage of large quantities of very dilute feed (50 to 90 g dry matter/kg). Studies were undertaken to develop and evaluate a low cost system for the collection, storage and feeding of silage effluent.

Drainage channels within the silo directed all effluent to a small collection tank. An electric mono-pump, operating on a float switch, transferred the effluent to an above-ground tank consisting of a butyl liner within a metal frame and covered with a canvas lid. Formalin was added at 3 litres/1000 litres effluent to aid preservation. During the feeding period the effluent was transferred to a header tank in the cattle shed using a submersible electric pump before being gravity fed into a butyl drinking trough controlled by a ballcock.

RESULTS

The system proved simple to operate and no practical problems were experienced. In-calf Friesian heifers offered unlimited access to the effluent consumed up to 45 litres/day.

The critical evaluation is, however, a cost/benefit analysis. The calculations below are based on unit costs for 1000 tonnes grass ensiled after a minimum wilt.

At December 1987 prices, the feeding value per thousand litres of effluent is £6.33 (based on an effluent composition of 50 g DM/kg, metabolisable energy value of 12.5 MJ/kg DM and crude protein content of 260 g/kg DM). The annual cost (after grant) of the collection, storage and feeding system described above is £6.57 per thousand litres effluent, resulting in a net deficit of 24 pence per thousand litres. In addition, the effluent is worth (net, after costs of spreading are charged) 48 pence per thousand litres as a fertiliser, raising the total deficit to 72 pence. A similar calculation based on unit costs for 500 tonnes of ensiled grass indicates an even higher deficit.

CONCLUSIONS

Where existing effluent handling facilities are adequate, it is uneconomic to store effluent, under this system, for winter feeding. However, where effluent facilities are presently unsatisfactory, requiring renewal or expenditure, this system is an option worth consideration.

FEEDING SILAGE EFFLUENT TO DAIRY COWS

O D Davies and S F Clench

MAFF/ADAS Trawsgoed/Pwllpeiran Experimental Farm,
Trawsgoed, Aberystwyth, Dyfed SY23 4HT

An increasing number of farmers are being prosecuted annually under Part 2 of the Control of Pollution Act 1974, which came into effect from 31 January 1985, for allowing silage effluent to contaminate water courses. Cheap but safe methods of disposal are sought, and one possible option is to feed the effluent to dairy cows. In a preliminary investigation at Trawsgoed EHF, where dry autumn-calving cows were offered silage effluent whilst at grass, intakes proved disappointing at only 1.58 litres of effluent per cow per day over a four week period.

A further investigation was subsequently undertaken using lactating cows.

PROCEDURE

Effluent, obtained from ensiling first and second cuts of perennial ryegrass treated with a formalin/formic acid additive at 4 litres/tonne, was offered in a fresh state to 30 October-calving cows for 75 minutes after each milking.

RESULTS

Over a total period of 31 days, the cows consumed an average of 26.5 litres of silage effluent/head/day (peak 40.7 litres), plus a further 8.5 litres of water whilst at grass. This compared to a similar group of 30 cows not offered effluent which consumed only 8.6 litres of water/head/day. For cows yielding about 14 kg milk/day this water intake was low, although the weather during the trial period was dull and overcast, with only 4 days without rain being recorded.

There were no significant effects from feeding silage effluent on milk yield, milk quality, liveweight change or animal health.

CONCLUSIONS

Lactating dairy cows will consume substantial quantities of fresh silage effluent when it is offered after milking. No change in water consumption or animal performance was seen in this study, thus the dairy cow appears to provide a cheap option for the disposal of this potentially dangerous pollutant.

Chapter 7

RESEARCH AND ADVISORY SERVICES

SMALL SCALE SILAGE RESEARCH—HIGH VALUE RESULTS

R Crawshaw

MAFF/ADAS, Block C, Government Buildings,
Brooklands Avenue, Cambridge CB2 2DR

Silage quality can be assessed properly only by feeding trials. The benefits of silage additives can best be measured in comparative trials on research stations, where animal performance can be quantified under controlled conditions.

Animal research facilities, however, are comparatively rare and animal studies are expensive. Fortunately, many aspects of silage making can be studied in small scale experiments. Just how useful such studies can be is evident from an experimental programme designed to investigate the effects of the addition of sugar beet pulp at ensiling.

Effects to be measured:

* Silage fermentation
* Silage digestibility
* Effluent flow
* Aerobic stability of silage

Several different levels of 'additive' can be compared, with replication to permit statistically valid interpretations. Different crops can be included so that advice can be given on the effects of additives at different dry matter levels, different stages of growth and a range of densities.

Such a programme would be impossible to carry out on a large scale. Not only would it be prohibitively expensive, but it would require the full facilities of a silage research centre to be committed for several years.

The final proof must be a farm scale study with animal feeding, although an exception to this general rule is aerobic decay. This important and unpredictable problem cannot be studied on a large scale. Thus the standard laboratory test developed by ADAS has been of significant help in identifying unstable silages, and in assessing the ability of various additives to prevent decay.

Research funds must be used to best effect. Small scale research enables the many options to be narrowed down to the most promising ones. ADAS will ensure you get value for money.

ADAS: SILAGE R & D FACILITIES

M Appleton

MAFF/ADAS Liscombe Experimental Husbandry Farm,
Dulverton, Somerset TA22 9PZ

Each year 17,500 tonnes of grass are ensiled on 8 of the ADAS Experimental Husbandry Farms (EHF). Most of the grass goes into the 60 clamps available, but both round and rectangular bales are also made. The clamps vary in size from 40 to 600 tonnes capacity, and there are 2 to 5 identical silos at each EHF. These can be filled simultaneously and are therefore suitable for comparative silage experiments. Big bales are stored individually (bagged or wrapped) or in packs. At 4 EHFs—Liscombe, Drayton, Rosemaund and Trawsgoed—effluent collection, monitoring and feeding is possible on an individual silo basis. Effluent production is measured either by tippler tanks or by total collection.

The sequence is completed by individual or group feeding facilities for dairy and beef cows, store and finishing beef cattle, store lambs and breeding ewes. Evaluation of complete or partial silage systems from field to feeding is possible, covering all aspects of silage fermentation, in-silo losses, effluent production, silage feeding value, silage intake and animal performance. Detailed investigations can be carried out by ADAS specialists—Nutrition Chemists, Analysts and Microbiologists.

Facilities for small scale work using mini-silos holding 4 kg of grass are available at the ADAS Slough Laboratory. Patterns of silage fermentation in both chemical and microbiological terms can be followed, together with effluent production.

ADAS specialists in Farm Buildings, Farm Waste, Mechanisation and Feed Evaluation may also be involved in the evaluation of silage systems.

These facilities are available for confidential contract investigations.

WATER AUTHORITIES ASSOCIATION

The water industry is managed by the water authorities whose areas are based on river basins; also 28 statutory water companies supply water. The water authorities are responsible to Parliament through the Secretaries of State for the Environment and for Wales. The Council of the Water Authorities Association is made up of the chairmen of the water authorities of England and Wales who are appointed by the Secretaries of State for the Environment and for Wales.

The ten water authorities are responsible for the development of water resources, water supply and distribution, river management and the control of pollution, sewage treatment and sewerage, land drainage, fisheries and water-based recreation, flood prevention and sea defences.

Water authorities recognise the need to protect the rivers against pollution of all kinds. In 1986/87 a total of 20,414 pollution incidents, including industrial and farm waste, were reported. In 1986 silage effluent was the cause of 592 pollution incidents, compared with 1,006 in 1985; this reduction occurred despite the increase in the total amount of silage produced. Although this improvement is welcomed, it is apparent that there are still inadequate sized effluent tanks and leaking stores or drains which need attention.

Silage liquor remains the most polluting waste produced on a farm; it is also extremely corrosive. Leakage through walls, floors or poor joints can go unnoticed and become a major cause of pollution particularly if the silage is not wilted—therefore silos need to be checked annually, repairs made and joints sealed before silage is made—"a stitch in time" increases the life of a silo. Liquor discharges cause the death of thousands of fish and render streams unfit for stock watering. Silage pollution can also affect public water supplies and those used by farmers for irrigation.

Farmers are asked to check their drainage systems regularly; advice and guidance on how to avoid pollution are available from water authority staff as well as from ADAS. Pollution of water courses by silage effluent can carry heavy fines for the polluter. In future these may be increased; recovery of costs for remedial action and replacement of fish stocks can add thousands of pounds to the fine.

Regional water authority contacts are as follows:

A Ferguson, Esq
Anglian Water
Ambury Road
Huntingdon
PE18 6NZ

0480 56181

G H Bielby, Esq
South West Water
Peninsula House
Rydon Lane
Exeter EX2 7HR

0392 219666

M Colley, Esq
Principal Pollution
  Control Officer
Northumbrian Water
Regent Centre
Gosforth
Newcastle upon Tyne
NE3 3PX
091 284315

C Davies, Esq
Thames Water
Nugent House
Vastern Road
Reading
RG1 8DB

0734 593333

C Clark, Esq
North West Water
New Town House
Buttermarket Street
Warrington
WA1 2QG

092 572 4321

R J Merriman, Esq
Welsh Water
Hawthorn Rise
Haverfordwest
Dyfed
SA61 2BH

0874 3181

J Wild, Esq
Severn Trent Water
Abelson House
2297 Coventry Road
Sheldon
Birmingham
B26 3PU

021 743 4222

Mrs Clare Dugdale
Wessex Water
Wessex House
Passage Street
Bristol
Avon BS2 0JQ

0272 290611

H Headworth, Esq
Southern Water
Guildbourne House
Worthing
West Sussex
BN11 1LD

0903 205252

L Beck, Esq
Yorkshire Water
West Riding House
67 Albion Street
Leeds
LS1 5AA

0532 448201

TRIDENT FEEDS
Molassed Sugar Beet Feed.

KEEPS THE GOODNESS IN YOUR SILAGE IN YOUR SILAGE.

Chapter 8

PRODUCTS FOR CONTROLLING SILAGE EFFLUENT

Name of product	MOLASSED SUGAR BEET FEED
Manufacturer	Trident Feeds
Description	Dried, molassed sugar beet shreds, nuts or pellets—all are equally effective absorbents.
Rate of inclusion with silage	Dependent upon silage dry matter (DM). Eg 15 to 85 kg/tonne for grass 25% to 15% DM.
Storage	Cool, dry conditions.
Manufacturer's claims for product	Reduces effluent loss from clamp—effluent is a potential pollutant. Absorbs 3 to 4 times its own weight of effluent. Silage/sugar beet mixture has improved nutritional value due to high energy and protein content of retained effluent, and high energy and digestible fibre content of sugar beet feed. Sugar content of molasses aids fermentation in clamp—reducing pH and improving palatability of silage. Sugar beet feed contains good quality protein to complement soluble protein of silage. For a 22% DM crop, added benefit of improved nutritional value of ensiled feed over cost of sugar beet inclusion is approximately 65 pence/tonne grass.
Cost (bulk) **per tonne product** **per tonne grass**	Approximately £100/tonne delivered £1.50 to £8.50 for grass of 25% to 15% DM
Further information	Dr G.D. Macleod Trident Feeds British Sugar plc PO Box 11 Oundle Road Peterborough PE2 9QX 0733 63171

Name of product   **MULTI-BACTERIAL INOCULANT (MBI)**

Manufacturer/agent   Multigerm (UK) Ltd

First launched in UK   1987

Ingredients (per gram)   4 strains of lactic $\left.\right\}$ 1×10^9
acid bacteria   viable organisms
on germinating
barley

Manufacturer's recommended application rate (grams/tonne)   3000 to 5000

Active ingredients applied (per gram crop) 3×10^6 to 5×10^6 viable organisms (at time of application)

Storage   Shelf-life 6 months in anaerobic conditions (i.e. sealed bag). Store in cool place

Manufacturer's claims for product   Lower effluent production than most silage additives. Ensures necessary lactic bacteria for good fermentation are both present and establish dominance within clamp. Initial nutrients for bacteria are supplied as germinating barley to allow successful ensilage of low sugar, low dry matter crops.

Cost (bulk)
per kg   £0.35
per tonne crop ensiled £1.05 to £1.75

Applicator   Multigerm applicator

Further information   L.O.A. Burt
Multigerm (UK) Ltd
Sandy Farm
The Sands
Farnham
Surrey GU10 1PX

02518 2654

Name of product   SILASWEET

Manufacturer   Pauls Agriculture Ltd

First launched in UK   1987

Description   A mixture of digestible feed ingredients, selected to achieve a high level of absorbency, and compressed into a high density pellet. Nutritional specifications similar to conventional dairy compound. Precise details available from local Pauls mills.

Rate of inclusion with silage   Dependent upon silage dry matter (DM). Recommended rates are from 40 kg per tonne for 15% DM grass to 15 kg per tonne for 24% DM grass.

Storage   Must be kept dry prior to use.

Manufacturer's claims for product   Retains grass juices in the silage and thereby reduces the nutrient losses of silage making. Controls effluent production. Can absorb up to 6 times its own weight of liquid; 3 kg of Silasweet can absorb 18 litres of grass juice, which has a feeding value equivalent to an extra 1.25 kg of compound feed. Since the product is a feedstuff itself, its value remains in the silage for subsequent feeding.

Application   No specialised equipment necessary. Silasweet should be mixed with grass at the time of ensilage to disperse the pellets throughout the clamp.

Cost (bulk)
per tonne of product   Approximately £130/tonne delivered
per tonne of grass   £1.95 to £5.20

Further information   M. Reece
                            Pauls Agriculture Ltd
                            PO Box 39
                            Key Street
                            Ipswich
                            Suffolk IP4 1BX

                            0473 232222

Name of product	SILO GUARD II
Manufacturer/agent	Holmen Feed Technology
First launched in UK	1986
Ingredients	α—1,4 glucan, 4 glucanohydrolase enzyme, α—1,4 glucan maltohydrolase enzyme sodium sulphate, potassium sulphate, sodium sulphite, dextrose
Manufacturer's recommended appication rate (grams/tonne)	Grass: 500 under normal conditions to 1000 for very high or low DM crops Maize: 250
Active ingredients applied (per gram of crop)	Total: 0.24 to 0.95 kg /tonne Enzymes: 825 to 3300 SKB units α—amylase activity/tonne
Storage	Shelf-life 2 years. Store in dry conditions, off floor
Manufacturer's claims for product	Reduces and can eliminate effluent. More efficient fermentation to conserve nutrients—on average, nearly 6 tonnes more DM recovered per 100 tonne crop DM ensiled compared to untreated silage. Enzyme-based fermentation enhancer to accelerate conversion of crop carbohydrates to sugars for naturally occurring lactic acid producing bacteria. Reduces fermentation time, temperature and heat damage. Improves palatability, intake and aerobic stability at feedout. Non-toxic.

Cost (bulk)
per kg   £2.80
per tonne crop ensiled £0.70 to £2.80

Applicator	Forage-harvester—any applicator for free-flowing powder, or as suspension using liquid applicator. Alternatively may be applied by hand at clamp. Dispensers supplied for hand application.
Further information	R J Steward Holmen Feed Technology PO Box 2 Basing View Basingstoke Hants RG21 2EB 0256 53442 Telex: 858371

Name of product	SWEET'N DRY
Manufacturer/agent	B. Dugdale & Son Ltd/ Kenneth Wilson Agriculture Ltd
First launched in UK	1986
Description	Free-flowing, dust-free, blowable mini-pellet. Rich in soluble sugars, proteins, organic acids and minerals. Crude protein content 9%. Energy content 12.7 MJ ME/kg DM.
Rate of inclusion with silage	Dependent upon silage dry matter (DM). Eg 15 to 45 kg/tonne for grass of 26% to 16% DM.
Storage	Cool, dry conditions
Manufacturer's claims for product	Retains and utilises silage effluent. Absorbs up to 6 times its own weight of water. Optimises fermentation conditions. Increases feed value of silage. Compensates for wilting in poor weather. Easily and safely incorporated. No further costs.
Application	Apply directly throughout the clamp, mixing as well as possible with silage. Various application methods including by hand, with fertiliser spreader or with buckrake.
Cost (bulk) per tonne product per tonne grass	Approximately £155/tonne delivered £2.33 to £6.98 for grass of 26% to 16% DM.
Further information	*Manufacturer* A. Sayle B. Dugdale & Son Ltd Bellman Mill, Salthill Clitheroe Lancs BB7 1QW 0200 27211
	Distributor T. Wilson Kenneth Wilson Agriculture Ltd Morwick Hall, York Road Leeds LS15 4NB 0532 737373 Telex: 556341

Name of products	**SUPER SILE PLUS (GRANULAR)** **SUPER SILE PLUS (PREMIX)**
Manufacturer/agent	Biotal Ltd/Axis Agricultural Ltd
First launched in UK	1984

Ingredients (per gram)

Granular:
Pediococcus pentosaceus $\left.\right\}$ 2.2×10^8
Lactobacillus plantarum $\left.\right\}$ viable organisms
Enzymes: amylases, hemicellulases, cellulases
Nutrients: to stimulate microbial growth
Carrier: maize prill
Premix:
  Sachet A (100 g)
  Pediococcus pentosaceus $\left.\right\}$ 2.5×10^{10}
  Lactobacillus plantarum $\left.\right\}$ viable organisms
Sachet B (400 g) Enzymes and nutrients as for granular product

Manufacturer's recommended application rate (grams/tonne)

Granular: 454
GPrefix: 20

Application procedure

Granular: applied as solid
Premix: solid applied as liquid suspension. Dissolve sachets A and B in 25 or 50 litres of water. Apply mixture at 1 or 2 litres/ tonne respectively.

Active ingredients applied (per gram of crop)

1×10^5 viable organisms

Storage

Shelf-life 18 months below 15°C. Store in original sealed packaging in cool, dry place. Reseal plastic liner tightly if contents not fully used (granular). Do not freeze (premix).

Manufacturer's claims for product

Reduces effluent flow. Controlled, replicated trials on 17% dry matter silage showed that Super Sile Plus- treated material produced 42% less effluent than acid-treated silage and 20% less effluent than silage treated with no additive.

Cost (bulk)
    per kg   **Granular**: £4
                       Premix: £90

    per tonne crop   £1.80
    ensiled

Applicator   **Granular**: any granular applicator
                       Premix: any liquid applicator

Further information   Axis Agricultural Ltd
                       36 High Street
                       Eccleshall
                       Staffordshire ST21 6BZ
                       0785 850941

Name of product	VITON-NIS
Manufacturer	Unitrition International Ltd
First launched in UK	1984
Description	Alkali-treated straw pellets, 8mm diameter.
Rate of inclusion	20 to 50 kg/tonne grass ensiled.
Storage	Cool, dry conditions
Manufacturer's claims for products	Nutritionally-improved straw. Retains nutrients in effluent. Absorbs 2 to 3 times its own weight of effluent in practical silo situations. Reduces need to wilt. Extends quality of available feed. Product is itself a valuable source of energy and digestible fibre.
Application	No specialised equipment required. Viton is tipped and spread evenly on clamp base as required, to depth of approximately 6 inches, and spread progressively as filling proceeds.
Cost (bulk)	
per tonne product	Approximately £68/tonne delivered
per tonne grass	£1.36 to £3.40/tonne
Further information	G. Jardine Unitrition International Ltd Basing View Basingstoke Hants RG21 2EQ 0256 841205 or Local Agricultural Merchant

PLYGENE ®

Tel:- 0298 812371 (4 lines)
(Write or telephone for more information)

Manufactured only by:-
H.D.SHARMAN LIMITED
CHAPEL-EN-LE-FRITH,
STOCKPORT CHESHIRE
SK12 6HW

PLYGENE® SILAGE PIT LINER is manufactured in our factory at Chapel-en-le-Frith. It is highly cost effective and carries a five year fair wear guarantee. It is a strong and durable material, acid and alkali resistant which not only protects the walls from the corrosive and damaging effects of silage effluent but which also helps stop wall side waste. When filling pits lined with Plygene® it is advisable to keep the wall sides well filled up because the shiny surface of the liner enables better wall side consolidation and compaction of the silage when rolling.

Plygene® Silage Pit Liner stops effluent seeping through the walls, even so, at the base of the wall, effluent can and often does seep out. (see fig.1.) Should this be a problem a simple drain channel in this area should be installed. (see fig.2.)

(fig.1.)

Plygene® Silage Pit Liner effectively seals pit walls. ➝

Effluent runs off concrete bases, a dpc membrane can be added as an extra safeguard

Effluent often seeps out into field drains below wall at the floor to wall junction in this area. Fitting a Rubberised Plygene Drain Channel is a relatively simple job.

(Not to scale)

Pre-creased Rubberised Plygene drain lining is manufactured in up to 20" widths and to any required length in roll form. Centrally positioned creases enable it to be snugly fitted into a small preformed channel. The wall side edge must be fitted at least 6" higher than the floor level and under the bottom edge of the wall liner. The opposite edge should be embedded in the floor concrete preferably just under the dpc membrane if used.

(fig.2.)

Plygene Silage Pit Liner
*Drill drain holes here.
Rubberised Plygene lined effluent drain channel.

dpc

To complete the job, the lower edge of the Plygene silage pit wall liner overlap should be drilled* to allow the outflow of liquid and prevent infill by silage. To fit drains in to existing pits it is necessary to break up the concrete floor at the wall side and embed a timber former into the new concrete creating a drainage fall in the process. To form a concrete floor edge to the drain a removable timber insert can be placed in the lined channel to act as a tamping edge. Obviously it is easier to fit drain channels into new pits or into existing pits where the floor is to be re-surfaced.

(We deliver direct to your farm)

QUOTATIONS AND SAMPLES OF PLYGENE® SILAGE PIT LINER & RUBBERISED PLYGENE® CREASED DRAIN CHANNEL LINER ARE AVAILABLE ON REQUEST:-

H.D.SHARMAN LTD. CHAPEL-EN-LE-FRITH. STOCKPORT CHESHIRE. SK12 6HW.

Name of product BALEFLEX

Manufacturer/agent DRG Plastic Films

First launched in UK 1988

Description Advanced cast extrusion process gives Baleflex superb elasticity, or 'memory', compared to many old-style films. The film can be stretched more tightly around the bale, giving improved cost-effectiveness. Less air is entrapped, therefore fermentation is better and quicker. The film continues to contract as the silage compacts, keeping the bale tight throughout fermentation. During wrapping, each layer is bonded securely to the one below by a natural cling built into its inner surface. This is part of the film's molecular structure and not a sticky adhesive. This adhesive is unaffected by weather and temperatures. Black, 24 micron (1 thou), 500 mm wide and 1800 metre per roll.

Cost Approximately £1.15 to £1.25 per average size bale.

Further information Ms L. Kalling
DRG Plastic Films
Carsons Road
Mangotsfield
Bristol BS17 3LN
0272 562525

Name of products	**BRITISH VISQUEEN BIG BALE SILAGE BAGS**
	BRITISH VISQUEEN SILAGE SHEETS
	BRITISH VISQUEEN POLISILE 3

Manufacturer/agent ICI Visqueen

First launched **Bags**: 1978
Sheets: 1960
Polisile 3: 1987

Description and **Bags**: Black polythene gussetted bags;
specification county green available in STD 4x4 ft size.

Bale size (ft)	Thickness (microns)	Bag width (mm)	Bag length (mm)
STD 4x4	110	2135	2800
Extra 4x4	110	2235	2800
STD 4x5	110	2135	3100
Extra 5x4	110	2235	3100
Green 4x4	110	2135	2800

All bags ex-stock in packs of 10 for easy handling. Rolls to order.

Sheets: Black, high grade polythene
Stock sizes (metres)
Natural (75µ, 300 gauge) 8x72
Black (75µ, 300 gauge) 8x72, 11x50
STD (125µ, 500 gauge) 2x100, 4x50,
   8x25, 8×50, 9.2x50, 11x25, 11x42,
   12.8x20, 12.8x25, 12.8x36, 14x30,
   16x30, 17.4x25, 18.3x25.
Black heavy duty (250µ, 1000 gauge) 8x28

Polisile 3: New concept in silage sheeting. Three inseparable layers of high quality raw material to give added strength, extra toughness, and increased resistance to punctures. One side is black, the other is white to reflect sun and prevent excessive heat build-up in the silo. 130µ, 520 gauge.
Stock sizes (metres)
8x25, 11x25, 11x42, 12.8x25, 14x25.

Cost Prices available on application

Further information A J Langford
British Visqueen Ltd
Yarm Road
Stockton-on-Tees
Cleveland TS18 3RD
0642 672288

Name of product	SILAWRAP
Manufacturer/agent	Volac Ltd
First launched in UK	1986

Description Two models.
Silawrap 90120 for maximum performance. Trailed with lifting arm. Bales can be picked up and placed on turntable for immediate wrapping in field. Possible to transport 2 bales over short distances—one on turntable, other on arm. After wrapping, bales are tipped hydraulically from machine.
Silawrap 90122 is lighter and mounted on tractor 3-point linkage. Turntable has tip cylinder similar to trailed model. Separate spool valves for operating turn and tip.
Pre-stretch unit, fitted as standard, stretches film 55% before wrapping on bale. Gives even, secure wrapping. Special support rollers are designed to keep badly shaped bales on turntable. Although both models are designed to wrap most bale sizes, best results are obtained with 4 ft x 4 ft (1.20m x 1.20m) bales. Lifting arm on Model 90120 can be adjusted to fit different bale sizes; will lift 700 kg bales and, with extra counterweights, 900 kg bales. Model 90120 is equipped with drawbar connected to tractor's link arms. Both models require only one oil outlet and one return. Spool valves on both models offer safe operation of machine.

Cost **Model 90120**: £5800 + VAT
Model 90122: £3625 + VAT

Further information Silawrap Department
Volac Ltd
Orwell
Royston
Herts SG8 5QX
0223 208021
Telex: 81622

Name of product   THE SILOSHIELD SYSTEM

Manufacturer   Surface Maintenance Systems Ltd—Burnley

First launched in UK   1983

Description   Range of products for structural repairs, sealing, surfacing and surface coating.

Products for sealing and surface coating

Supergrade RS: effective heavy-duty wall and structural coating for silage pits and similar structures. Bonds permanently to most surfaces. Applied by brush in 1 coat, with no priming on sound surfaces.

Supaflex SR: seamless rubber lining for 'tanking-out' structures for effluent and silage. Offers high resilience and excellent flexibility for maximum life. Supplied in liquid form for easy application.

Trowelplus: simple, economical joint-filling and sealing compound. Requires no mixing, priming or equipment. Used extensively on silage pit walls and floor joints, and to repair cracks. Used in conjunction with Supergrade RS.

Hydroprime: primes, seals and increases adhesion on problem surfaces, prior to applying other Siloshield compounds. Provides economical, acid- resistant finish to new or sound silage pit floors when applied as 2-coat treatment (can be completed in 1 day).

Tufprufe: tough 2-pack epoxy pitch coating for steelwork or equipment in contact with highly corrosive farm effluents. Totally impervious, and resistant to chemicals and abrasion. Applied by brush. No priming and minimal preparation.

HS Scrim: Rot-free flexible woven mesh, cut to size with scissors. Used with Supergrade RS for large gaps, where excessive movements occur or to reinforce joints where filler has been used.

Products for structural repair and surfacing

Supafoam: filling, fixing, sealing foam in large piston-type aerosol. Used for filling awkward joints prior to sealing or coating.

Surfajoint-2: advanced 2-pack epoxy urethane floor joint sealer, particularly suitable for expansion joints and large construction joints in silage clamps, slurry pits etc. Effluent-resistant. Quick-setting without shrinkage. Applied by trowel or may be poured.

Blockprime: alternative to cement render on porous building blocks. Fast drying. Fills, primes, seals and skims in one application. Suitable for pits, tanks, channels etc. Applied with brush.

Creetex rapid: quick setting, resin-based cementitious compound for tough repairs to localised damage, spalling, joint erosion and machine damage on silage pits etc. Applied by trowel.

Creetex: polymeric grano resurfacing compound for 12 mm screed to eroded floors. Forms effluent-proof new floor over old. More resilient than concrete. Applied by trowel/float/tamp in a quarter the time of concrete.

Contiscreed: very strong epoxy resin mixed with special aggregates to form surface to withstand heavy abrasion and chemical attack by effluents. Most cost-effective with new and sound surfaces to increase their life span. Applied by trowel.

Approximate costs

Supergrade RS £2/m^2 Supafoam £15/unit Supaflex SR £4/m^2 Surfajoint-2 £4/m run Trowelplus £1/m run Blockprime £2.50/m^2 Hydroprime £0.50/m^2 Creetex rapid £11.25/10 kg pack Tufprufe £2/m^2 Creetex £6/m^2 HS Scrim £0.50/m run Contiscreed £18/m^2

Further information

R.J. Taylor
Surface Maintenance Systems Ltd
Grisedale House
PO Box 42
Burnley
Lancs BB11 5SU
0282 33280

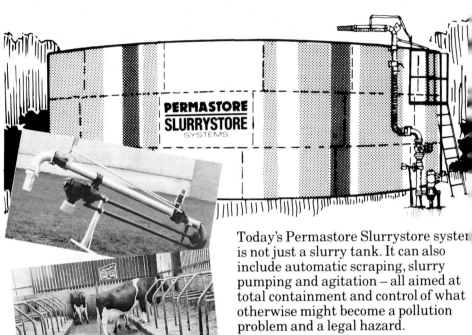

Name of product   SYLOGUARD SYSTEMS

Manufacturer/agent   Tremco Ltd/Livestock Systems UK Ltd

First launched in UK 1987

Description   New brush-on sealer to protect silo walls and
                          floors against silage effluent. Highly resistant to
                          aggressive acids. Acts within minutes of applica-
                          tion, to coat and impregnate concrete surfaces.
                          Effluent attack on new concrete is minimised.
                          Further deterioration of existing damage is
                          arrested. If properly applied should be effective
                          for at least 5 years.

Approximate cost   Price on application

Further information   M. Snow
                          Livestock Systems UK Ltd
                          Rose Hill
                          Chulmleigh
                          Devon EX18 7DD
                          0769 80600

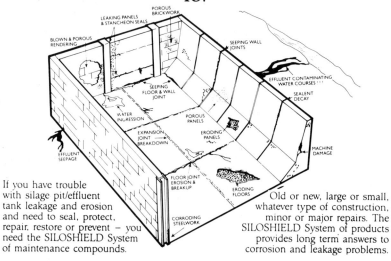

Name of product	PERMASTORE TOTAL CONTAINMENT TANKS
Manufacturer/agent	Permastore Ltd
Description	Above ground storage tanks made from glass fused to steel panels, bolted and mastic sealed, making them completely water-tight. Resistant to weather corrosion outside and chemical corrosion inside. All tanks built to BS449, and, where appropriate, CP3 Chapter V, part 2. Sizes range from 10,683 litres (2,350 gallons) to 2,750,330 litres (605,000 gallons). Range of extendable and non-extendable models available. Tanks can be dismantled and moved to another site, giving them a resale value. Above ground tanks, with Permastore long life glass coating, keep a new look for years, making them less unsightly than other similar structures. They are safer both to children and to animals than below ground tanks or lagoons. Tanks are supplied through a nation-wide dealer network.
Approximate cost	Prices on application
Further information	Permastore Ltd Eye Suffolk 0379 870723 Telex: 975228

Name of products	SILAGE EFFLUENT TANK DRAINAGE UNITS GRC SAMPLING & GAUGING CHAMBER

Manufacturer BCM (Contracts) Ltd

First launched in UK Effluent tank 1983 Drainage units 1981
GRC sampling chamber 1986

Description **Silage effluent tank**
Two sizes: 3150 litres (700 gallons) and 4500 litres (1000 gallons). Weights 400 kg and 500 kg. Constructed in Glass Reinforced Cement (GRC). Weather-proof, shatter-proof and frost resistant when installed below ground. Does not require concrete back-fill in a fenced area. Ready to take reinforced concrete in a traffic area. Fitted with lifting handles. Interior coated specifically to withstand silage acids. Supplied with 6 inch inlet pipe and an access man-hole cover.

Drainage units—headwalls, inspection chamber, inlets
Made from GRC. Light to man-handle. Shatter-proof, enabling installer to create exact pipe position on site. Large and small headwalls for pipe outfalls in banks; extension splash plates for deeper banks. Chambers at underground junctions for inspection and cleaning. Inlets, including galvanised grids, prevent bank erosion and blockages due to stream rubbish.

GRC sampling and gauging chamber
Chamber used by Severn Trent Water Authority. GRC has several advantages over alternative materials. It is a rigid cementatious material, enabling holes and waterproof joints to be made *in situ*, using only hammer, chisel and trowel. Its rigidity reduces installation costs as back-fill and lifting equipment are not needed. Rigid baffles can be cut to reduce effluent flow. Rigid shape of chamber allows accurate gauging. Easy sampling from front of baffle.

Approximate costs **Effluent tank** £500 to £600 delivered
Drainage units £10 to £149
GRC sampling chamber £210

Further information Mrs S Carden/D K Clark
BCM (Contracts) Ltd
Highgate Centre, Liverpool Road,
Whitchurch, Shropshire SY13 1SW
0948 5321

Name of product   **SPEL SILAGE EFFLUENT TANKS**

Manufacturer/agent   Spel Products

First launched in UK   1977

Description   Specifically designed underground tanks for silage effluent. Acid-resistant lining. Completely sealed, strong, lightweight unit. Simple and quick to install with unskilled labour. Detailed advice on siting and instructions given for installation. Capacities of tanks range from 2750 to 50,000 litres. Capacity should be based on volume of effluent produced in one day (at the minimum) under adverse weather conditions. Eg. 1000 tonne silo requires tank of 25 m^3 (25,000 litres) to contain peak daily effluent production after ensilage of 20% dry matter grass. With wetter grass, a 40 m^3 (40,000 litre) tank would be needed.

Approximate cost   Sample prices 2700 litres (600 gallons) £298
                        9000 litres (2000 gallons) £860

Further information   Spel Products
Lancaster Road
Shrewsbury SY1 3NQ

0743 235200

Error

Glitch, Noise, and Jam in
New Media Cultures

Error

Glitch, Noise, and Jam in
New Media Cultures

Edited by
Mark Nunes

B L O O M S B U R Y

NEW YORK • LONDON • NEW DELHI • SYDNEY

Bloomsbury Academic
An imprint of Bloomsbury Publishing Plc

175 Fifth Avenue   50 Bedford Square
New York   London
NY 10010   WC1B 3DP
USA   UK

www.bloomsbury.com

First published by Continuum International Publishing Group 2011
Paperback edition first published 2012

Library of Congress Cataloging-in-Publication Data
Error : glitch, noise, and jam in new media cultures / edited by Mark Nunes.
p. cm.
Includes bibliographical references.
ISBN-13: 978-1-4411-2120-2 (hbk. : alk. paper)
ISBN-10: 1-4411-2120-X (hbk. : alk. paper)
1. Information technology–Social aspects. 2. Errors–Social aspects.
I. Nunes, Mark, 1965–II. Title.
HM851.E76 2010
303.48'33–dc22
2010011328

ISBN: HB: 978-1-4411-2120-2
PB: 978-1-4411-1021-3

Typeset by Newgen Imaging Systems Pvt Ltd, Chennai, India
Printed and bound in the United States of America

Contents

List of Figures

List of Contributors

Susan Ballard is a writer and curator whose research focuses on media art histories with a particular emphasis on sound and utopia in contemporary digital and time-based installation art. Su is the Principal Lecturer in Electronic Arts, Dunedin School of Art at Otago Polytechnic, New Zealand. She is a trustee of the Aoteoroa Digital Arts Network (ADA) and co-editor of *The Aotearoa Digital Arts Reader* (Clouds, 2008).

Tim Barker is a Research Fellow at the iCinema Research Centre, College of Fine Arts, University of New South Wales. His research is focused upon an investigation of the philosophy of time and process in relation to contemporary art and new media, as well as the development of an aesthetic philosophy that accounts for the ongoing creative role of an artwork's technological architecture.

xtine burrough is a media artist and educator, and the co-author of *Digital Foundations: Intro to Media Design with the Adobe Creative Suite* (New Riders/AIGA, 2009). She is informed by the history of conceptual art and practices in the era of social-networking. Using tools common to consumer web practices, such as databases, search engines, blogs, and applications—sometimes in combination with popular sites like Facebook, YouTube, or Mechanical Turk, she creates web projects and communities that foster interpretation and autonomy. xtine believes that art can shape social experiences by mediating consumer culture and envisioning imaginary practices. As an educator, she is interested in the art of instruction. She teaches at California State University, Fullerton.

Michael Dieter is completing a PhD in the School of Culture and Communication at the University of Melbourne. His research is focused on new media art, software design, and political philosophy.

Ted Gournelos is an Assistant Professor of Critical Media and Cultural Studies at Rollins College. His publications range from applications of trauma theory in contemporary U.S. film to interrogations of irony and politics in popular culture. He is the author of *Popular Culture and the Future of Politics* (Lexington Books, 2009), and has two edited collections forthcoming: the first on humor's role in post-9/11 politics, and the second on transgression in digital media.

Benjamin Mako Hill is a PhD candidate in a joint program between the Sloan School of Management and the Media Lab at MIT, and a Fellow at the MIT Center for Future Civic Media. Hill has worked as a consultant, programmer, and advocate for free and open source software. He is a contributor to the Debian GNU/Linux project, a founding member of the Ubuntu project, and a director of the Free Software Foundation.

Peter Krapp is Associate Professor of Film & Media Studies, English, and Informatics at the University of California, Irvine, where he also served as Director of the PhD. Program in Visual Studies, and as a Program Faculty Member in Art-Computing-Engineering. His research interests include digital culture, media history, and cultural memory. In 2007, he served as Visiting Professor in the International Human Rights Exchange (University of the Witwatersrand, Johannesburg). He is the author of *Déjà Vu: Aberrations of Cultural Memory* (Minnesota, 2004) and editor of *Medium Cool*, a special issue of *South Atlantic Quarterly* (Duke, 2002).

Elizabeth Losh is currently the Writing Director of the Humanities Core Course at University of California, Irvine, where she also teaches digital rhetoric courses. She is author of *Virtualpolitik: An Electronic History of Government Media-Making in a Time of War, Scandal, Disaster, Miscommunication, and Mistakes* (MIT Press, 2009). Her daily online column by the same name won the John Lovas Award for best academic weblog in 2007. Her current book project, *Early Adopters: The Instructional Technology Movement and the Myth of the Digital Generation*, looks at a range of digital projects in higher education and the conflicts between regulation and content-creation that universities must negotiate.

Stuart Moulthrop is Professor of English at the University of Wisconsin-Milwaukee. He is an award-winning digital artist and writer, author of *Victory Garden*, "Reagan Library," and other important works of electronic literature and interactive multimedia. His essays on digital culture have been widely republished and translated, and have appeared, among other places, in the *Norton Anthology of Theory and Criticism*.

Mark Nunes is Associate Professor of English and Media Studies, and Chair for the Department of English, Technical Communication, and Media Arts at Southern Polytechnic State University. He is author of *Cyberspaces of Everyday Life* (Minnesota, 2006). He has written extensively on the role and place of new media technologies in a network society.

Chad Parkhill is working on a doctorate at the University of Queensland, where he also teaches. His dissertation is an interdisciplinary historical analysis of constructions of heterosexuality from the late nineteenth century to the present. His published academic works include a queer reading of Stephen King, an analysis of Daft Punk's film *Electroma* as an existentialist critique of transhumanism, an intellectual history of the early development of the concept of heterosexuality, and an article entitled "What Do Heterosexual Men Get Out of Girl-Girl Pornography?" In his spare time, he works as a music critic and amateur DJ.

Jessica Rodgers is a PhD student in the School of Media, Communication and Journalism at the Queensland University of Technology. Her dissertation, *Australian Queer Student Activists' Media Representations of Queer*, investigates understandings of queer identity and queer activism in Australian queer student activist media.

Tony D. Sampson is a London-based academic and writer. He lectures on network culture, new media, usability studies, and interactive design at the University of East London. Tony received his PhD from the Department of Sociology at the University of Essex. He is the coeditor (with Jussi Parikka) of *The Spam Book: On Viruses, Porn, and Other Anomalies From the Dark Side of Digital Culture* (Hampton Press, 2009), and is currently finishing work on his next book, *Virality: Contagion Theory in the Age of Networks* (Minnesota, forthcoming).

Christopher Grant Ward is a professional media designer, artist, and writer from San Francisco, California. For over twelve years, he has developed media campaigns and interactive web experiences for firms and companies in San Francisco, Los Angeles, Portland, and Madrid. He is also a frequent speaker on the themes of design, media discourse, and visual/verbal communication. He holds an MA in Rhetoric from Carnegie Mellon University and has studied Graduate Media Design at the Art Center College of Design in Pasadena, California.

Acknowledgments

I would like to thank the contributors to this collection for their patience and dedication in seeing this book through from its initial conception to completion. I would also like to thank Katie Gallof at Continuum for her guidance and support through the review and publication process. Special thanks also goes to Axel Bruns at *M/C Journal* who gave me the opportunity to coordinate a special issue on the topic of error, which planted the seeds for this book.

Chapter 7: Losh, Elizabeth. "Artificial Intelligence: Media Illiteracy and the SonicJihad Debacle in Congress." *M/C Journal* 10, no. 5 (Oct. 2007). http://journal.media-culture.org.au/0710/08-losh.php.

Chapter 10: Dieter, Michael. "Amazon Noir: Piracy, Distribution, Control.' *M/C Journal* 10, no. 5 (Oct. 2007). http://journal.media-culture. org.au/0710/07-dieter.php.

Introduction

Error, Noise, and Potential: The Outside of Purpose

Mark Nunes
Southern Polytechnic State University

We need no longer speak of a "coming age" of the network society. Today, a majority of our social interactions assume the proper, efficient functioning of global, digital networks. How we conduct business, exchange ideas, entertain ourselves, and participate in politics relies upon dependable flows of information. This network society is governed by what Jean-Francois Lyotard defined more than thirty years ago as a "logic of maximum performance":[1] a cybernetic ideology driven by dreams of an error-free world of 100 percent efficiency, accuracy, and predictability.

Given the growing dominance of this ideology of informatic control, *error* provides us with an important critical lens for understanding what it means to live within a network society. Error reveals not only a system's failure, but also its operational logic. This logic of maximum performance demands that error is either contained as predictable deviation, "captured," for example, by the all-too-familiar error messages of everyday life, or nullified as an outlying and asignifying event. Occasionally, though, error slips through. In these moments, error calls attention to its etymological roots: a going astray, a wandering from intended destinations. In its "failure to communicate," error signals a path of escape from the predictable confines of informatic control: an opening, a virtuality, a *poiesis*.

Error gives expression to the *out of bounds* of systematic control. When error communicates, it does so as noise: abject information and aberrant signal within an otherwise orderly system of communication. While often cast as a passive, yet pernicious deviation from intended results, error can also signal a potential for a strategy of misdirection, one that invokes a logic of control to create an opening for variance, play, and unintended outcomes. Error, as errant heading, suggests ways in which failure, glitch, and miscommunication provide creative openings and lines of flight that allow for a reconceptualization of what can (or cannot) be realized within existing social and cultural practices.

This book explores the ways in which an ideology of information defines and delimits the dominant social and cultural structures in a network society, and how error and noise mark a destabilizing moment within the same system that attempts to capture or banish these errant expressions. This book explicitly examines the role of error as *errant communication* within a culture increasingly dominated by a logic of maximum performance. While not the first book to explore a cybernetic understanding of noise in the context of literary and artistic production, this collection attempts to examine more directly how one might understand a "poetics of noise" within a network society predicated upon the control of information, and how strategies of misdirection serve as both cultural and artistic interventions.

Control

Increasingly, the networks that underpin our economic, political, and social interactions require informatic control systems to minimize error and maximize performance. Our information age utopia is an error-free world of efficiency, accuracy, and predictability. In effect, in the growing dominance of "the network" as social space, we are witnessing the transcendence of a social and cultural system that must suppress at all costs *the failure to communicate*. This system operates within a paradoxical moment of maximum flow and maximum control. On one hand, these global networks are celebrated for their democratizing, grassroots potential—as an expression of "the power of organizing without organizations," to quote from the subtitle of Clay Shirky's highly popular and highly influential book, *Here Comes Everybody.*[2] But at the same time, it is clear that these networks of a participatory, new media culture operate within a dominant structure of transnational markets that, while perhaps drawing immense capital from the free flow of information across national boundaries, can only do so within a framework that insists upon dependable transmission. All Web 2.0 (or 3.0 or 4.0) hype aside, underlying these interactive networks is a cybernetic logic, a logic that ultimately depends upon information control. This balance between total flow and total control parallels Deleuze and Guattari's discussion of a regime of signs in which anything that resists systematic incorporation is cast out as asignifying scapegoat "condemned as that which exceeds the signifying regime's power of deterritorialization."[3] This free-flowing system ultimately depends upon a control logic in which everything that circulates communicates . . . or is cast aside as abject.

As a control system, "communication" reduces to a binary act of signal detection, a Maxwell's demon for circulating signal and casting out noise. This forced binary imposes a kind of violence, one that demands a rationalization of all singularities of expression within a totalizing system. To the extent that systems of information imply systems of control, the violence is less metaphor than metonym, as it calls into high relief the differend of the scapegoat, the abject remainder whose silenced line of flight marks the trajectory of the unclean. The violence of information is, then, the violence of *silencing or making to speak that which cannot communicate.* None of this violence of exclusion is visible within the dominant ideologies exemplified by a logic of maximum performance implicit in the new media practices of everyday life. Who, after all, would champion the cause of "bad results"? We anticipate consistency in our mass-produced products, dependability in our 24/7 connectivity, and quality assurance in our customer service. Within this dominant ideology, the unfortunate outcomes of error serve only one purpose: to remind us of the need for greater control.

"Control" is by no means a new concept in the developed world—in fact, we might go as far as to say that what marks the developed world as developed (and the developing nations as developing) is the integral relationship between systems of production, distribution, and consumption and systems of control. In his classic *The Control Revolution,* James Beniger explores the "crisis of control" that emerges in the nineteenth century, when increased production speed through rapid industrialization led to the need for increased speed in distribution and consumption.[4] Key to his analysis of this crisis of control is the rise of what today we would call the network society, namely an increasingly complex and system-oriented arrangement of production, distribution, and consumption within a society. This need for systematic control foregrounds information processing as a central concern in manufacturing, allowing for the development of "programs" for production (the Jacquard loom and the Hollerith punch card), distribution (coordination of telegraph and rail), and ultimately consumption (brand marketing and mass advertising).[5] In this regard, Beniger sees our current information society as a contemporary expression of a control society that is over a century old, with ultimately the same goal: process management. What has shifted, he argues, from the rise of bureaucracy in the late nineteenth century to the dominance of information and communication technologies (ICTs) in the present is the degree to which the crisis of production in the nineteenth century—how to accommodate a speed-up in process through industrialization—has increasingly become a crisis of control. Furthermore, as process control increasingly operates at a level abstracted from production itself, information flow—transmitted, received, and acted upon as feedback—has become the central feature of operational control.

If we think of process in terms of programs, "control" describes that set of actions that directs behavior toward intended outcomes. Beniger, in fact, defines control as "purposive influence toward a predetermined goal."[6] This relation between "purpose" and control emerges as an area of significant concern along multiple fronts in the mid-twentieth century. While Beniger draws on systems theory and von Neumann's distinction between *order* and *organization* to associate control with "end-directedness,"[7] it is really within Norbert Wiener's cybernetics that we first see a fully developed concept of control as purpose-directed behavior. The connection between purpose and control, and the role of error in articulating this relationship, merits further discussion.

In his popularized account of cybernetics in *The Human Use of Human Beings,* Wiener defines this emerging field of study as an (inter)discipline designed "to attack the problem of control and communication in general."[8] Key to Wiener's discussion of control and communication is the role of information in providing feedback to a system. Feedback serves as a measure of the gap between "actual performance" and "expected performance."[9] While given full form in the late 1940s, Wiener's interest in feedback and control emerges much earlier in his work during World War II with the National Defense Research Committee on antiaircraft artillery, specifically in relation to voluntary action, target tracking, and error correction.[10] This attempt to develop an effective "self-correcting" antiaircraft device, along with follow-up research on "purpose tremors" in individuals with cerebellum damage, led to his seminal ideas on feedback, which would eventually appear in a coauthored publication entitled "Behavior, Purpose, and Teleology" in 1943. It is in this piece that Wiener first explicitly discusses (to a non-military audience, at least) the relationship between "purpose," "error," and voluntary control:

> Positive feed-back adds to the input signals, it does not correct them. The term feed-back is also employed in a more restricted sense to

signify that the behavior of an object is controlled by the *margin of error* at which the object stands at a given time with reference to a relatively specified goal. The feed-back is then negative, that is, the signals from the goal are used to restrict outputs which would otherwise go beyond the goal. It is this second meaning of the term feed-back that is used here (emphasis added).[11]

Error, in effect, serves its purpose as a *corrective*—what keeps purpose on purpose and tasks on goal.

Error marks a deviation from a predetermined outcome. Control, then, must account for deviation in order to correct it and contain it within a program or system. This approach to error correction and purposive control, of course, need not be limited to artillery shells and moving objects. Beniger notes that in the control revolution of the early twentieth century, the "scientific management" of labor, most notably through the figure of Frederick Winslow Taylor, provides an exemplary instance of feedback control in mass production by rationalizing labor to a flow process, one that could be measured, quantified, and fine tuned to maximum efficiency.[12] In his time studies, the efficiency expert breaks down activity into "elementary operations and motions . . . Eliminat[ing] all false movements, slow movements, and useless movements" and then compiling the "quickest and best movements" as a replacement for the "inferior series which were formerly in use."[13] Beniger points out that Taylor's time studies really amount to a "reprogramming of even the most basic human movements to conform to system-level rationality" as an implementation of control.[14] In a similar vein, George Radford's work on quality control in the 1920s marks quality by way of uniformity "within tolerance limits" for both mass production and end consumption.[15] Likewise, Walter Shewhart's statistical quality control and the concept of "maximum acceptable defects" provided a program for processes that established levels of tolerable deviation in output before correction became necessary.[16] In each instance, purpose drives process, with error serving as—by way of feedback loops—a structural limit that assures expected outcomes. Error is tolerable to the degree that deviation remains systematically contained within a program of control.

Six Sigma is more than the latest corporate trend. Driven by two concepts—the ability to identify and maximize any action as process, and a shrinking of tolerable deviation to six standard deviations (fewer than four in a million)—it stands for the systematic reduction of all activity to measurable—and therefore predictable and controllable—processes.

As Six Sigma standards migrate from the world of manufacturing to a wide range of institutions, we find a standard of maximum predictability and minimum error as the latest coin of the realm. What is one to do, then, with error? How does this ideology address the place of the error that escapes—or exploits—systems of process control? As error is increasingly programmed out of (and therefore, ironically *into*, by way of error handling messages and feedback responses) social and market processes, how does error offer an opportunity for unintended trajectories? Consider, for example, Taylor's classification of "false," "slow," and "useless" movements as deviations from an optimal performance toward a purposive end. These deviant, wandering movements fall outside of efficiency, and as such, mark "error" in need of corrective feedback. If, as Beniger argues, this rationalization replaces the individual worker with a rationalized process, one can also read the *singularity* of the individual as an errancy in its own right. To the extent that individuals become "dividuals" within societies of control,[17] do these singular moments—no matter how useless or without purpose—suggest a form of asystematic resistance?

A Genealogy of Error

As mentioned above, "error" has an etymological root that emphasizes wandering. In his study of error, the Enlightenment, and revolution in eighteenth-century France, David Bates notes: "There is, etymologically, an ambivalence at the heart of error, a tension, that is, between merely aimless wandering and a more specific aberration from some path."[18] Whereas Wiener's understanding of error is explicitly a deviation of actual course from an intended (or ideal) path, Bates calls attention to error, in the sixteenth-century usage of the term, as a path in its own right, but one without a heading or a purpose. By the seventeenth century, the use of the term "error" to denote a deviation from "true" judgment gains common usage, but one can still find traces of wandering at work in its connotative field—for example, as "a vagabondage of the imagination, of the mind that is not subject to any rule."[19] Bates argues that while the emphasis on error as a deviation from the truth would dominate in the Enlightenment, "the hidden duality of error, a duality that persists into the modern period, hinges on this crucial distinction between 'wandering' and 'deviation.'"[20]

Deleuze picks up on a similar line of argument in *Difference and Repetition* (and in other texts as well), in which he explores "the image of thought" in modern philosophy.[21] In Enlightenment thought, error marks a move away from an increasingly accurate conception of the world. While knowledge may be imperfect, reason and rational thought could reveal—and therefore correct—error. Deleuze discusses this orthodox image of thought as based on an assumption that "thought has an affinity with the true; it formally possesses the true and materially wants the true."[22] To think, in this orthodox image of thought, Deleuze argues, is to "recognize" the true and thereby enact a repetition of the true. Deviation, wandering—in short, any "misadventure of thought"—is defined as error.[23] Within the Enlightenment, Deleuze argues, "error" functions as a delimiting concept to categorize any image of thought that does not coordinate with the orthodox image of thought as a "recognition" of truth.[24] He writes:

> Who says, "good morning, Theodorus" when Theatetus passes, "It is three o'clock" when it is three-thirty, and that 7+5=13? Answer: the myopic, the distracted, and the young child at school. These are the effective examples of errors, but examples which . . . refer to thoroughly artificial or puerile situations and offer a grotesque image of thought because they relate it to very simple questions to which one can and must respond by independent propositions. Error acquires a sense only once the play of thought ceases to be speculative and becomes a kind of radio quiz.[25]

Error, for Deleuze, is always defined within systematic closure: "as though error were a kind of failure of good sense within the form of a common sense which remains integral and intact."[26] As a "misadventure of thought," error assumes the possibility (and necessity) of proper thought, and therefore as a concept remains "what is in principle negative in thought."[27] As such, Deleuze sees limited use in pursuing what error might reveal as an image of thought. But what if, in allowing error to stand as what is negative in thought, we might also gain a sense for the parameters of the programmatic itself? Would we not find in error its double sense: not only the predefined misstep, but also a going astray that poses a challenge to purposive intent? In doing so, we would be establishing *the place of error*—within and exterior to the programmatic—as a kind of real, yet excluded possibility, much as Deleuze elsewhere discusses the relation between the virtual and the actual.

A number of chapters in this book will return to these speculations, but for the sake of this genealogy of error, we must leave Continental philosophy aside and turn to mathematics and logic, where the concept of error gains increasing importance in the nineteenth century. In particular, it is Carl Friedrich Gauss who, in addition to his work on normal distributions of data, introduces a means for understanding "true" measurement in terms of error, namely through the least square method for fitting observational data to a presumed distribution. For Gauss, actual measurements are seen as deviations from idealized or expected results as dictated by an underlying equation. And with Gaussian mathematics arises the possibility of a statistical view of the world, in which probability of outcomes marks a "truth" that supersedes individual measurements. As such, we need no longer see error as something that must be corrected; rather, error is the stuff of empirical measurement itself—singular instances within a pattern of deviation that *reveals* true results.

This mathematical understanding of error emerges as a full-fledged philosophical concept later in the nineteenth century through the figure of Charles Sanders Peirce. While maintaining a belief in an absolute truth, Peirce insists that knowledge advances as the result of the collective action of a "community of inquirers" engaged in partial approximations at truth. We see this idea as early as 1870, when Peirce conducted a series of experiments on errors in multiple, relative measurements for the U.S. Coast Survey in an attempt to advance a "Theory of Errors of Observation."[28] In later essays such as "Fallibilism, Continuity, and Evolution," Peirce maintains what he calls a Doctrine of Fallibilism, insisting that while an absolute answer to an inquiry exists at a theoretical horizon, at any given moment, answers are always partial, incomplete, and, in short, fraught with error.[29] What advances knowledge, then, is not authority and certainty, but rather *doubt.* Unlike an Enlightenment view of error as a failure to recognize the truth, error serves a purpose in Peirce's thought in ways that foreshadow Wiener's faith in underlying patterns of order. As Deborah Mayo notes, Peirce claims that "inductive inference" is "self-correcting," based on the ability of error to create doubt not merely in findings but in process and method as well. As a result, inductive inference "not only corrects its conclusion, *it even corrects its premises.*"[30] In Peirce's theory of errors, individual acts of observation are to some degree inconsequential; it is the underlying function that governs how multiple observations deviate from one another that gives us the best account of the process and method of observation itself. For Peirce, the fallibility of individual measures revealed that "truth" was an abstraction beyond human measure. Rather than treating error as a deviant instance in need of correction, Peirce maintained that the *equation* (and *not* the instance) contained the potential to reveal truth, a kind of real, yet unactualized pattern that transcended the missteps of error in actual human execution.

As probabilistic studies advance in number and complexity from the late nineteenth century into the early twentieth century, a changing perception of the world begins to arise, one driven by statistical analysis and an understanding of error in terms of standard deviation. As Beniger notes, the rise of "information" in the late nineteenth century as a dominant organizing principle emerges with the growing systematization of corporate and governmental processes through programmatic control, first under rationalization and bureaucracy, and eventually through ICTs.[31] In this regard, the shift toward "information" over knowledge as the driving force in process control leads to a recasting of "error" as a *deviation* from an intended preprogrammed outcome. While Wiener never cites Peirce as an intellectual forefather, this vision of pattern, error, and deviation is clearly an essential, underlying concept in all of his work in cybernetics.[32] In fact, it is his use of mean square approximations to correct for gun position in his work with the NDRC, much as Peirce did for the U.S. Coast Survey, that would provide him with a mathematical basis for his concept of the relation between feedback systems and purposive control.[33] As with Peirce, what transcended individual observation was the equation that accounted for this deviation, with error serving its function in delimiting an underlying pattern (and hence purpose). For all of the randomness associated with this probabilistic view of the world, Wiener's cybernetics attempted to demonstrate through these same statistical principles that pattern played itself out as a backdrop to all orderly and purpose-driven endeavors; animals and machines alike, and the systems they develop, provide feedback control to correct for error and approximate goals. Wiener's faith in control would lead to a definition of information as a creation of order in the face of entropy. It is important to note the degree to which order for Wiener is tied to a purpose-driven prediction of results, in which error provides corrective feedback. As Wiener writes, describing the legal system in cybernetic terms: "The technique of the interpretation of past judgments must be such that a lawyer should know, not only *what* a court has said, but even with high probability what the court is *going to* say."[34] What Wiener describes here, I would argue, provides us with a contemporary view of error within the dominant ideology of a network society. Error, as captured, predictable deviation *serves* order through feedback and systematic control.

Except in those moments when error, in its wanderings, evades prediction, program, and protocol. In those moments, an interstitial gap opens, an outside *within* the logic of the system that threatens "the good" of the system itself.

Noise and the Seduction of Error

It is worth noting the degree to which these gaps in systematic control consistently mark an area of anxiety for Wiener. In this same passage on the predictability of the law, Wiener describes the "interstices" between systems of interpretation as a place occupied by malevolent potential: "a no-man's land in which dishonest men prey on the differences in possible interpretation of the statues."[35] I will return to this discussion in my own contribution to this collection, but for the time being, suffice it to say that for Wiener, the containment of error within the framework of purpose-driven behavior through feedback is the essence of order and control. Remember as well that for Wiener, increases in information signal a *decrease* in entropy and an increase in order. We can contrast this view with Claude Shannon's statistical analysis of communication as a system, which pairs an increase in information with an *increase* in entropy. The uncertainty that Wiener associates with an opportunity for malevolent manipulation becomes in Shannon's account a measure of the total information in a system. This is an important shift in terms (what Katherine Hayles refers to as "Shannon's Choice"[36]), and one that repositions error as something other than a negation of intent. In theorizing "errors in transmission" as information in its own right, Shannon's communication theory provides a critical link in establishing a technical and theoretical relationship between "error" and "noise."[37] Noise, in effect, is errant signal, or as termed by Warren Weaver, "spurious information."[38] What occurs in place of Wiener's vision of error as deviation from intended results is Shannon's concept of "equivocation," what Weaver describes as "an undesirable uncertainty" that the message received was the message sent.[39]

This shift in understanding the relationship between entropy and information allows error to communicate not only as system feedback, but within the system itself as errant signal. Purposive behavior still plays a role here, but with Shannon's information theory, it is not only the relationship between information and error that alters, but the role that error itself plays within the communicative system. Error, in effect, communicates as information *without a purpose*—or at *cross-purposes* to programmatic control. As Shannon discusses error correction through feedback processes, his discussion of noise as spurious information places more emphasis on the role of noise in creating uncertainty in selection, rather than the degree to which a given signal "strays" from its intent. In effect, while Wiener's emphasis on order defines control as purpose-driven behavior, Shannon's discussion of equivocation explores the impact of noise, error, and other errant signals on the proliferation of choice. While for Wiener error serves its purpose as input for feedback control, Shannon, in allowing errors in transmission to communicate within a system, creates the possibility of a kind of informational virtuality—a "potential of potential," in the words of McKenzie Wark, that signals an indetermination that exceeds systematic control.[40] In other words, in Shannon's work entropy provides a measure of the variance between a communication system's virtuality (all possible messages) and its actualized performance (the message sent). If Wiener's cybernetics foregrounds purposive behavior as a key feature in feedback systems, then the spurious information of noise—from the pops and glitches of transmission error to the hacks and jams of counter-agents—functions as a kind of information that exceeds programmatic control by *widening* the gap between the actual and the possible, rather than narrowing the deviation between the intended and the actual. Control provides a system for guiding communication from intention to intention. In contrast, the error of noise marks a potential to throw off systems of control by deferring the actual (message received) and sustaining the virtuality of equivocation.

Error also reveals the degree to which everyday life plays itself out within this space of equivocation. Errors come in many kinds, but increasingly, our errors arrive prepackaged. Think of the last time you mistyped a web site URL and received a "404" error. The failure notices that are a part of our networked everyday life correspond to a specific category of error—a *potential* error that the system must predict before it has *actually* occurred. While the error notice signals failure, it does so within the successful, efficient operation of a system. To this extent, then, in a majority of the new media practices of everyday life, every error is a "fatal error" in that Baudrillardian sense[41]—a failure that is subsumed in the successful operation of a closed system. Such systems are *all-actualizing*. Control systems do not deal in the singularities of instances, but with fields of the possible, what is called in statistics the event space. Pushed further, such systems are often only interested in events "worth" looking at: the sample space. This logic of the normal—that one is often best off ignoring the outlier—suggests that something very risky occurs when an event or an instance falls outside the scope of predicable results.

But as Stuart Moulthrop addressed some ten years ago, "Error 404" can also highlight "the importance of not-finding":[42] that error marks a path in its own right, and not merely a misstep. Such a critical perspective runs contrary to a dominant, cybernetic ideology of efficiency and control. While we may find indulgence for errors, glitches, and noise in art, music, or literature, such erratic behavior finds little favor in a world increasingly defined by protocol and predictable results.[43] But is there not something seductive in error precisely because it draws us off our path of intention, interrupting the course of goals, objectives, and outcomes and pulling us toward the unintended and unforeseen? Uncaptured error refuses to signify within a system of feedback control, and as such, threatens to disrupt the cybernetic regime of efficiency and maximum performance. These outlying signals—the statistical abject—fall outside of the sample space, a singularity that deviates "out of range." Yet it is this materially and informatically abject form, failing to be captured (and hence signify) within a system, that marks the potential of potential. Error is virtual only to the extent that it exceeds (or falls short of) the possibility of existing as an actual feedback event. It is a "failure" that at the same time marks a potential, an opening in what we may hesitantly call a culture of noise. This asignifying poetics of noise, marked by these moments of errant information, simultaneously refuses and exceeds the cybernetic imperative to communicate.

In this tension between capture and escape marked by the scapegoats of error and noise stands a potential for aesthetic, political, and social insinuations within an increasingly programmatic network society. As such, it marks an increasingly fertile territory for critical and tactical interventions. Recent focus on "hacker culture" serves as an instance in which gaps and fissures in systems of control provide for possibilities of creative, unintended consequences. Alexander Galloway and Eugene Thacker's *The Exploit: A Theory of Networks*, for example, attempts to expose how networks structure systems of power and social control, and how the "hack" functions as a mode of resistance to control to the extent that it "exploits" control protocols to its own ends.[44] Likewise, McKenzie Wark's *A Hacker Manifesto* holds forth the political and social potential of a hacker class to liberate "the possibility of production" from the commodifying interests of a dominant vectoral class, and to do so by exploiting the virtual as an "inexhaustible domain of what is real but not actual, what is not but which may become."[45] Any "hack" in this regard reveals and exploits error to the extent that it leads to an outcome unintended by a system's purposive organization. Error, in its excesses and its failures, signals a real, yet virtual locus for social and cultural interventions.

And there is certainly precedent for considering the artistic potentials of error and noise, going back at least as far as the Futurist and Dadaist engagements with "high art" in the early twentieth century, as discussed by xtine burrough in this collection. Likewise, error in the form of unintended sound—from John Cage's experiments with environmental noise to more recent *glitch* music—has provided a basis for what Kim Cascone has called "the aesthetics of failure" in the contemporary computer music scene.[46] This glitch aesthetic has over the past decade worked its way into the visual arts as well, as discussed by Tim Barker in his contribution to this collection—and given gallery book treatment in the recently published *Glitch: Designing Imperfection*.[47] In literature, books such as William Paulson's *The Noise of Culture: Literary Texts in a World of Information*[48] and N. Katherine Hayles's *Chaos Bound: Orderly Disorder in Contemporary Literature and Science* have drawn upon information theory, and in particular, Shannon's general model of communication, to provide a critical framework for discussing the relationship between noise and literary production. In all of these works, noise rather than serving as an impediment to otherwise clear communication, creates an opportunity for alternate modes of expression. This book continues work in this vein by exploring the creative potential of errors, glitches, and noise, through what Galloway and Thacker call "counterprotocological practice"[49] and what we might also refer to as a poetics of noise. In highlighting poetics, I do so within a broader framework of *poeisis* as creative opening and as touched upon in Umberto Eco's discussion of information theory in *The Open Work*.[50] Eco elaborates on Shannon and Weaver's information theory by distinguishing between *actual* communication (the message sent) and its virtuality (the possible messages received). Eco argues, in effect, that communication *reduces* potential in its desire to actualize signal at the expense of noise. In contrast, poetics *generates* potential by sustaining the equivocation of the text.[51] Eco is intrigued by Shannon and Weaver's insight that communication is always haunted by equivocation—the uncertainty that the message received was the signal sent.[52] Likewise, a poetics of noise foregrounds the creative potential of the errant and the unintended outcome—a purpose that has no purpose within an existing (actual) system of meaning or order.

In marking these interstitial openings between "purpose" and outcome, error also suggests a strategy of misdirection, the potential of getting back a result other than what one expected. Until recently, typing the word "failure" into the search engine Google, for example, produced as a top response George W. Bush's web page at www.whitehouse.gov.[53] By building web pages in which the text "failure" linked to the former President's page, users managed to "game" Google's search algorithm to provide an errant heading. "Google bombing," as the practice is called, creates an informatic structure that plays off of the creative potential of equivocation, a form of sabotage in which the cybernetic imperative is turned against itself. From within the framework of programmatic control, these are the tactics of "bad actors," who game the system in order to divert the rules toward their own ends. The recuperation of the "hack" by Wark and others as a term associated with a liberatory potential does not take away from the fact that for a majority of daily users of new media technologies, the hacker is a figure of malevolent potential. The place of the bad actor and the errant practices associated with this behavior figure prominently in several contributions to this collection, my own included. In this regard, an extended study of error and the poetics of noise as presented in this collection might fall under what Matthew Fuller and Andrew Goffey have recently called an "evil media studies," in its attempt to bypass a moralizing approach to new media culture (good vs. bad media practices), and to instead foreground those "agonisms" that are both inherent in and abject to a network society.[54] Error and noise function as "strategies of the object" when *seduction* exceeds a system's ability to capture the errant in the service of purpose.[55] From the perspective of cybernetics, we are indeed speaking the language of evil; Wiener for one certainly did not shy away from a rhetoric of good and evil when discussing order, entropy, and control, be that the Augustinian evil of a world that passively resists our ability to reverse entropy, or a more determined Manichean evil of exploited uncertainty.

But perhaps to best understand the "evil" of error as a strategy of the object, we should focus our attention not on the usual suspects of trolls, scam artists, and cyberterrorists, but rather on the unexpected evil of none other than the *Sesame Street* Muppet, Bert.

I am referring, of course, to an event in 2001, chronicled most notably by Henry Jenkins and Mark Poster, resulting from a Photoshopped image created by a high school student named Dino Ignacio, depicting the Muppet Bert with Osama Bin Laden—part of his humorous web site project entitled "Bert is Evil." A Pakistani-based publisher scanning the web for images of Bin Laden came across Ignacio's image and, perhaps not recognizing the *Sesame Street* character, incorporated it into a series of anti-American posters. According to Henry Jenkins's account, in the weeks that followed, "CNN reporters recorded the unlikely sight of a mob of angry protestors marching through the streets chanting anti-American slogans and waving signs depicting Bert and Bin Laden."[56] As the story of the Bert-sighting spread, new "Bert is evil" web sites sprang up, and Ignacio found himself the unwitting center of a full-blown Internet phenomenon. Jenkins finds in this story a fascinating example of what he calls convergence culture, the blurring of the line between consumer and producer within participatory networks.[57] For Poster, the tale of "evil Bert" reveals not only the "aberrant decoding" of the Pakistani protesters, but their inability to filter out the very same American/Western pop cultural influences they were protesting against, suggesting the degree to which in a network society "transmissions . . . can be both noiseless *and* incoherent."[58] Of particular concern to this collection, however, is how this Bert/Bin Laden error, rather than resulting in a failure to communicate, instead created its own viral network in a kind of "error-contagion," as Tony D. Sampson discusses in his chapter. Error here *opens* a channel for communication, a strategy of the object beyond the intention of Ignacio, the anti-Western protesters, or any other agent. While in this instance unintentional, here is the essence of "gaming the system," in which a gap in protocol allows for errant practices that lead to unintended results—or more accurately, results that are "outside of purpose." The evil of Bert the Muppet showing up at an anti-American rally in Pakistan marks such a moment of slippage, in which an error communicates a creative potential that escapes purpose and systemic control.

In the moment of equivocation—not so much beyond good and evil, but rather in the interstitial spaces between signal and noise—slippage, error, and misdirection suggest an opening onto a potential outside of purpose and control. The violence of the cybernetic imperative expresses itself in this moment of insistence that whatever circulates signifies, and that which cannot communicate must be silenced. In such an environment, we would do well to examine system failures, as well as the ways in which error and noise seduce us off course.

Overview

This book attempts to sketch out as an area of critical interest these moments in which error makes visible both the dominant logic of all-actualizing systems as well as those interstitial spaces that escape control. The chapters are arranged in three sections, clustered around metaphors for a set of practices that call attention to the potential of potential inherent in error. In the first section, "Hack," contributors explore the ways in which errors, glitches, and failure provide an opening into what I have described above as a poetics of noise. In addressing how error and noise provide an opportunity for both critical and aesthetic interventions, these authors explore the ways in which the errant and the unintended, rather than impeding performance, provide for a productive aesthetic in its own right. In the first chapter, "Revealing Errors," Benjamin Mako Hill explores how media theorists would benefit from closer attention to error as a "reveal" for the role of new media in everyday life. By allowing uncaptured errors to *communicate*, he argues, we gain a perspective that makes invisible technologies all the more visible. Tim Barker's chapter, "Aesthetics of the Error: Media Art, the Machine, the Unforeseen, and the Errant" examines the function of productive error in digital art, both as an expression of artistic "intent" and as an emergent, unintended expression within the digital medium. This "glitch aesthetic" foregrounds the errant and the uncontrollable in media art. In doing so, Barker argues, error hacks art, offering a field of potential for the media artist in the form of the unforeseen. In a similar vein, Susan Ballard's chapter, "Information, Noise, et al.," explores the productive error of noise in the digital installation art of the New Zealand artists' collective et al. through a reading of Shannon's concept of equivocation. Rather than carefully controlling the viewer's experience, et al.'s installations place the viewer within a signal/noise dilemma, in effect encouraging misreadings and unintended insertions. xtine burrough's "*Add-Art* and *Your Neighbors' Biz: A Tactical Manipulation of Noise*" looks to parallels between the junk aesthetic of early twentieth-century dada and two contemporary, tactical uses of new media as hacks within dominant cultural practices. Drawing in particular on Eco's reading of Shannon, she explores a kind of poetics of noise that exploits commercial new media practices for counter-commercial ends. In contrast, Christopher Grant Ward takes this reading of a poetics of noise in the opposite direction in "Stock Imagery, Filler Content, Semantic Ambiguity" in a reading of how the stock image industry attempts to encourage equivocation rather than control semiotic signal in corporate communications. Here it is corporate interest that exploits the virtual to rhetorical ends, increasing entropy through visual fillers and granting consumers greater selection within a range of possible meanings, thereby creating a more ambiguous corporate ethos.

In the second section, "Game," the contributors explore how error marks interstitial spaces outside or between the rules, allowing for intentional and accidental co-opting of protocol toward other ends. Here, "gaming the system" reveals both systems of control within a network society and opportunities to exploit these failures to control errant practices. In "Gaming the Glitch: Room for Error" Peter Krapp compares the human-computer interface designers' attempt to control and normalize user error with game design, in which failure is highlighted as a measure of user performance, suggesting that much of new media culture is located between these two parameters for interaction. Turning to the work of new media artists, he explores how these works exploit failures in user interaction in a way that calls attention to the limits—and possibilities—for "free play" in a network society. In her chapter, "The Seven Million Dollar PowerPoint and Its Aftermath: What Happens When the House Intelligence Committee Sees 'Terrorist Use of the Internet' in a *Battlefield 2* Fan Film," Elizabeth Losh focuses on a highly public misreading of a fan video by experts from the Science Applications International Corporation in their testimony before Congress on digital terrorism. Losh argues that Congress's willingness to give audience to this misreading is a revealing error in its own right, as it calls attention to the ability of experts and power brokers to game new media networks. At the same time, she notes, appropriation and remix, as presented in this instance in fan videos, offer an opportunity for a range of errant practices that exploit media to their own ends. Following along similar lines, Ted Gournelos's chapter, "Disrupting the Public Sphere: Mediated Noise and Oppositional Politics" explores three case studies of media "pranks" that game mass media to impact political affect. Key to these moments of subverting the media are the networks of exchange in which these pranks circulate, in effect creating an oppositional politics within the same networks that support and maintain the dominant culture.

And finally, in my own contribution, "Wikipedia, Error, and Fear of the Bad Actor," I examine the popular and institutional anxieties surrounding the activity of bad faith actors within the collective intelligence. By paying attention to the "scandals" surrounding highly publicized errors, misuses, and acts of vandalism in Wikipedia, I hope to reveal both the implicit faith in communal activity that underlies much of the rhetoric concerning a new media collective intelligence, and the degree to which the errant practices of "bad actors" force us to reconsider these assumptions.

As Fuller and Goffey note, these errant practices need not be seen as "intentional";[59] in fact, a resistance to intentionality and purposive behavior is, in some instances, a strategy of the object well in keeping with error's exploits and misadventures. In the final section, "Jam," the contributors explore error as a resistance to informatic control through both intentional and unintended misdirections of the cybernetic imperative. While touching upon themes addressed in earlier sections, these chapters are explicitly interested in "counterprotocological practices" as either inherent agonisms within systems of control, or as tactical interventions by counteragents. In "Contingent Operations: Transduction, Reticular Aesthetics, and the EKMRZ Trilogy," Michael Dieter examines three projects by the group Ubermorgen.com that exploit productive networks toward ulterior outcomes. Dieter demonstrates how these tactical interventions challenge a cybernetic system of control by opening it up to new and ambiguous creative processes. Chad Parkhill and Jessica Rodgers's piece, "Queer/Error: Gay Media Systems and Processes of Abjection," turns to the ways in which gay normative identity establishes itself within gay media, and how these media networks route and contain the abject headings posed by queer identity. In particular, they explore the HIV+ body as abject within gay media systems; as infected by an "unrepresentable" disease that exploits boundary errors within the immune system, the HIV+ body threatens to jam identity networks if not treated as excluded other. Tony D. Sampson examines a different form of the viral in "Error-Contagion: Network Hypnosis and Collective Culpability." Drawing on the work of Gabriel Tarde, he looks at recent financial crises and instances of "botnet" cybercrime as examples of error-contagion that undermine our rationalist assumptions of a logic of maximum performance in a network society. Error-contagion does more than jam network function through DoS attacks and the like; it introduces a structure for a collective ethics that does not resolve easily to the rationalist ideology of a dominant society. In the final piece in the collection, Stuart Moulthrop's "Error 1337" returns to considerations of error's path in a network society. Looking toward the multilinearity of networks, and the role of "paths" in hypertext, he explores how the rhetoric of "literacy" enters a kind of "halting state" as we attempt to translate it to the activities that make up new media practices of everyday life in a network society. Offering "L337" as a self-ironizing counter-term, he introduces a jamming signal within this discussion of new media practices that attempts to open a new frame of discourse for media studies.

The chapters in this collection approach error from a range of critical and cultural perspectives, each exploring through differing methodologies and case studies the place of error in a society increasingly dominated by information structures and a logic of maximum performance. In many ways, this volume provides a countermeasure to what can be seen as a second wave of network euphoria, this time expressed as a celebration of the participatory potential of new media cultures. In an era of failed banks, collapsed markets, and spiraling debt, perhaps a book on error, noise, and the limits of systematic control is well timed. This is not to say that the authors in this book celebrate failure "in its own right." But as a critical intervention, and as a set of artistic and political tactics, this collection argues that error provides a critical path worth pursuing, one that suggests a potential immanent in—yet excluded from—systematic control.

Notes

1. Jean-Francois Lyotard, *The Postmodern Condition: A Report on Knowledge* (Minneapolis: University of Minnesota Press, 1984), xxiv.
2. Clay Shirky, *Here Comes Everybody* (New York: Penguin, 2008).
3. Gilles Deleuze and Felix Guattari, *A Thousand Plateaus: Capitalism and Schizophrenia* (Minneapolis: University of Minnesota Press, 1987), 117.
4. James Beniger, *The Control Revolution* (Cambridge, MA: Harvard University Press, 1989).
5. Ibid., 241–278.
6. Ibid., 7.
7. Ibid., 35.
8. Norbert Wiener, *The Human Use of Human Beings: Cybernetics and Society* (New York: Da Capo, 1988), 17.
9. Ibid., 24.
10. Norbert Wiener, *Cybernetics: Or Control and Communication in the Animal and the Machine* (Cambridge, MA: MIT Press, 1961), 5–8. See also Flo Conway and Jim Siegelman, *Dark Hero of the Information Age* (New York: Basic Books, 2005), 107–136. Wiener's work on artillery path prediction actually dates to an even

earlier moment in his life, in his brief stint as a trajectory tabulator during World War I at the Aberdeen Proving Grounds. See Conway and Siegelman, 42–43.

[11] Arturo Rosenblueth, Norbert Wiener, and Julian Bigelow, "Behavior, Purpose, and Teleology," 2, http://pespmc1.vub.ac.be/Books/Wiener-teleology.pdf. Originally published in *Philosophy of Science* 10 (1943): 18–24.

[12] Beniger, 294.

[13] Quoted in ibid., 294.

[14] Ibid., 295.

[15] Ibid., 308.

[16] Ibid., 309.

[17] Gille Deleuze, "Postscript on the Societies of Control," *October* 59 (Winter 1992): 5.

[18] David W. Bates, *Enlightenment Aberrations: Error and Revolution in France* (Ithaca: Cornell University Press, 2002), 20.

[19] Quoted in ibid., 20.

[20] Ibid., 21.

[21] Gilles Deleuze, *Difference and Repetition* (New York: Columbia University Press, 1994). See also Gilles Deleuze and Felix Guattari, *What is Philosophy* (New York, Columbia University Press, 1994).

[22] Ibid., 131.

[23] Ibid., 148.

[24] Ibid., 133–135.

[25] Ibid., 150.

[26] Ibid., 149.

[27] Ibid., 52.

[28] Charles Sanders Peirce, "On the Theory of Errors of Observation," *Report of the Superintendent of the United States Coast Survey Showing the Progress of the Survey During the Year 1870* (Washington, DC: Government Printing Office, 1873), 200–224, http://docs.lib.noaa.gov/rescue/cgs/001_pdf/CSC-0019.PDF.

[29] Charles Sander Peirce, "Fallibilism, Continuity, and Evolution," *The Collected Papers of Charles Sanders Peirce* (Cambridge, MA: Harvard University Press, 1931), 1.141–1.175, http://www.scribd.com/doc/7221904/PIERCEs-Collected-Papers.

[30] Quoted in Deborah Mayo, "Peircean Induction and the Error-Correcting Thesis," *Transactions of the Charles S. Peirce Society*, 42, no. 2 (Spring, 2005): 312.

[31] Beniger, 13–21.

[32] Interestingly enough, Wiener suggests ideas not too far removed from Peirce's theory of Fallibilism in a juvenile publication entitled "A Theory of Ignorance." See Conway and Siegelman, 16.

[33] Wiener, *Cybernetics*, 9. Wiener employs similar methods in his work on the use of Fourier transformations to determine "conditional probability" within random systems, most notably in Brownian motion. See Wiener, *Cybernetics*, 69–76.

[34] Wiener, *The Human Use of Human Beings*, 110.

[35] Ibid., 108.

[36] N. Katherine Hayles, *Chaos Bound: Orderly Disorder in Contemporary Literature and Science* (Ithaca: Cornell University Press, 1990), 31–60.

37 Claude E. Shannon and Warren Weaver, *The Mathematical Theory of Communication* (Urbana: University of Illinois Press, 1949).

38 Ibid., 19.

39 Ibid., 21.

40 McKenzie Wark, *A Hacker Manifesto* (Cambridge: Harvard University Press, 2004), par. 14.

41 Jean Baudrillard, "From the System of Objects to the Destiny of Objects," *The Ecstasy of Communication* (New York: Semiotext(e), 1988), 77–95.

42 Stuart Moulthrop, "Error 404: Doubting the Web," http://iat.ubalt.edu/moulthrop/essays/404.html.

43 It is worth noting, however, that even within the dominant ideology of maximum performativity, there is still room for the error that slips past our programs of control—as long as it is still willing to play by the rules of the game. Does not that same world of business that insists on maximum efficiency mythologize the creative moment that allows error to turn a profit? Post-It Notes and Silly Putty present two classic instances in which happenstance, mistake, and error mark a moment in which "thinking outside of the box" saves the day.

44 Alexander R. Galloway and Eugene Thacker, *The Exploit: A Theory of Networks* (Minneapolis: University of Minnesota Press, 2007).

45 Wark, par. 74, 77.

46 Kim Cascone, "The Asetetics of Failure: 'Post-Digital' Tendencies in Contemporary Computer Music," *Computer Music Journal*, 24, no. 4 (Winter 2000): 12–18.

47 Iman Moradi, Ant Scott, Joe Gilmore, and Christopher Murphy, *Glitch: Designing Imperfection* (New York: Mark Batty, 2009).

48 William Paulson, *The Noise of Culture: Literary Texts in a World of Information* (Ithaca: Cornell University Press, 1988).

49 Galloway and Thacker, 97–101.

50 Umberto Eco, *The Open Work* (Cambridge: Harvard University Press, 1989).

51 Ibid., 50–58.

52 Roland Barthes also picks up on this idea in *S/Z*, as Katherine Hayles notes in her discussion of information theory and poststructuralism. See Roland Barthes, *S/Z* (New York: Hill & Wang, 1974), 144–147. See also Hayles, 186–196.

53 "Googlebombing Failure," *The Official Google Blog* September 16, 2005, http://googleblog.blogspot.com/2005/09/googlebombing-failure.html.

54 Matthew Fuller and Andrew Goffey, "Toward an Evil Media Studies," in *The Spam Book: On Viruses, Porn, and Other Anomalies from the Dark Side of Digital Culture*, ed. Jussi Parikka and Tony D. Sampson (Cresskill, NJ: Hampton Press, 2009), 141–159.

55 Ibid., 143.

56 Henry Jenkins, *Convergence Culture* (New York: New York University Press, 2006), 1–2.

57 Ibid., 3.

58 Mark Poster, *Information Please* (Durham: Duke University Press, 2006), 21, 24.

59 Fuller and Goffey, 143–144.

Hack

Chapter 1

Revealing Errors

Benjamin Mako Hill
Massachusetts Institute of Technology

Introduction

In "The World is Not a Desktop," Marc Weisner, the principal scientist and manager of the computer science laboratory at Xerox PARC, states that, "a good tool is an invisible tool."[1] Weisner cites eyeglasses as an ideal technology because with spectacles, he argued, "you look at the world, not the eyeglasses." Although Weisner's work at PARC played an important role in the creation of the field of "ubiquitous computing," his ideal is widespread in many areas of technology design. Through repetition, and by design, technologies blend into our lives. While technologies, and communications technologies in particular, have a powerful mediating impact, many of the most pervasive effects are taken for granted by most users. When technology works smoothly, its nature and effects are invisible. But technologies do not always work smoothly. A tiny fracture or a smudge on a lens renders glasses *quite* visible to the wearer.

Anyone who has seen a famous "Blue Screen of Death"—the iconic signal of a Microsoft Windows crash—on a public screen or terminal knows how errors can thrust the technical details of previously invisible systems into view. Nobody knows that their ATM runs Windows until the system crashes. Of course, the operating system chosen for a sign or bank machine has important implications for its users. Windows, or an alternative operating system, creates affordances and imposes limitations. Faced with a crashed ATM, a consumer might ask herself if, with its history of rampant viruses and security holes, she should *really* trust an ATM running Windows.

Technology is powerful for a number of reasons. First, technologies make previously impossible actions possible and many actions easier. In the process, they frame and constrain possible actions. For example, communication technologies allow users to communicate in new ways but constrain communication in the process; in a very fundamental way, these technologies define what their users can say, to whom they say it, and how they can say it—and what, to whom, and how they cannot. Second, in a related sense, technology mediates. Acting as an intermediary, technology stands between any given user and an other, the users' goals, or information. As a function of its operation, technology transforms information as it passes it on. In the process, technology gives its operators and providers the immense power associated with the ability to monitor or change users' messages—often without users' knowledge. Third and finally, in a very broad sense, technology hides and abstracts by presenting users with a series of "black boxes." By intentionally controlling what users can see and understand, technology designers' and providers' implicit power becomes more fully entrenched and firmly enforced. In all three cases, the values of technology designers and providers play an important role in shaping users' experience.

Humanities scholars understand the power, importance, and limitations of technology and technological mediation. Weisner hypothesized that, "to understand invisibility the humanities and social sciences are especially valuable, because they specialize in exposing the otherwise invisible."[2] Technology activists, like those at the Free Software Foundation (FSF) and the Electronic Frontier Foundation (EFF), understand this power of technology as well. Largely constituted by technical members, both organizations, like humanists studying technology, have struggled to communicate their messages to a less technologically savvy public. Before one can argue for the importance of individual control over who owns technology, as both FSF and EFF do, an audience must first appreciate the power and effect that their technology and its designers have. To understand the power that technology has on its users, users must first *see* the technology in question. Most users do not. Both the EFF and the FSF have struggled in their appeals to technology users who are not also technologists and developers—the communities both organizations are explicitly dedicated to serve.

Errors are underappreciated and underutilized in their ability to reveal technology around us. By painting a picture of how certain technologies facilitate certain mistakes, one can better show how technology mediates. By revealing errors, scholars and activists can reveal previously invisible technologies and their effects more generally. Errors can reveal technologies and their power and can do so in ways that users of technologies confront daily and understand intimately.

Affordances

Errors can reveal distinct and overlapping aspects of the technologies that mediate our lives and the designers of those technologies. First, and perhaps most fundamentally, errors can reveal the affordances and constraints of technology that are often invisible to users. Through these affordances and constraints, technologies make it easier to do some things, rather than others, and either easier or more difficult to communicate certain messages. Errors can help reveal these hidden constraints and the power that technology imposes.

The misprinted word

Catalyzed by Elizabeth Eisenstein, the last 35 years of print history scholarship provides both a richly described example of technological change and an analysis of its effects.[3] Unemphasized in discussions of the revolutionary social, economic, and political impact of printing technologies is the fact that, especially in the early days of a major technological change, the artifacts of print are often quite similar to those produced by a new printing technology's predecessors. From a reader's purely material perspective, books are books; the press that created the book is invisible or irrelevant. Yet, while the specifics of print technologies are often hidden, the affordances of any particular print technology has important effects on what is printed and how, effects that are often exposed by errors.

While the shift from scribal to print culture revolutionized culture, politics, and economics in early modern Europe, it was near-invisible to early readers. Early printed books were the same books printed in the same way; the early press was conceived as a "mechanical scriptorium."[4] Gutenberg's black-letter Gothic typeface closely reproduced a scribal hand. Of course, handwriting and type were easily distinguishable; errors and irregularities were inherent in relatively unsteady human hands.

Printing, of course, introduced its own errors. As pages were produced *en masse* from a single block of type, so were mistakes. While scribes would reread and correct errors as they transcribed a second copy, no printing press could. More revealingly, print opened the door to whole new categories of errors. For example, printers setting type might confuse an inverted *n* with a *u*—and many did. Of course, no scribe made this mistake. An inverted *u* is only confused with an *n* due to the technological possibility of letter flipping in movable type. As print

moved from Monotype and Linotype machines, to computerized type-
setting, and eventually to desktop publishing, an accidentally flipped *u*
retreated back into the realm of impossibility.[5]

Most readers do not know how their books are printed. The output
of letter presses, Monotypes, and laser printers are carefully designed
to produce near-uniform output. To the degree that they succeed, the
technologies themselves, and the specific nature of the mediation,
becomes invisible to readers. But each technology is revealed in errors
like the upside-down *u*, the output of a mispoured slug of Monotype, or
streaks of toner from a laser printer.

Changes in printing technologies after the press have also had
profound effects. The creation of hot-metal Monotype and Linotype,
for example, affected decisions to print and reprint and changed how
and when it is done. New mass printing technologies allowed for the
printing of works that, for economic reasons, would not have been
published before. While personal computers, desktop publishing software, and laser printers make publishing accessible in new ways, it also
places real limits on what can be printed. Print runs of a single copy—
unheard of before the invention of the type-writer—are commonplace.
But computers, like Linotypes, introduce their own affordances and
constraints and render certain formatting and presentation difficult
and impossible.

Errors provide a space where the particulars of printing make technologies visible in their products. An inverted *u* exposes a human typesetter, a
letterpress, and a hasty error in judgment. Encoding errors and botched
smart quotation marks—a *?* in place of a *"*—are only possible with a
computer. Streaks of toner are only produced by malfunctioning laser
printers. Dust can reveal the photocopied provenance of a document.

Few readers reflect on the power or importance of the particulars of
the technologies that produced their books. In part, this is because the
technologies are so hidden behind their products. Through errors, these
technologies and the power they have on the "what" and "how" of printing
are exposed. For scholars and activists attempting to expose exactly this,
errors are an underexploited opportunity.

Typing mistyping

While errors have a profound effect on media consumption, their effect
is perhaps more strongly felt during media creation. Like all mediating
technologies, input technologies make it easier or more difficult to

create certain messages. It is, for example, much easier to write a letter with a keyboard than it is to type a picture. It is much more difficult to write in languages with frequent use of accents on an English language keyboard than it is on a European keyboard. But while input systems like keyboards have a powerful effect on the nature of the messages they produce, they are invisible to recipients of messages. Except when the messages contain errors. Typists are much more likely to confuse letters in close proximity on a keyboard than people writing by hand or setting type. As keyboard layouts switch between countries and languages, new errors appear. The following is from a personal email:

> *hez*, if there's not a subversion server *handz*, can i at least have the root password for one of our machines? I read through the instructions for setting one up and i think i could do it. [emphasis added]

The email was quickly typed and, in two places, confuses the characters *y* with *z*. Separated by five characters on QWERTY keyboards, these two letters are not easily mistaken or mistyped. However, their positions are swapped on German and English keyboards. In fact, the author was an American typing in a Viennese Internet cafe. The source of his repeated error was his false expectations—his familiarity with one keyboard layout in the context of another. The error revealed the context, both keyboard layouts, and his dependence on a particular keyboard. With the error, the keyboard, previously invisible, was exposed as an inter-mediator with its own particularities and effects.

This effect does not change in mobile devices where new input methods have introduced powerful new ways of communicating. SMS messages on mobile phones are constrained in length to 160 characters. The result has been new styles of communication using SMS that some have gone so far as to call a new language or dialect called TXTSPK.[6] Yet while they are obvious to social scientists, the profound effects of text message technologies on communication is unfelt by most users who simply see the messages themselves. More visible is the fact that input from a phone keypad has opened the door to errors which reveal input technology and its effects.

In the standard method of SMS input, users press or hold buttons to cycle through the letters associated with numbers on a numeric keyboard (e.g., *2* represents *A*, *B*, and *C*; to produce a single *C*, a user presses *2* three times). This system makes it easy to confuse characters based on a shared association with a single number. Tegic's popular T9 software allows users to type in words by pressing the number associated with each letter of each word in quick succession. T9 uses a database to pick the most likely word that maps to that sequence of numbers. While the system allows for quick input of words and phrases on a phone keypad, it also allows for the creation of new types of errors. A user trying to type *me* might accidentally write *of* because both words are mapped to the combination of 6 and 3 and because *of* is a more common word in English. T9 might confuse *snow* and *pony* while no human, and no other input method, would.

Users composing SMSes are constrained by the affordances of SMS technology. The fact that text messages must be short and the difficult nature of phone-based input methods has led to unique and highly constrained forms of communication like TXTSPK.[7] Yet, while the influence of these technological affordances is profound, users are rarely aware of them. Errors can expose the particularities of a technology and, in doing so, provide an opportunity for users to connect with scholars speaking to the power of technology and to activists arguing for increased user control.

Of course, affordances and constraints are far from arbitrary. In the case of SMS length restrictions, there are historical technical reasons for the limitation. In the case of T9, constraints imposed by small handhelds with a small number of buttons informed that technology's design. But affordances and constraints are ultimately designed and implemented by human designers who have enormous power to impose their own values in their design. This power, and the important design decisions through which it is enacted, is also revealed through errors.

Any T9 user attempting to input profanity onto their phone will run into such an affordance. To avoid the impropriety of suggesting a four-letter word as a possible completion, Tegic removes all profanity from its T9 wordlist. The result, lampooned at one point by the British comedy duo Armstrong and Miller on the BBC, is improbable suggestions like "shiv" and "ducking" (both first suggestions for their respective key combinations) followed by decreasingly plausible options.

As ugly and offensive as they may be to some, "shit" and "fucking" are English words and the inability to type them on a T9 system is an error from the perspective of any user unable to complete a profane message using T9's software. Built from word frequency counts of SMS messages and existing databases, T9's inability to compose profanity is an intentional decision on the part its designers. Not only does the error reflect Tegic's corporate values, it codifies them and forces them onto its users.

The result is that it is easier to write a non-profane message using T9 than it is to write a profane one. It is easier to write "darn" than it is to write "damn." In a small but important way, T9 forces the values of its designers on its users every time to every user who tries to swear using their technology. But in that the result is, from the perspective of the user, in error, it also makes those values visible. In the process, it reveals one of the affordances of a normally invisible technology, points toward a plausible description of "why," and reveals the power of designers, in at least one small way, to determine what users can and cannot say.

Hidden Intermediaries

While affordances constrain what users *can* do or say, the role of technology as an intermediary gives technology designers the ability to change what users say after the effect. In that it transmits, repeats, and copies users messages, communication technologies in particular intermediate as a fundamental function. Especially in online environments, any message is invisibly passed through a barrage of technological intermediaries that can monitor, censor, and even alter its content. To the degree that these technologies are invisible, this intermediation is invisible as well. Of course, errors can reveal the presence of this hidden mediation.

Clbuttic

Tegic's desire to avoid profanity, as described in the previous section, is hardly unique. Their method however, of making swearing possible, but more difficult, than not swearing, is less heavy handed than some other approaches. Another large class of errors results from intermediaries that monitor, censor, and even change user input to avoid profanity. A now famous mistake, repeated several times, has become known as the "clbuttic" mistake.

The term "cbluttic" is the result of a computer program that monitored user input to an online discussion forum and swapped out instances of profanity with less offensive synonyms. For example, "ass" might become "butt," "shit" might become "poop" or "feces," etc. To work correctly, the script must look for instances of profanity between word boundaries— that is, profanity surrounded on both sides by spaces or punctuation.

On a number of occasions, programs did not. As a result, not only was "ass" changed to "butt," but any word that contained the letters "ass" were transformed as well. The word "classic" was mangled and left as "clbuttic." Today, web searches for "clbuttic" turn up thousands of hits on dozens of independent web sites. In the same vein, one can find references to a "mbuttive music quiz" a "mbuttive multiplayer online game," references to the average consumer is a "pbutterby," a "transit pbuttenger executed by Singapore," "Fermin Toro Jimenez (Ambbuttador of Venezuela)," "the correct way to deal with an buttailant armed with a banana," and a reference to how "Hinckley tried to buttbuttinate Ronald Reagan."[8]

For any reader, each error reveals the presence of an intermediary in the form of an anti-profanity script; obviously, no human would accidentally misspell or mistake the words in question in any other situation.

What is perhaps more impressive than this error is the fact that most programmers do not make this mistake when implementing similar systems. On thousands of web sites, posts and messages and interactions are "cleaned-up" and edited without authors' consent or knowledge. As a matter of routine, their words are silently and invisibly changed. Few authors, and even fewer of their readers, ever know the difference. Each error is a stark reminder of the power that technology gives the designers of technical systems to force their own values on users and to frame—and perhaps to substantively change—the messages that their technologies communicate.

Tyson Homosexual

In the lead up to the 2008 Olympic games in Beijing, One News Now, a news web site run by the conservative Christian American Family Association published an Associated Press article with the headline, "Homosexual eases into the 100 final at the Olympic trials." The AP published no article with such a headline. The first paragraph revealed the source of the error and a hidden intermediary running on the ONN web site in its explanation that, "Tyson Homosexual easily won his semifinal for the 100 meters at the U.S. Olympic track and field trials and seemed to save something for the final later Sunday." Of course, there is no U.S. sprinter named Tyson Homosexual; but there is one named Tyson Gay.

ONN provides Christian conservative news and commentary. One of the things they do is offer a version of the industry standard Associated Press news feed. Rather than just republishing it verbatim, however, AFA runs software to modify the feed's language so it more accurately reflects their organization's values and choice of terminology. They do so with a hidden intermediary in the form of a computer program.

Indeed, the error is similar to the "clbuttic effect" described above—an errant text filter attempting to "clean up" text by replacing offensive terms with theoretically more appropriate ones. Among other substitutions, AFA replaced the term "gay" with "homosexual"—many conservative Christian groups prefer the term homosexual which they argue sounds more clinical and pathological than "gay." In this case, their software changed the name of champion sprinter and U.S. Olympic hopeful Tyson Gay to "Tyson Homosexual."

AFA never advertised the fact that it changed words in its AP feed, and it did so silently. One must assume that most of ONN's readers never realized that the messages they received and the terminology used was being intentionally manipulated by AFA. AFA's substitution, and the error it created, revealed the presence of a hidden script downloading news articles and processing them before they were published. And yet, ONN had been publishing its edited feed for years before anyone realized the presence of the intermediary. The reason, of course, is that unlike the clbuttic example, most of the time, AFA's intermediary worked correctly. It was not until the system produced an error was AFA's invisible mediator thrust into view.

Although few edit the content of the article quite like ONN, every news feed uses technology to parse and edit articles before publishing to some degree. Errors can reveal the presence of these hidden intermediaries. In doing so, these errors can highlight the power that technological designers and service providers have over our communication. Nearly every message we send or receive is as vulnerable to the type of manipulation that the AP's article is.

However, the Tyson Gay error also reveals a set of values that AFA and ONN have about the terminology around homosexuality and the way that these values invisibly frame users' experience of reading an AP article through their system. It is possible that many of ONN's readers know and support AFA's values and their choice of terminology changes. What they did not know is the ways in which the media they're consuming are being automatically and silently manipulated and mediated. AFA's error thrust hidden technology into view, giving a clear image of the power that designers and service providers have over their users.

Opening Black Boxes

As technologies become more complex, they often become more mysterious to their users. While not invisible, users know little about the way that complex technologies work both because they become accustomed to them and because the technological specifics are hidden inside companies, behind web interfaces, within compiled software, and in "black boxes."[9] Errors can help reveal these technologies and expose their nature and effects. As technology becomes complex, the purpose of technology is to hide this complexity. As a result, the explicit creation of black boxes becomes an important function of technological design processes and a source of power. Once again, errors that break open these boxes can reveal hidden technology and its power.

Google News denuded

Google's News aggregates news stories and is designed to make it easy to read multiple stories on the same topic. The system works with "topic clusters" that attempt to group articles covering the same news event. The more items in a news cluster (especially from popular sources) and the closer together they appear in time, the higher confidence Google's algorithms have in the "importance" of a story and the higher the likelihood that the cluster of stories will be listed on the Google News page. While the decision to include or remove individual sources is made by humans, the act of clustering is left to Google's software.

Because computers cannot "understand" the text of the articles being aggregated, clustering happens less intelligently. We know that clustering is primarily based on comparison of shared text and keywords—especially proper nouns. This process is aided by the widespread use of wire services like the Associated Press and Reuters which provide article text used, at least in part, by large numbers of news sources. Google has been reticent to divulge the implementation details of its clustering engine but users have been able to deduce the description above, and much more, by watching how Google News works and, more importantly, how it fails.

For example, we know that Google News looks for shared text and keywords because text that deviates heavily from other articles is not "clustered" appropriately—even if it is extremely similar semantically. In this vein, blogger Philipp Lenssen gives advice to news sites who want to stand out in Google News:

> Of course, stories don't have to be exactly the same to be matched—but if they are too different, they'll also not appear in the same group.

If you want to stand out in Google News search results, make your article be original, or else you'll be collapsed into a cluster where you may or may not appear on the first results page.[10]

While a human editor has no trouble understanding that an article using different terms (and different, but equally appropriate, proper nouns) is discussing the same issue, the software behind Google News is more fragile. As a result, Google News fails to connect linked stories that no human editor would miss.

But just as importantly, Google News can connect stories that most human editors will not. Google News's clustering in April of 2006, for example, of two stories—one by *Al Jazeera* on how "Iran offers to share nuclear technology," and one by the *Guardian* on how "Iran threatens to hide nuclear program," seem at first glance to be a mistake. Hiding and sharing are diametrically opposed and mutually exclusive.

But while it is true that most human editors would not cluster these stories, it is less clear that it is, in fact, an error. Investigation shows that the two articles are about the release of a single statement by the government of Iran on the same day. The spin is significant enough, and significantly different, that it could be argued that the aggregation of those stories was incorrect—or not.

The "error" reveals details about the way that Google News works and about its limitations. It reminds readers of Google News of the technological nature of their news's meditation and gives them a taste of the type of selection—and mis-selection—that goes on out of view. Users of Google News might be prompted to compare the system to other, more human methods. Ultimately it can remind them of the power that Google News (and humans in similar roles) has over our understanding of news and the world around us. These are all familiar arguments to social scientists of technology and echo the arguments of technology activists. By focusing on similar errors, both groups can connect to users less used to thinking in these terms.

Show Me The Code

A while ago, software engineer Mark Pilgrim wrote about being prompted with the license agreement shown in Figure 1.1.[11]

Most people have difficulty parsing the agreement because it is not the *text* of the license agreement being presented but the "marked up" XHTML code. Of course, users are only supposed to see the processed

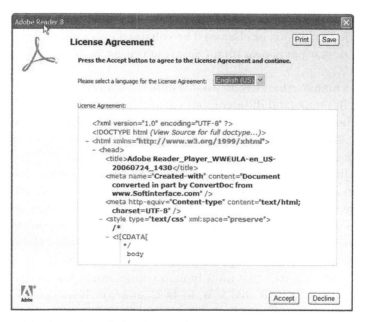

FIGURE 1.1   An unintended code reveal, as captured by Mark Pilgrim, http://diveintomark.org/archives/2007/11/26/wtf-adobe-reader-8

output of the code and not the code itself. Due to an error, Pilgrim was shown everything. The result is useless.

Conceptually, computer science might be described as a process of abstraction. In an introductory undergraduate computer science course, students are first taught syntax or the mechanics of writing code that computers can understand (e.g., SICP). After that, they are taught abstraction. They will continue to be taught abstraction, in one form or another, until they finish their careers. In this sense, programming is just a process of taking complex tasks and then hiding—abstracting—that complexity behind a simplified set of interfaces. Then, programmers build increasingly complex tools on top of these interfaces and the whole cycle repeats. Through this process of abstracting abstractions, programmers build up systems of incredible complexity. The work of any individual programmer is like a tiny cog in a massive, intricate machine.

The error Mark Pilgrim encountered is interesting because it shows a ruptured black box—an acute failure of abstraction. Of course, many errors tell us very little about the software we're using. With errors like Pilgrim's, however, users are quite literally presented with a view of parts of the system that a programmer was explicitly trying to hide. While the error Pilgrim showcases is embarrassing for authors of the software that caused it, it is reasonably harmless. Users can understand how a technology works in more detail but learn little that the technology designer would be hesitant to share. In other cases, the view into a broken black box can be shocking.

On two occasions, Facebook accidentally configured their web server to publish source code to the software that runs the Facebook network service—essentially all of the code that, at the time, ran Facebook. Reports at the time show that people looking at the code found little pieces including these code snippets (comments, written by Facebook's authors, are bolded):

```
$monitor = array( '42107457' => 1, '9359890' => 1);
//*Put baddies (hotties?) in here

/* Monitoring these people's profile viewage.
Stored in central db on profile_views.
Helpful for law enforcement to monitor stalkers and stalkees. */[12]
```

The first block describes a list of "baddies" and "hotties" represented by user ID numbers that Facebook's authors have singled out for monitoring. The second stanza is simply a comment that should be self-explanatory but that raised important concerns for privacy cautious users skeptical about their viewing habits being monitored in collaboration with law enforcement.

Facebook has since published a "source protector" plugin to their web server which will help them, and others who use it, avoid errors like these in the future. As a result, users are less likely to get further views of this type into the Facebook code. Of course, we have every reason to believe that this code, and perhaps other codes like it, still runs on the active version of Facebook. But as long as Facebook's black box works better than it has in the past, users will never know exactly what Facebook is doing with their data.

Like Facebook's authors, many technologists do not want users knowing what their technology is doing. Very often, designers use black boxes to create a better usability experience or a more manageable and modular implementation. Sometimes, like Facebook, technology is designed to do things that users are shocked and unhappy to learn about. Errors that provide clear views into black boxes provide a view into some of what we might be missing and reasons to be discomforted by the fact that many technologists go to extreme lengths to keep users in the dark.

Conclusion

Reflecting on the role of the humanities in a world of increasingly invisible technology for the blog, "Humanities, Arts, Science and Technology Advanced Collaboratory," Duke English professor Cathy Davidson writes:

> When technology is accepted, when it becomes invisible, [humanists] really need to be paying attention. This is one reason why the humanities are more important than ever. Analysis—qualitative, deep, interpretive analysis—of social relations, social conditions, in a historical and philo-sophical perspective is what we do so well. The more technology is part of our lives, the less we think about it, the more we need rigorous humanistic thinking that reminds us that our behaviors are not natural but social, cultural, economic, and with consequences for us all.[13]

Davidson concisely points out the strength and importance of the humanities in evaluating technology. She is correct; users of technolo-gies do not frequently analyze the social relations, conditions, and effects of the technology they use. Activists at the EFF and FSF argue that this lack of critical perspective leads to exploitation of users.[14] But users, and the technology they use, are only susceptible to this type of analysis when they understand the applicability of these analyses to their tech-nologies. Davidson leaves open the more fundamental question: How will humanists first reveal technology so that they can reveal its effects?

Scholars and activists must do more than contextualize and describe technology. They must first render invisible technologies visible. As the revealing nature of errors in printing systems, input systems, online forums, news feeds, and social networks show, errors represent a point where invisible technology is already visible to users. These errors can reveal several important features of technologies connected to the power that it, and its designers, have over users. In particular, these examples can speak to the power of technological affordance constraints, tech-nologies that act as intermediaries, and the technology that uses "black boxes" in explicit attempts to hide the technology in question. In all three cases, errors can also reveal the values of the technologies' design-ers. As such, these errors, and countless others like them, can be treated as the tip of an iceberg. They represent an important opportunity for humanists and activists to further expose technologies and the beginning of a process that aims to reveal much more.

Notes

[1] Marc Weisner, "The World is not a Desktop," *Interactions* 1, no. 1 (January 1994): 7.

[2] Ibid., 8.

[3] Elisabeth L. Eisenstein, *The Printing Press as an Agent of Change: Communications and Cultural Transformations in Early-Modern Europe* (Cambridge: Cambridge University Press, 1979).

[4] Ibid.

[5] For further discussion, see, Ottmar Mergenthaler, *The Biography of Ottmar Mergenthaler, Inventor of the Linotype* (New Castle, DE: Oak Knoll Books, 1989); Alvin Garfield Swank and United Typothetae of America, *Linotype Mechanism* (Chicago: Deptartment of Education, United Typothetæ of America, 1926).

[6] Crispin Thurlow. "Generation Txt? The Sociolinguistics of Young People's Text-Messaging," *Discourse Analysis Online* 1, no. 1 (2003), http://extra.shu.ac.uk/daol/articles/v1/n1/a3/thurlow2002003.html.

[7] John Sutherland, "Cn u txt?" *Guardian Unlimited* November 11, 2002, http://www.guardian.co.uk/technology/2002/nov/11/mobilephones2.

[8] Links to web pages that include each of these examples are in a blog post on my *Revealing Errors* blog here: http://revealingerrors.com/clbuttic

[9] Bruno Latour, *Pandora's Hope: Essays on the Reality of Science Studies* (Cambridge, MA: Harvard University Press, 1999).

[10] Philipp Lenssen, "How Google News Indexes, "*Google Blogscoped* July 28, 2006, http://blogoscoped.com/archive/2006-07-28-n49.html.

[11] Mark Pilgrim, "WTF: Adobe Reader 8," *Dive into Mark* November 26, 2007, http://diveintomark.org/archives/2007/11/26/wtf-adobe-reader-8.

[12] Although Facebook has successfully stopped the source code from being distributed, several news articles and blog posts are still available online and on the Internet Archive. Some of these are linked from an article describing this incident on my *Revealing Errors* blog here: http://revealingerrors.com/show_me_the_code.

[13] Cathy Davidson, "When Technology Is Invisible, Humanists Better Get Busy," *HASTAC* June 7, 2007, http://www.hastac.org/node/779.

[14] Richard M. Stallman, in *Free Software, Free Society: Selected Essays of Richard M. Stallman*, ed. Joshua Gay. (Boston: Free Software Foundation, 2002).

Chapter 2

Aesthetics of the Error: Media Art, the Machine, the Unforeseen, and the Errant

Tim Barker
University of New South Wales

The machine must become a work of art! We will discover the art of the machine.
Bruno Munari,
Manifesto del Macchinismo, *1952*

Bruno Munari's comments published in 1952 come from a manifesto aimed at reconciling art with the machine.[1] Founded upon Futurist ideals and promoting the art and design movement known as "Movimento d'Arte Concreta" (MAC), in this manifesto Munari aimed to compel artists to abandon their "romantic" oil paints and embrace the "anatomy" of the machine. The creative expression of this attitude can be seen in his series of *Useless Machines*, produced prior to the publication of the manifesto, between the 1930s and 40s. The *Useless Machines* were abstract forms made of lightweight materials strung together by thin threads and designed to be suspended in mid-air. These "machines," which could be termed kinetic sculptures similar to the hanging mobiles of Alexander Calder, were designed to interact with their surrounding environment.[2] They have no internal power source to drive them. Rather, they rely on external forces, such as the wind, to set them in a gentle undulating motion. As such, we could say that the movements of the *Useless Machines* are not pre-scripted or programmed by Munari; the artist does not define the machine's routine. Instead, Munari designs them with only one purpose in mind: to allow them to find their own creative force. Of course he builds them, designing their weight and shape and thus directing the way they will behave in the wind, setting limitations on what they can achieve, setting the degrees of freedom in which they must operate, but the machine itself works with its external forces, which are unpredictable, to actuate a particular function. The art of the machine here is an art in which the machine, after being built by human hands, is itself creative.

We can bring this concept of creativity to bear on recent digital art and the aesthetics of the error by thinking of the art of the machine as an art outside the machine's pre-programmed routine. It is an art outside of the errorless algorithm, for this would merely amount to an art of the computer programmer. Rather the art of the machine is the art of open systems, relationships, and importantly the capacity to actualize process, which may lead to unexpected and errant outputs.

In order to understand the aesthetics of the error in digital art I would like to separate this argument into two sections. The first, titled "The Art of the Machine/The Art of the Error" examines artworks in which the artist's role is to set up situations in which errors manifest, and to exploit these errors in the art making process. In this section I also tie this practice to art history, namely to the aesthetic experiments that were undertaken in the early to mid-twentieth century, specifically those that relied on outside and largely unpredictable forces. In the second section, titled "Errors/Potentials/Virtuality," I put forward an understanding of the error as an outcome of particular conditions and potentials embedded in digital technologies. Throughout this argument I am interested in what Gilles Deleuze describes as the actualization of the virtual, a process by which novel unforeseen and unformed events are made actual. The term "virtual" as Deleuze uses it, and as will be explained later in the argument, signifies a conditioning that directs the way that the present moment actualizes, which, in our case, involves the becoming of an error.[3] As such, both sections focus on process: the first focuses on the creativity of an error that arises from a set of conditions established by an artist; the second then focuses on the process by which an error comes into being, attempting to understand both philosophically and aesthetically how errors, and in fact any unforeseen information, manifests in our interactions with technology.

The Art of the Machine/The Art of the Error

The condition that marks the post-digital age may be precisely the condition for error. In the condition where machinic systems seek the unforeseen and the emergent, there is also a possibility for the unforeseen error to slip into existence. This condition can be seen in the tradition of artists using the error, just as Munari used the wind, as a creative tool. The difference though is that whereas the *Useless Machines* rely on outside forces, the error is something internal to the machine, it is something that is immanent to the machine's process. In his paper

"The Aesthetics of Failure: 'Post-Digital' Tendencies in Contemporary Music," Kim Cascone points to the way in which composers, using digital means, exploit the inadequacies of a particular compositional or performative technology.[4] Cascone cites composers such as Ryoji Ikeda who create minimalist electronic compositions using media as both their form and theme. In these compositions, the errors, imperfections, and limitations of the particular compositional media are the central constituting elements of the piece. In addition to music, this glitch aesthetic is also exploited in the visual arts. Artists such as Tony Scott set up situations in which errors are able to emerge and be exploited in the art making process. Scott's work in his *Glitch* series consists of brightly colored geometric forms, assembled in rhythmic compositions, that on the surface appear to be in the mold of Frank Stella's linear works or Bridget Riley's Op Art. However, these digital prints are assembled from the visual outputs of computer crashes and software errors caused by Scott. In these types of work the artist's role is to prompt a glitch or an error to arise in a specific system, then to reconfigure and exploit the generative qualities of the unforeseen error.

The approach to art making as an emergent process, realized as an interaction between the artist and the limitations of the machine, is also seen in the live VJ performances of Jorge Castro. In the first few minutes of Castro's video performance *Messy* (2005), flashes of a moving image, which slowly moves through a landscape, erratically comes to view through various effects of visual noise. This error, intentionally sought out by the artist and given form by digital means, appears to be the product of an analogue tape head sporadically losing contact with the videotape. Here, flashes of an intelligible image—a picturesque Argentinian landscape—are seen through the unintelligible patterns caused by the loss of video signal. Similarly, in his work *Witness* (2006) (Figure 2.1) Castro uses error as an aesthetic device. In this work the digital error is similarly used to *re*-present the original source material. In this video work a large man sits in front of a building, head down, asking for money. Only one woman, from the steady stream of passers by, stops to acknowledge the man on the corner, dropping money into his jar. The video work continuously loops back in staccato jump-cuts to the interaction between this woman and the heavy set man, seeming to skip, as though a glitch in the code, then returning to the stream of passers by. Apart from the man and the woman, all figures are distorted by either a vertical extension of the upper parts of their bodies, continuing upward out of frame, or a horizontal extension of their lower bodies

FIGURE 2.1   Jorge Castro, *Witness* (still), 2006. Reproduced with permission of the artist

out of frame. This gives the impression that there has been an augmentation to the digital signal, a mistranslation from the code of the computer to the image on the screen. A distinctly digital aesthetic is created here as the generative qualities and idiosyncrasies of an error make obvious the processes of the computer.

The aesthetics of the error or the aesthetics of the glitch, however, are not necessarily something new and are not necessarily confined to the realm of the digital. For instance, at the beginning of the 1960s Franz Erhard Walther, experimenting with art "informel," utilized errors and chance occurrences as generative tools. The story goes that Walther was using a bucket full of water to weigh down a collage when the bucket unexpectedly sprang a leak and drenched the paper, seemingly ruining the collage. However, as the paper dried, it took on new forms, unforeseen by the artist.[5] From here Walther's role as an artist changed: he now saw his role to set up situations in which transformational processes could occur. The emphasis of the creative act is now on process, and in particular opening up this process to outside forces. This includes his famous *Werkstücke* (Work Pieces), works made of fabric that the viewer is invited to wear or to actively manipulate in order to "activate" the art-work. Prior to this, artists such as Jean Arp, as he dropped cut-outs onto a canvas, allowed the creative process to be directed by the outside forces of gravity and wind. Similarly, the Japanese artist Fujiko Nakaya, in her fog sculptures from the 1970s onward, allowed the wind and the landscape to guide the shape and intensity of a mist, generated by water forced through small nozzles. Also, the group that have become known as "process artists," artists such as Bruce Nauman, allowed the act of making to take precedence over the finished form. The artist's job was to initiate process, not to control the outcome. All of these aesthetic experiments are involved in a process of actuating an event from a field of potential; the works take on a particular form because of activities and interactions that occur between their generative elements.

In a sense the artist sets up particular degrees of freedom as a set of internal limitations in which the system must function, directing, but not producing, the end product. The idea of *degrees of freedom* comes from Manual DeLanda's work on the philosophy of science. This concept, which DeLanda takes from the discipline of mathematics, refers to the ways in which an object may change.[6] For instance, he gives the example of a pendulum, which, as it can only ever change its velocity and its position, has two degrees of freedom. He also points out that a bicycle, because it has – for the sake of the argument – five moving parts, (handle bars, front wheel, crank-chain-rear-wheel assembly, and two pedals) has ten degrees of freedom, as each part may change its position and its velocity.[7] The degrees of freedom are thus the limits in which a system unfolds; they are the boundaries that direct the process of the system. Transplanting this thinking to the aesthetics of the machine and the aesthetics of the error, we see that aesthetic processes are an output of a particular condition that is set up by the artist. Artists working with digital media such as Scott and Castro and artists using more traditional media such as Munari and Walther set these degrees of freedom upon the creative systems. They restrict or design the conditions of the machine. In Munari's case this involves designing the wind resistance of his mobiles, and for Walther, the mechanics of his works involves bringing certain materials together, such as paper and water or a sculpture made out of a piece of fabric and a viewer, and allowing them to work on one another. For Scott and Castro the aesthetics of the machine, and the subsequent aesthetics of the error, involve setting up situations that might cause computer crashes, software glitches, and errors in information processing.

Scott and Castro thus direct the machine toward a particular operation, setting up the conditions for an error to emerge, but not actually solely designing the aesthetics of the error. Instead, these aesthetic outputs are emergent based on the initial conditions that the artist sets and the machine's operation within these degrees of freedom. For instance, in *Witness*, as Castro uses a particular digital effect to simulate an error across the digital image, he initiates a process by which digital information is filtered through an effects channel. This is a process that alters the image; importantly though, this alteration is not completely foreseen by Castro. He has a general idea of the type of error that the effect will cause, due to his experience with the digital medium. But the actual appearance of the error is something that is generated as the original digital information comes into contact with the particular conditions of transmission established by Castro.

This can also be seen in the works of Joan Heemskerk and Dirk Paesmans, who together make up Jodi. In these Internet based works the user's computer seems to be running into errors at every instant of interaction. For instance, in *404.jodi.org* it is as though the user and her computer are stuck in a looping error. After accessing the work's web site the user encounters a brightly colored screen with the large error code "404": an error code that usually designates that the user's server could not find the requested file or web page. After clicking on the error code the user links to another page, which displays a list of seemingly random numbers. The user may send an email response on this page, which just seems to generate more meaningless numbers. The user may then click on any particular entry within the list of numbers, taking them to yet another page displaying the 404 error code. Throughout the interaction the user activates a selection of different pages, all of which display what seem to be various errors, triggered by the email responses, and all of which link to a new iteration of the 404 page, each time a different color. As the user moves through this network of linked web pages, what emerges from interaction is not that which is expected or at all controlled by the user. The user may interact with this work, sending an email response, but this information, which causes the seemingly meaningless and unrelated outputs, is passed through a filter that reduces the message to gibberish. This filtering renders the system unusable, in the traditional sense of Human Computer Interaction (HCI). In this work the user is unable to exploit the system; the system does not work *for* the user. Instead errors are continually unfolded from a system. By this, Jodi's formalist investigation of the digital medium exploits the limitations of the digital network and the errors that are enfolded in the system.

In relation to the technological mediations that occur in these works, we can understand the generative capabilities of error through Lev

Manovich's cultural communication model. Manovich explains that a "pre-media" or "pre-digital" cultural communication model represents the transmission of a signal as SENDER—MESSAGE—RECEIVER.[8] In this original model the sender encodes and transmits a message over a communication channel; as Manovich indicates, in the course of transmission the message is affected by any noise that exists along the communication channel. The receiver then decodes the message. Here the message is susceptible to error in two ways. First, the noise that originates in the communication channel may alter the message. Second, there may be discrepancies between the sender and receiver's code.[9] In order to propose a post-digital consideration of transmission, Manovich develops this model by including the sender and receiver's software. Post-digital cultural communication can now be considered as SENDER—SOFTWARE—MESSAGE—SOFTWARE—RECEIVER.[10] In this model the cultural significance of software is emphasized. The software, much more than the noise introduced by the communication channel, may change the message. Significantly, the software may introduce an error into the message.

For instance, this can be seen in *404.jodi.org*. In this work the user sends a message, using the dialogue box at the bottom of the web page. The message is composed as the user activates various basic computational processes in order to input her message into the dialogue box. The message is then passed through the web page's software, which in this case involves various filters that convert the message into an unrecognizable and unintelligible mess. For instance one filter appears to remove all the consonants from the message, as well as replacing certain characters with images of white rabbits. The software here is creative; it works with the user, translating her original message, in order to actualize an error.

The cultural role that Manovich ascribes to software also becomes elucidated in the emerging tradition of artists, such as the already mentioned Jodi, as well as Mark Daggert and the German artists group Rolux.org, that are producing alternative web browsers, in a sense interrogating the way in which most of us receive information from the web. One such project is Dimitre Lima, Iman Morandi, and Tony Scott's *Glitchbrowser*. Rather than attempting to assist user navigation, this browser creates errors when displaying the pages that it accesses. The images of any page accessed by *Glitchbrowser* are distorted or glitched through color saturation and abstraction. In this work, following Manovich's cultural communication model, the software that intervenes between sender and receiver alters the content of the message. Thus in *Glitchbrowser* as well as *404.jodi.org*, the artists remind us that the information we receive is largely reconstituted by the system that it travels through. In a sense the machine reveals itself, rather than creating the illusion of a transparent interface to information. In the application of *Glitchbrowser* the user witnesses the way that messages are transmitted and altered by the interface, with the machine reminding the user of its existence.[11]

In *Glitchbrowser* as well as Jodi's *404.jodi.org* we see Munari's aesthetics of the machine, perhaps updated to include an aesthetics of software. In both these works the machine's process that actuates an error is positioned as the creative force of the work. In both works information is filtered through a layer of software prior to its visualization on the screen interface. It is this process that creates the error, and it is this process where the "arthood" of the work is realized. This is a process in which software works with information designed either by an artist or a user. Just as the wind works with objects designed by Munari in order to set the *Useless Machines* in motion, so does the software work on information in order to realize the art of the error.

The art of the error is a type of art that is articulated by unforeseen processes, and it is this process-based aesthetic that links the art of the error to various domains of art history. There is a kind of Duchampian legacy, to borrow David Hopkin's term, emergent in these works. In his book *After Modern Art*, Hopkins traces a legacy from Marcel Duchamp, through Robert Rauschenberg, John Cage, Jasper Johns, and Ed Keinholz that positions art as a dematerialized concept that awaits actualization by a spectator.[12] To continue this pursuit we could situate the error or glitch aesthetic inside this paradigm. As has already been argued we can certainly think of the error in a system in the same manner as artists such as Walther and Arp think of chance as a creative tool. Just as Dada works such as *Collage Arranged According to the Laws of Chance* (1916–17) exploit the chance event as a creative force and hence move into the realm of the potential, works such as Jodi's and Lima, Morandi, and Scott's move into the unforeseen as errors direct the aesthetic. Also, we can see similarities between the art of the error and Rauschenberg's *White Paintings* (1951). These large rectangular white canvases, following Hopkins, are "passive receptors, awaiting events rather than prescribing sensations."[13] The canvases exist not as art objects in themselves, but await an audience to initiate their *transformation* into art. The works exist as empty spaces that are to be filled by the audience and all those peripheral events that occur around them. Art is not inside the *White Paintings*, but rather outside them. In this sense the *White Paintings*, as they exist as open potentiality, are similar to John Cage's famous piano composition *4'33"*. As the concert hall is filled with the silence of the performance of Cage's composition, the background noises and activities of an audience are allowed to fill the empty space. Similarly, works that utilize error involve setting up computer situations in which software fulfills the potential for error, just as the audience fills the potential of Cage's silence and Rauschenberg's whiteness. In this sense, in Cage's and Rauschenberg's work, as with the art of the error, both the artist and the audience find themselves in the field of the emergent. The artist must provide the condition for the emergent and unforeseen, and the audience (in Cage and Rauschenberg's case) or the software (for our concerns with error) must bring this condition to satisfaction.

Errors/Potentials/Virtuality

In order to further understand the way in which an error, as a potential event, may slip into existence, and the creativity that might arise via this process, we may turn to Deleuze's philosophy of the virtual. This philosophy is not a philosophy of events that have actually taken place, but rather a philosophy of process and the conditionings from which these events have emerged. As Steven Shaviro explains, "the virtual is like a field of energies that have not yet been expended, or a reservoir of potentialities that have not yet been tapped."[14] The virtual is the impelling force that drives a becoming; it is a conditioning that allows the production of something new or novel.[15] Understanding questions of aesthetics and technology through the virtual is thus not an investigation of the events that actually occurred in a system, but rather understanding the system based on the events that could have potentially taken place, if certain circumstances had been different.[16]

It must be remembered though that Deleuze's virtual is not some kind of transcendental essence or "ideal" that the actual is yet to become.[17] Deleuze's philosophy of the virtual moves past this type of transcendental idealism, instead positing what has been described as a transcendental *empiricism*: an approach that focuses on the *conditions* from which experience emerges. In other words, this approach privileges the field from which events, objects, and "things" come into being.[18] For our concerns a Deleuzian approach would privilege the *processes* that give form to the aesthetics of the error. This approach can be described as empiricism as it focuses on experience; however, for Deleuze, as different from traditional empiricism, the *conditions* and *potential* for experience are positioned as just as real as actual experience. These qualities are positioned as a real, but not actual, virtuality. They are the virtual qualities of every actual entity, and the elements that give actual events their character. Following in this line of thought we can understand diverse things like cities, societies, and people as well as technology and art not by their appearance or their role but rather by the invisible set of organizational structures, rules, laws, and protocols and their interaction with other individuals that directs their becoming.[19] For Deleuze there exists two planes of events developing simultaneously: on one level is actual events, as real events that are the solutions to particular problems, and on the other level is the virtual, as a set of ideal events embedded in the condition of the problem.[20] This would be the virtual that surrounds every actual event.

In these terms, the virtual, in respect to interactive art, may be thought of as a field of conditions, imposed by both the internal programming and limitations of the computer as well as the processes initiated by a human user. Following Deleuze, we may say that the software may articulate a link to the field of potential—in this case a field of potential errors—in order to generate unforeseen, and perhaps unwanted, information. This is because the virtual that Deleuze theorizes is a mode of reality that is articulated in the emergence of new potentials; to understand how any collective, assemblage, or machine, including the digital machine, is able to produce new or novel information we need to understand the virtual as a field of emergence, a field or grounding that conditions the manner in which novelties actualize. To change into something new the machine must enter into the field of the potential, and within this reality of change Deleuze's virtual, as the entirely real but not actual conditions of these potentialities, is always implicated.[21]

We can think of an error as just this potential that may or may not become actualized. The system that seeks the actualization of unforeseen potential is also a system that has the capacity to become errant; it is a system that is surrounded by a cloud of potential errors, or, as Deleuze would put it, a cloud of the virtual.[22] In other words, at any moment, any system that seeks the unforeseen, the novel, or the new is involved in the process of actualizing potential information. At any moment this system is traversing a field of potential. Within this field exists the virtual error, waiting to be actualized by an errant system. At any point in its process, a system is traversing potential errors and at any point, one may become actualized.

For instance, in traditional computer use, such as our everyday Internet searches, the computer's software activates particular pieces of information, navigating its way through an archive of data. But this is unstable; at any moment the machine may return an errant response, at any moment there may emerge a bug in the system. The technological mediation of data, while for the most part attempting to facilitate the clear exchange of information, may actually give rise to a greater potential for miscommunication. We can see this in its extreme in another of Jodi's Internet-based works, *Blogspot.jodi.org.* When accessing Jodi's blog, the web page is filled with error messages that appear to indicate that the blog has failed to load. Upon clicking on any one of the links presented within the error messages, the user activates other web pages that seem to be errant. It seems that the user is navigating a system that at every turn runs into an error.

Rather than thinking of a digital event as the process by which pre-formed or pre-conceived *possible* information becomes *realized,* we can only think of an error as coming into being as unformed and unforeseen *potential* is *actualized.* The error is potential in the sense that it is not pre-formed or pre-programmed by the artist. It can only be described as potential, which is inherent in the machine. This potential emerges from unique activities that occur in the process of a system, processes that deterritorialize the system, removing it from its usual functioning, which open the system so that unforeseen information may emerge.[23] If a system runs through its sequence of procedures without the potential for error it is essentially closed. In this stable, neat, and predictable transition from cause to effect there is no potentiality for the emergent or the unforeseen; there are no lines of escape. The system may, however, be destabilized or deterritorialized, a process by which parts of the assemblage of the machine are made to function against itself, moving the assemblage from its usual operation into a new regime.[24] It is only by allowing the capacity for potential errors, by moving away from the territory of the preconceived aesthetics of errorless machines, that we may provide the opportunity to think the unthought, to allow digital technologies to become-other.[25]

In a sense, when there is potential for an error to emerge in a system, the system cannot be regarded as a preformed linear progression. Rather, it can only be thought as a divergent process that actualizes elements of the virtual. In terms of error, the condition for an errant event is inbuilt into the digital encounter. When we interact with technology there is always the potential that we will activate an error. We might use the technology in the wrong way, or the technology itself may have a fault; either way the potential for interaction to return something other than what we ask for is always present. In other words, in any event of human-computer interaction that seeks to actualize the unforeseen, there is enfolded in the event the potential for error. This error may not become actualized, but it is there immanent to the system, waiting to be unfolded from the system, virtual.

Yann Le Guennec's *Le Catalogue* (2003-ongoing) (Figure 2.2) is an example of artist designed software causing unforeseen errors.[26] This Internet-based work allows public access to a catalogue of images and installations created between 1990 and 1996. Every time a page is accessed from the archive, an intended error is activated in the form of an intersecting horizontal and vertical line, generated at random points over the image. The more that the page is viewed, the greater its deterioration by the obscuring intersecting lines and the closer the

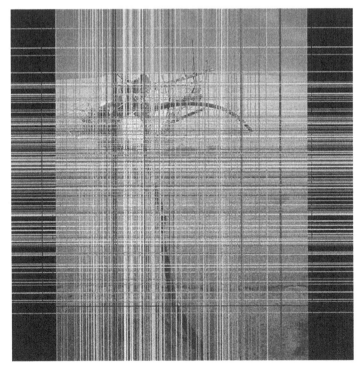

FIGURE 2.2 Yann Le Guennec, *Le Catalogue* (detail), 2003-ongoing. Reproduced with permission of the artist

image comes to abstraction. As Eduardo Navas states, "the archive is similar to analogue vinyl records losing their fidelity and being slightly deteriorated every time the needle passes through the groove."[27] In Le Guennec's catalogue the act of accessing and consulting the information of the archive in essence causes an internal error to the information. This is an error that is inbuilt; it is an error that we cause by the act of looking at or accessing any of the images. As we access the image we allow a potential error to become actual. Eventually the error will take over the original information, and the image will be more about error than it ever was about its referent.

Just as in Cascone's glitch music, the form and the theme of *Le Catalogue* is error. In the work we see the potential for error whenever information is mediated; *Le Catalogue* becomes a reflection on the act of looking, but looking through a particular paradigm, looking through the interface. The works can only be viewed by the act of accessing them in the archive, viewing them through processes associated with database management and computer software. But this act of looking destroys the images; they can only be preserved by allowing them to exist invisible and un-accessed, behind the interface. But this work is not about preservation. It is ultimately about the ephemeral and its uniqueness. Each error caused by the user—that, directed by a set of conditionings, becomes actual from a field of potential—is unique, and each time the archive is accessed it is differentiated from its past. Every time an image is accessed, it becomes its own original; every time an error from the field of the virtual is actualized, the unforeseen emerges. As Pierre Lévy states, "the virtual is that which has potential rather than actual existence . . . The tree is virtually present in the seed."[28] The seed does not know what shape the tree will take; instead it must actualize the tree as it enacts a process of negotiation between its internal limitations and the environmental circumstances that it encounters along the way. Likewise, the errant system does not know the errors that it may actualize; it must rather actualize these errors as it explores its degrees of freedom and the outside circumstances that might allow the emergence of error.

We can further see this process of the error in Cory Arcangel's *Data Diaries* (2002) (Figure 2.3).[29] In this work Arcangel extracts the data file of the computer's memory and inputs this information into QuickTime, directing the system to treat the data file as a video file, destabilizing the system and causing it to work against the clear and faultless transmission of information. The result is a visualization of data in abstract hard edge

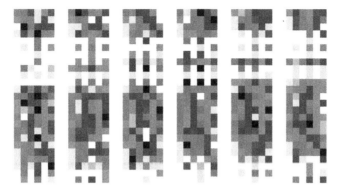

Figure 2.3 Cory Arcangel, *Data Diaries* (still), 2002. Reproduced with permission of the artist

patterns. As Alexander Galloway states in the introduction to this work, "every so often an artist makes a work of art by doing almost nothing. No hours of torturous labor, no deep emotional expression, just a simple discovery and out it pops. What did Cory Arcangel do in this piece? Next to nothing. The computer did the work, and he just gave it a form."[30] These images take on a distinct digital aesthetic, as they are primarily conjured by the machine, as it works within the degrees of freedom set by the artist. They are the result of computer programming, but a type of programming that is designed to produce an unforeseen error by asking a particular system to operate outside of its intended function. Neither Arcangel nor the system knows what shape the data-video will take, just as Levy's seed does not know the shape of the tree. Rather these images are actualized as the system operates within its internal programming, reading the information from the data file that Arcangel has inputted. Here, as with the artists discussed previously, the error has become an aesthetic tool that is exploited in the art making process. This is an art of the found object; an art practice in which the artist gives new meaning to an object, in this case, the error. It is an art that deterritorializes itself as the outputs of the QuickTime software is made unstable, and it is by this that Arcangel's work can produce the unforeseen. Galloway and Eugene Thacker,[31] as well as McKenzie Wark,[32] have previously explicated this kind of practice in terms of network culture. For these thinkers it is the act of hacking, of exploiting glitches in the network, that can be a deterritorializing or destabilizing force to the social organization implicit in the network. For both Arcangel's work and the hacking practices theorized by Wark and Galloway and Thacker, systems are allowed to become other, to operate outside of their intended function, to turn back against themselves, and to actualize emergent forms as they exploit the potentiality of the virtual.

As the potential for error marks the potential for the new and the unforeseen, we can see that an error in itself may be creative. An error may be utilized. It may be sought out and used to create the unforeseen within traditional systems, such as routine computer use, musical compositions, or visual art practice. In these instances, as the unique generative qualities of error are actualized, the artist can no longer be thought of as the sole creative force. Rather it is now the artist's role to provide the circumstances for an error to emerge. The error fills the potentiality of a system with meaning, whether intended or unintended by the designer. When an error occurs, unforeseen to the artist, the work is affected and possibilities are created for new meanings to emerge.

As Adrian Mackenzie has already pointed out in his book *Cutting Code*, and as is seen quite clearly in the aesthetics of the error that can be unfolded from digital systems, any contemplation of the reception of the image of the interface must also consider the aesthetics of the machine and its particular software.[33] The media event and its aesthetic are thus articulated not just by the image of the interface but also by the digitality of the system, including the work's computational processes as well as its interactivity. Manovich, as already discussed, also points out the cultural role played by software by positioning it as a creative force that not only operates as a tool, meant to be invisible to the user, but also plays an important role in cultural communication.[34] This has also been pointed out previously by Mark Hansen, who states, "the image can no longer be restricted to the level of surface appearance, but must be extended to encompass the entire process by which information is made perceivable."[35] As such any understanding of the aesthetic event of digital art needs always to take into account the processes that occur along the hidden layers of software.

We have seen that the aesthetics of the error involves an unfolding or becoming-situated within particular pre-scripted for preprogrammed conditions. As such, the works examined here, and the creativity of the error immanent to digital aesthetics, are at their core works that are performed over time by a set of processes. These processes—whether they be the transmission of digital information through an effects channel, as is the case with Castro's work, Arcangel's imposition of a data file into a system that reads video files, or the actions of a user who interacts with a system to cause errors—are all creative gestures that traverse a field of potential; they are all processes of destabilization or deterritorialization, processes that shake the system free of its precise or pre-programmed functioning. From this, the processes actualize unforeseen errors from the system, operating within the conditions set up by the artist, giving form to previously unformed information and generating a distinct aesthetics of the machine. As the wind blew Munari's *Useless Machine's*, as chance arranged Arp's collage, and as the audience actuated both Cage's and Rauschenberg's works, so too does the usually unseen levels of software reveal the creativity of the error.

Notes

[1] According to Pierpaolo Antonello's recent work on Munari, although published in 1952, the manifesto was actually written in 1938.
[2] Henri Gabriel, "The Hanging Mobile: A Historical Review." *Leonardo* 18, no.1 (1985): 40.
[3] Gilles Deleuze, *Cinema 2: The Time Image* (London: Continuum, 2005), 45.
[4] Kim Cascone, "The Aesthetics of Failure: 'Post-Digital' Tendencies in Contemporary Computer Music," *Computer Music Journal* 24, no.4 (2000): 13.
[5] Michael Linger, "From Being to Seeing: Michael Linger on Franz Erhard Walther," *FORUM International* 4 (1990): 50–55, http://ask23.hfbk-hamburg. de/draft/archiv/ml_publikationen/kt90–3_en.html.
[6] Manuel DeLanda, *Intensive Science and Virtual Philosophy* (London: Continuum, 2002), 13.
[7] Ibid.
[8] Lev Manovich, "Post-Media Aesthetics," in *(Dis)Locations*, ed. Astrid Sommer (Karlsruhe: ZKM, 2001), 18.
[9] Ibid.
[10] Ibid., 17–18.
[11] Lev Manovich, *The Language of New Media* (Cambridge, MA: The MIT Press, 2001), 206.
[12] David Hopkins, *After Modern Art 1945–2000* (Oxford: Oxford University Press, 2000), 41.
[13] Ibid., 42.
[14] Steven Shaviro, *Without Criteria: Kant, Whitehead, Deleuze and Aesthetics* (Cambridge, MA: The MIT Press, 2009), 34.
[15] Ibid.
[16] DeLanda, 35.
[17] Claire Colebrook, *Philosophy and Poststructural Theory: From Kant to Deleuze* (Edinburgh: Edinburgh University Press, 2005), 244.
[18] Constasis V. Boundas, *Deleuze and Philosophy* (Edinburgh: Edinburgh University Press, 2006), 9.

[19] Elizabeth Grosz, *Chaos, Territory, Art: Deleuze and the Framing of the Earth* (New York: Columbia University Press, 2008), 42–48.

[20] Gilles Deleuze, *Difference and Repetition*, trans. Paul Patton (London: Continuum, 1997), 237.

[21] Brian Massumi, "Sensing the Virtual, Building the Insensible," *Architectural Design* 68, no. 5/6 (1998).

[22] Gilles Deleuze and Claire Parnet, "The Actual and the Virtual," in *Dialogues 2*, ed. Eliot Ross Albert (London and New York: Continuum, 1987), 148.

[23] DeLanda, 36–37.

[24] Gilles Deleuze and Felix Guattari, *A Thousand Plateaus: Capitalism and Schizophrenia*, trans. Brian Masumi (London: Continuum, 2004), 156–158.

[25] D. N. Rodowick, *Reading the Figural, or, Philosophy after New Media* (Durham: Duke University Press, 2001).

[26] Available online at http://www.datapainting.com/catalogue/.

[27] Eduardo Navas, "Net Art Review November 30–December 6, 2003," http://www.netartreview.net/featarchv/11_30_03.html.

[28] Pierre Levy, *Becoming Virtual: Reality in the Digital Age* (New York: Plenum Trade, 1998), 23.

[29] Available online at http://turbulence.org/Works/arcangel/.

[30] Alexander Galloway, "Introduction to *Data Diaries*", in *Data Diaries*, http://www.turbulence.org/Works/arcangel/alex.php

[31] Alexander Galloway and Eugene Thacker, *The Exploit: A Theory of Networks* (Minneapolis: University of Minnesota Press, 2007).

[32] McKenzie Wark, *The Hacker Manifesto* (Cambridge, MA: Harvard University Press, 2004).

[33] Adrian Mackenzie, *Cutting Code: Software and Sociality* (New York: Peter Lang Publishing, 2006).

[34] Lev Manovich, "Avant-Garde as Software", in *Ostranenie*, ed. S. Kovats (Frankfurt and New York: Campus Verlag, 1999).

[35] Mark Hansen, *The New Philosophy of New Media* (Cambridge, MA: The MIT Press, 2004), 10.

Chapter 3

Information, Noise, et al.

Susan Ballard
Dunedin School of Art

The two companions scurry off when they hear a noise at the door. It was only a noise, but it was also a message, a bit of information producing panic: an interruption, a corruption, a rupture of information. Was the noise really a message? Wasn't it, rather, static, a parasite?

Michel Serres, 1982[1]

Since, ordinarily, channels have a certain amount of noise, and therefore a finite capacity, exact transmission is impossible.

Claude Shannon, 1948[2]

There is much truth in much of our knowledge, but little certainty. We must approach our hypothesis critically; we must test them as severely as we can, in order to find out whether they cannot be shown to be false after all.

et al., 2005[3]

Recording Information

Over the past ten years the New Zealand artist's collective et al. have deployed a series of gallery installations that operate at the edges of information and noise. These installations challenge viewers to engage with aesthetic tools that blur the disciplines, media, expectations, materials, and models through which an artwork is viewed. The possibilities suggested are at once confusing and didactic. The experience becomes one of chasing a series of constantly changing informatic moments. In each of these installations noise infects the gallery either through the sonic intonations of large mechanical structures, or through the layering and erasure of objects within space. This chapter weaves together a discussion of the viewer experience of three of et al.'s recent installations with a consideration of the histories of informatics on which they draw.

In every model of information there is noise. This is because information travels through technology. Technology cannot exist without movement, and without movement there is no information. Movement no matter how imperceptible introduces noise. As soon as something moves it picks up traces of dust and dirt, glitches, mistakes, and error. Without movement there is no information, and without noise there is nothing to hear. One seizes the other. Exact transmission is impossible. Nevertheless, we continue to construct technologies to shift information around. These methods occupy two modes—analog and digital. Analog and digital are like the two companions in Serres's tale; each suffers the relationship of noise to information as internal rupture and external interference. Analog information relationships are keyed as continuous data sets represented by variable, exchangeable, and interchangeable modes of measuring, and made up of sometimes physical but at other times sub-physical quantities such as length, width, voltage, or pressure. The digital information relationship is discrete rather than interchangeable and results in segmented, one-to-one correspondences to movement; a bit like moving one finger at a time. Both digital and analog are informational, both operate through movement and both have to deal with noise. Transmission is messy, never direct and always imperfect. Given the above it should not matter if we utilize either digital or analog information transfer technologies, as any engagement with information must contend with noise. However, the subtle shifts introduced by each technological mode are significant for this discussion of et al.'s gallery installations. Art practices engage media as source, concept, and material. Any slight movement transforms the artwork. When informational media are introduced to gallery spaces we need to pay attention to the materials used to occupy the gallery space. The means of transmission affect the methods of transmission and thus impact the manner in which the corresponding relationships of viewer to artwork can be construed. How we experience the glitch, error, and noise of informational media and how we describe what we see and hear is crucial for media art histories. With this in mind this chapter will look at how et al. employ a range of material technologies to demarcate the space, rupture, and tension of information transmission. Simultaneously I will engage the history of some models of informational technology that feed into these works. Histories of cybernetics offer a background to the kinds of controls, feedbacks, and systems present in these installations.

Voice, Information, and Noise

When Edison first recorded a ghostly "hello" in 1877 it was the direct result of vibrations scratched onto a tin cylinder. The spoken voice produced vibrations in the form of sound waves that were captured by a diaphragm and directly and mechanically inscribed onto the cylinder.[4] For playback the process was reversed; the vibrations of a needle on the cylinder produced vibrations on the diaphragm and sound was heard. Edison's work presented a method for inscribed sound that continues to be used today. In this system it is difficult to exclude any frequencies. Once captured sound and noise are indistinguishable as both are made from vibration; they are continuous. Their movement depends on analogies and the continual set of the analog is realized by the interchangeability of relational qualities in the recorded information. When played back needles move across the inscribed surface, and noise within and without contributes its own information.

These problems of noise and information came to the fore towards the end of the World War II when groups of mathematicians, engineers, and physicists were seeking to perfect command, communication, and control systems. In their examination of the historical construction of information, Hobart and Schiffman locate the source of noise within physical materials: "all analogue machines harbour a certain amount of vagueness, known technically as 'noise'. Which describes the disturbing influences of the machine's physical materials on its calculations."[5] These certain amounts of vagueness were essential to Claude Shannon's mid-century articulation of a theory for information transfer. In 1865 James Clerk Maxwell had described an innovation heralded by James Watt in steam engine design. Through a process of negative feedback, a power amp device or governor could interrupt a signal so that when the engine accelerated the steam supply reduced, resulting in speed or velocity becoming stabilized.[6] All of this occurred independent of "energetic considerations"; that is, the information sent by the signal was abstracted.[7] This influential idea defined information transmission as the changing of abstract signals with time. These changing signals formed the basis for Shannon and Weaver's book *The Mathematical Theory of Communication* (1948) in which they suggest a new model for communications technologies.[8] Shannon's initial brief was to develop efficient telephone lines, and his approach involved breaking down a system into subsystems in order to evaluate the efficiency of the channels and codes. The concern was with signal and the emphasis was on the consistency of transmission and reception of information through any given medium. To increase efficiency Shannon insisted that the message be separated from its components; in particular, those aspects that were predictable were not to be considered information. Furthermore, Shannon was adamant that information must not be confused with meaning.[9] Information was not a message to be understood, but something that induced feedback. The problem that Shannon had to contend with was noise. As Weaver writes in the introduction to their book:

> In the process of being transmitted, it is unfortunately characteristic that certain things are added to the signal which were not intended by the information source. These unwanted additions may be distortions of sound (in telephony, for example) or static (in radio), or distortions in shape or shading of picture (television), or errors in transmission (telegraphy or facsimile), etc. All of these changes in the transmitted signal are called noise.[10]

Noise is both the material from which information is constructed as well as the matter which information resists. Unwanted and disruptive, noise became symbolic of the struggle to control the growth of systems. The more complex the system, the more noise needed to be addressed. To enable more efficient message transmission, Shannon designed systems that repressed as much noise as possible, while also acknowledging that without some noise, information could not be transmitted. Shannon's conception of information meant that information would not change if the context changed. This was crucial if a general theory of information transmission was to be plausible, and meant that a methodology for noise management could be foregrounded.[11] Without meaning, information became a quantity—a yes or a no decision that Shannon called a "bit." His emphasis on separating the message from predictability and external noise appeared to give information an identity where it could float free of a material substance and be treated independent of context. The fluidity of information and the possibilities offered for encoding it meant that information, although measurable, did not have a finite form. Within the equally material confines of message and signal, this was a definition of information as probability function, dependant "both on how probable an element is and on how improbable it is."[12] By situating noise as the material of information, Shannon opened up the possibility for operations within informatic environments that are not merely technical, but that embrace shifting qualities and quantities of any kind of materials. Thanks to Shannon information became quantifiable. Moreover, like light and energy, the transformations that it made on material objects could be mapped.

Shannon's productive dual positioning of noise and information opens a space to address the impure, uncontrolled, and troublesome experience of viewing art installations such as those by et al. An engagement with these works involves an assessment of the relationships of materiality and media, in order to examine patterns and modes of operation as well as highlight the way these works operate within complex gallery and informational systems.

Material Practices

We walk into an art installation and expect certain things. In the gallery something is seen, identified, viewed, or presented as a series of relationships that might be established between individuals, groups, environments, and sensations. Understood this way art is an aesthetic relationship between differing material bodies, images, representations, and spaces. It is experienced as an event rather than via a search for meaning. Et al. is the working name of an artists' collective, operating within New Zealand for around twenty years. Previously et al. have used numerous monikers including l. budd, blanche readymade, popular productions, lionel b., and p. Mule among others. Since 2000 et al. have encompassed most but not all previous incarnations of the collective (e.g. p. mule is still active and a senior lecturer at the University of Auckland, and l. budd has redefined itself as an estate).[13] As a non-name or placeholder et al. suggests a positioning of the artists as critics outside the structures of the cultural object they might create. Et al. explain that the Latin abbreviation "and others" is neither masculine nor feminine, and is always used in the plural to refer to a number of people. The multiplicity of the name and the artists' refusal to speak publicly themselves (although they do have a number of publicists) continues to cause some consternation in New Zealand.[14] Furthermore, et al.'s installations are not easily accessible through the common vernacular of the gallery space that, as Brian O'Doherty explains, is constructed to become an invisible white context freed from traces of history and time.[15] Sidestepping the rules of engagement usually established by the "white cube," et al. enfold information and noise with the structures of the gallery exhibition and raise the possibility that in transforming the structures and materials through which they travel, information and noise have left important traces that can be harnessed by art installation. These traces are located in the artwork's systems and materials. The fractured spaces of et al.'s installations are haunted by histories of information.

Like art, information science remains concerned with the material spaces of transmission—whether conceptual, social, or critical. Discussions of information transmission through phone lines occurred alongside the development of technology capable of computing multiple discrete and variable packets of information; that is, the digital computer. Pulling together all these threads the computer became a central operational space within which art could be produced, constructed, distributed, and viewed. This resulted in concepts, materials, and media behaving in similar ways. In a critique of this convergent consensus, et al. overlay the discrete iterations of fundamentalism with the continual mechanistic movements of politics. In these installations the space is always permeated by noise from outside the suspended space and time of the gallery. History, power, and the visual and sonic control of networked technologies all impact the viewing experience. The viewer is immersed in a series of engagements with political ideologies, cold-war control mechanisms, truth, information, theory, and technology. The concern is with people tempered by the politics of the machine.

maintenance of social solidarity–instance 5 was first exhibited in 2006 as part of the Scape Biennal of Art in Public Space (Figure 3.1).[16] The installation occupies a small alcove partially blocked by a military-style portable table stacked with newspapers. Inside the alcove are gray wooden chairs, some headphones, and a large modified projection of data from Google Earth. Everything feels temporary, provisional. The space is dim, lit by the glow of desk lamps. Electrical and data cables are strung across the space. It is not immediately clear if the viewer is allowed within the alcove to listen to the headphones; anything we might hear would be cancelled out by the monotonous voices that fill the room intoning what seem to be political, social, and religious platitudes. Are the headphones a tool to block out the noise? In the installation it appears that multiple messages have been sent but their source, channel, and transmitter are unintelligible to the receiver. All that is left is information divorced from meaning. Everything is set up for a briefing yet despite the provision of copious media the viewer cannot unpack much of the information contained within the environment.

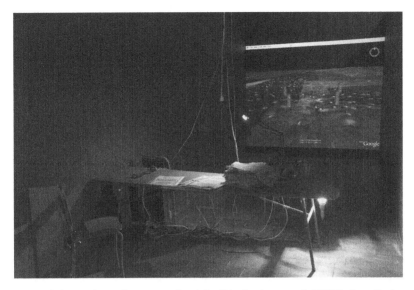

FIGURE 3.1   et al., *maintenance of social solidarity–instance 5* (2006), installation. Exhibited in Scape Biennal of Art in Public Space, Christchurch Art Gallery, Christchurch, New Zealand, September 30–November 12, 2006. Reproduced with permission of blanche readymade trust. Photographer: Lee Cunliffe

To enter the installation we slip around the table only to find ourselves extremely close to the projection screen. It is not clear if the chairs are sculptural items, props, or an invitation to sit. On screen, and apparently integral to the Google Earth imagery, are five animated and imposing black monolithic forms. These are the artists' Autonomous Purification Units or APU, perhaps the source of all this noise. Because of their connection to the voices around us, the APU begin to map an imposition of narrative, power, and force in various disputed territories. This is not a location for interrogation, but a record of it. Like their sudden arrival in Kubrick's *2001: A Space Odyssey* (1968) it is the contradiction of the visibility and improbability of the monoliths that renders them believable. On the video landscape the APU seem to house the dispassionate voices of many different media and political authorities. Their presence modulates the layering of media forces, and in between error slips in.

Errors are usually accidents—the kind of noise that Shannon found within his communications channel. In his discussion of virtualized space-time relationships, Paul Virilio locates two kinds of "accident."[17] Prior to the domination of instantaneous telecommunications, we only experienced specific localized accidents; now, Virilio argues, we are subject to the delocalized "general accident."[18] This is a very real shift in which the only consistency is accident, noise, and error. There is an "accidental transfer" as the error mediates between real and virtual.[19] Virilio further contextualizes his discussion of the accident within image cultures where real and virtual are not opposed. By mapping three phases of the image, Virilio suggests a movement from reality (the local) through actuality, to virtuality (the delocalized) that manifests as a shift in the representational authority of digital images. Virilio says that as viewers "we still cannot seem to get a grip on the virtualities of the paradoxical logic of the videogram, the hologram, or digital imagery."[20] And, despite our awareness of the noise of the image and the accident it encompasses, it is our inability to understand the multiple modes of virtuality that result in repeated attempts to capture meaning in order to understand the information before us. In *maintenance of social solidarity–instance 5* there is a mixing of real-time presence with this particular, yet accidental, constructed virtuality. The APU occupy a fixed spatial ground, imposed over the shifting navigation of Google Earth (this is not to mistake Google Earth with the "real" earth). Together they form a visual counterpoint to the texts reciting in the viewer's ears, which themselves might present as real, but again, they are not. Virilio's argument parallels Shannon's. Information cannot be tied to meaning. Instead, in the race for authority and thus authenticity we find interlopers; noisy digital images that suggest the presence of real-time perception. The spaces of *maintenance of social solidarity–instance 5* meld representation and information together through the materiality of noise. And across the different modalities employed, the appearance of noise is not through formation, but through error, accident, and surprise.

The driving impetus for the development of digital information technologies is a desire for the elimination of noise or the replacement of accident with accuracy. For example, in digital photography the push for greater and greater megapixel resolution is tied to the illusion of a pure image; a digital version of reality. However photographic reality results from a viewer's inability to see digital noise. We don't know how noisy a three megapixel image is until we see a ten megapixel image. We can only identify the noise in hindsight. The reason for this is now apparent. Digital noise is not located in the specific breakdown of media but in the materializing forces of the accident. This is not a new sort of noise but is part of the very material it disrupts. It is a delocalized accident, not the result of the analogical movements of its parts. Digital noise is always already present as glitch, error, a misread numeral, or a gap in the waveform. In looking at the shifting ground of representation and virtuality, Virilio mapped a movement from particular to generalized accidents. In *maintenance of social solidarity–instance 5* we are forced to align real-time with virtual spaces and begin to believe in the temporal truths that we see on the screen before us. The installation is a site for testing. The noise of the voices blends with the noise of the image and a realization of the impossibility of a single reading of the installation.

Two Types of Noise

In Shannon's communication model, information is not only complicit with noise; it is dependent upon it for elucidation. Without noise, either encoded within the original message or present from sources outside the channel, information cannot get through. Shannon's model traces a path from sender through encoder to signal, decoder, and receiver. This path is interrupted by noise.[21] Visually and schematically this noise is a disruption inserted in the nice clean lines of the message. This does not mean, however, that noise is a last minute consideration.

In his discussion Shannon locates noise in two crucial places. The first position accorded to noise is external and marked by an arrow that demonstrates the introduction of noise into the message channel while in transit. External noise confuses the purity of the message, while equivocally adding new information. External noise has a particular materiality and enters the equation as unexplained variation and random error. This is disruptive presence rather than entropic coded pattern. Weaver writes:

> If noise is introduced, then the received message contains distortions, certain errors, certain extraneous material, that would certainly lead one to say that the received message exhibits, because of the effects of the noise, an increased uncertainty. But if the uncertainty is increased, the information is increased, and this sounds as though the noise were beneficial![22]

Noise as interference is not necessarily audible. For example, we understand a noisy picture on television to result from some kind of interruption in the broadcast waves. In Shannon's context interference in the telephone line is usually due to channel overload—not the source, but the channel capacity—a limitation of the material infrastructure.

Shannon offers this equivocal definition of noise to be everything that is outside the linear model of sender-channel-receiver. Hence, anything can be noise if it enters a channel as an unwelcome or unintended addition to the message.

Secondly, noise was defined as unpredictability or entropy found and encoded within the message itself. This for Shannon was an essential and, in some ways, positive role. Entropic forces invited continual reorganization and assisted with the removal of repetition enabling faster message transmission. Shannon entered difficult terrain here. In lifting the term entropy from thermodynamics he reversed its meaning. In thermodynamics entropy refers to information loss and the qualitative degradation of system functioning, whereas in Shannon's model of information entropy is the very measure of information (its quantity). For Shannon, entropic noise was essential to the functioning of information. This is because entropy could enable calculation of relative probabilities. If it were possible to calculate the entropy of the signal, it would be possible to measure the capacity of the channel necessary to transmit the message. But external noise meant this level of certainty was often elusive. Weaver called the measurement of this shifting relationship between entropy and message "equivocation."[23]

Weaver identified equivocation as central to the manner in which noise and information operated. A process of equivocation identified the receiver's knowledge. Equivocation thus mediated between useful information and noise, as Shannon measured both through the binary unit or "bit."[24] This meant that there was no such thing as an unmediated message. Crucial here is that noise is not abstract but encompasses material processes of reorganization, disruption, and distraction all of which can be representational. Hobart and Schiffman observe that "what we call a piece of fixed information represents only an arbitrary point at which we have halted the movement of information processing."[25] Furthermore, Shannon and Weaver's intentionally polyvalent use of the term "entropy" has been responsible for many misreadings of their arguments.[26] Following Shannon, most models of information and transmission seek to equivocate the presence of noise, at the same time acknowledging its necessary presence. Shannon's multiple positioning of information and noise gives both a dynamic form. As a message strives for differentiation—to perhaps become encoded content—noise saturates its boundaries. Information understood in this way is also about relationships between differing material bodies—representations and spaces, connected together for the purposes of transmission. Information, like art, is an event.

Material Correlations

Thinking about information and art as events appears to suggest a correlation between the processes of information transmission and viewing in galleries. This is unintended. Although the contemporary information channel is essentially a tube with fixed walls (constrained by physical properties, bandwidth and so on), and despite the implicit spatialization of information models, I am not proposing a direct correspondence between information channels and gallery installation spaces. This is because I am not interested in "reading" the information of either environment. What I am suggesting is that both environments share the material of noise. In each instance, noise is present in four places: noise is within the media errors of transmission; it is within the media of the installation, (neither of which are one way flows); the viewer or listener introduces noise as interference; and lastly, it is present in the very materials through which it travels. Noise layered on noise.

In et al.'s *the fundamental practice*, first exhibited at the 51ˢᵗ Biennale di Venezia in 2005, noise quickly grows out of proportion, and information does not travel in straight lines (Figure 3.2).[27] Occupying an old orphanage located behind Santa Maria della Pieta Church *the fundamental practice* seems at first glance to be in the process of being constructed from discarded technologies. Barriers and signs limit access and five large caged but moving APU noisily roll back and forth in irregular patterns. Confronted with the mobile APU (these are the original steel constructions that were rendered as virtual forms in *maintenance of social solidarity–instance 5*) the viewer is left attempting to decode the un-encoded. According to a cue known only to themselves the APU emit long bursts of didactic speech, in the manner of either a call to war or a call to prayer. These figures suggest answers: that perhaps somewhere within this installation are methodologies to confront the materializing force of digital noise. By wiring up conductors, antenna, desktop (mainframe) computers, and receivers in a seemingly random feedback loop, et al. introduce the possibility that such a system relies on the unseen operations of electromagnetic mathematical ratios of feedback. The system dominates. The large steel barriers control and determine only certain

FIGURE 3.2   et al., *the fundamental practice* (2005), installation. Exhibited at New Zealand's National Pavilion at the 51st Biennale di Venezia Palazzo Gritti, Calle della Picta, Venice, Italy, June 12–November 6, 2005. Reproduced with permission of the artists. Photographer: Jennifer French

movements and despite the use of sophisticated sound engineering tools noise leakage is everywhere. At the same time that it highlights the impossibility of fixed media *the fundamental practice* addresses the noise of fundamentalism as discovered in between information and noise. This is not a fundamentalism with directed positions and singular leaders. The complexity here is in the fundamentals themselves. The steel gray caging of the technology generates resonances, more noises that imply that as a viewer enters the spaces of *the fundamental practice* she becomes complicit (that is, complicity is thrust upon her). But complicit with what? Does the viewer's presence within the installation and before the APU imply some kind of agreement? As a receiver of messages, is it the viewer's job to feed back, to reciprocate?

When discussing an earlier installation by et al. Gregory Burke wrote of "the absence of subjects [being] combined with the persistence of the apparatus."[28] It appears that this technique is being repeated in *the fundamental practice*. But how can an apparatus remain without its subjects? Maybe what is actually happening is that so-called "subjects" are being constructed differently. The viewer is not necessarily the subject of, or subject to, a message. After all, it is the shifting subject called the viewer who is able to simultaneously process ruptures, discontinuities, and the soaring exhilaration of being within these spaces. The viewing subject is not absent in *the fundamental practice* but begins construction as soon as he or she enters the machinic spaces of fundamentalism. As with any fundamental practice (this might be a religion, society, or culture) the first step is to stop, listen, and pay attention. As soon as a threshold is passed and attention is captured, a new modality begins to be produced. In *the fundamental practice* the production of subjectivity occurs by way of the noise or disturbances generated by multiple authorities; each metal box shouts or recites different contradictory instructions or truisms which are repeatedly interrupted by the static sounds of transmission. Because there is no single position offered which is subject forming, it is easy to understand how the subject might appear to be absent. However, the installation could not exist without a viewer. Fundamental training cannot continue without our presence.

The fragments of noise manifest in *the fundamental practice* are observational, meaning that to approach the work is to invite an unanswerable relationship: a location that operates as a filtering of power. The parameters for this power are in a constant state of flux. As well as mapping these forces et al. invite resistance. Can we be both complicit and resistant? How long are we expected to stay within this space? Should we be resistant to the assemblage within which sound is found and constructed? *The fundamental practice* acknowledges the viewing subject as another (potentially noisy) material force as his or her presence impacts on, manipulates, and transforms the materials and spaces of the installation. *The fundamental practice* highlights how the shifting open nature of installation means that subject-formation and thus feedback is both external and immanent to the informational system. Despite resistance, it is the feedback of the viewer that shifts the work and its relations in unpredictable ways. The viewer leaves with more questions than answers, and it is the traces of noise that are left to counter the drone of fundamentalism.

Sonic Logic

If it is measured as a quantity, noise is productive; it adds information. But not always. In thinking through the logics of information transfer von Neumann comments: "Error as a matter of normal operation and not solely . . . as an accident attributable to some definite breakdown, nevertheless creeps in."[29]

Error creeps in. In *the fundamental practice* et al. disrupts signal transmission by layering ambiguities into the installation. Gaps are left for viewers to introduce misreadings of scale, of space, and of apprehension. Rather than selecting meaning out of information within different contexts, a viewer finds herself in the same sphere as information. Noise imbricates both information and viewer within a larger open system. When in conversation about the structures of viewing that have pervaded their work since 2001, et al. collaborator p. mule explains how the group stretch and reconfigure sometimes very small excerpts of found data:

These days we talk generally about knowledge being reduced down to components and digital information—it's always reconfigured. Hypertext is never linear either; it's what cut-and-paste has become. There's the whole interest in interweaving texts and the way society is compelled forward by differing levels of information and the paranoia that surrounds this.[30]

The question remains: when both content and transmission are blurred to differing levels how do we receive information? Paranoia is an appropriate response to noise. Et al.'s installations constantly turn to representational and meaning-making systems. As one instance, *the fundamental practice* demonstrates both how Shannon and Weaver foretold of the digital and how the arrival of digital materials has challenged the traditions of information theory. Digital logics are modulated by redundancies and accidents. And equally, neither information nor noise is positive or negative but both are placeholders for pattern and randomness. In *the fundamental practice* it is not possible to have information without noise. If et al. operate between noise and information in an ongoing disruption of the legacies of representation, immersion, and interaction, their works open up new material languages for art installation. The space and structure of the gallery shifts as it is permeated by noise. Furthermore, an engagement with noise and error results in an obscuring of the structures of information, generating a position from which we can discuss the viewer immersed within the system. All the same, it seems too neat. What happens when spaces and subjects don't fit inside one another? In many ways, this is the case in the works discussed here. *maintenance of social solidarity–instance 5* defines and generates the space of the gallery, rather than the other way around. It is not a comfortable fit for a viewer, who can easily be disoriented by the motion of the screens and the volume of the sound. In *the fundamental practice* the disorientation of multiple sounds is further exaggerated because of the layering of an informational and machinic environment within the preexistent social and cultural environments of the Venice Biennale. Because of the building-site aesthetic of the installation, many viewers developed what might be termed a figure-ground problem: the work was not seen, and presuming the work was simply not-finished—or even, not-there—viewers turned away at the door. The space of the installation leaked physically, but also into other modes, as it moved outside of the boundary of the gallery. This has more than economic or critical implications. It is about the implicit contract that viewers make with the artwork within a certain space. And hence it is also about the subject of art. In what way, then, do et al.'s installations unmake a viewer's subjection to the artwork and make the artwork itself the starting point for subject-formation? Or is this another red herring? An error introduced by over-reading the situation?

Both *maintenance of social solidarity–instance 5* and *the fundamental practice* demand an investment from viewers. In both works, it is the role of sound that disrupts any neat formation of subjects, and it is noise that makes the very space of the installation appear. The sounds documented shift modes. We hear noise one minute, and a fragment of information the next. It is impossible to listen to, and difficult to attempt to isolate the information from the noise. Standing at the entrance of *the fundamental practice* listening to the rumblings and rustlings picked up by the machinic devices, it is not hard to be affected by an aggressive feedback loop of metal, text, fundamentals, lessons, and violence. The feedback, techno, and biorhythms of the movements of the technology itself, as it rolls about within the space, makes the presence of each individual element less fixed; they move, or shift ground. The work becomes a pattern of sounds that have been captured and tuned into aural noise. We cannot help but connect these materials made from noise back to some fragment of information.

Although *the fundamental practice* seems to offer an insight into the presence of techno and biorhythms, its sonic pattern suggests randomness.

At a broader level it makes us aware of the dialectic between the different meaning-making systems of technoscience and art. Because it presents a feedback torsion, there is a doubling that forces the installation away from itself, and away from any neat resolution in the eyes and ears of the viewer. Shannon understood that information needed to be replicable, but that repetition did not add anything to the information content of a message and was in fact an impediment to smooth transmission. In Shannon's formulation, repetition is redundancy. Too much information is redundant and theoretically not essential to the transmission of a message. To constitute a message, a transmission must contain a mixture of pattern and noise with a minimum of repetition. If noise is the materiality of information, then we understand Shannon's model by what information is. If we add to this the way that the information, as noise and materiality, operates we begin to understand what it does. The structures of representation and repetition as encountered in *the fundamental practice* are *not* informational and, despite all the racket, they *exclude* noise. The viewer is rendered passive and unable to introduce his or her own information. Noise between the elements of the installation presents the viewer with an unfinished environment where sound becomes integral to the work and its experience. Enmeshed in these informational flows we are forced to adapt our viewing techniques and methods to get out of unresolved loops. The installation is present—like all works operating within the circumstances of the gallery—in a state of unfinish, within a situation designed to be always incomplete and where new information and new noise continuously flow.

Insoluble

It is this idea of control, whether immersive or interactive, that et al.'s installations question, and which viewers in effect resist. If a relationship with installation is one of entering complex preexistent controlled environments, there is the implicit assumption that the installation is static, that it will remain the same for each viewer. However, the experience of these works is one of process rather than achieving goals. And nothing ever stays the same. When Shannon was constructing models for information transfer others in the building were developing systems for weapons control that would eventually become the command, communication, and control systems of networked technologies. Noise gained a level of importance whereby information transfer became a life

FIGURE 3.3 et al., *That's obvious! That's right! That's true!* (2009), installation. Exhibited at Christchurch Art Gallery, Te Puna o Waiwhetu, Christchurch, New Zealand, July 23–November 22, 2009. Reproduced with permission of the artists. Photographer: David Watkins

or death decision. Information and noise remained a process of movement, but knowledge of movement was formed from direct engagement with overwhelming materials and technologies. Noise became dangerous. In their 2009 installation *That's obvious! That's right! That's true!* at the Christchurch Art Gallery et al. make these political connections with information theory even more explicit (Figure 3.3).[31] *That's obvious! That's right! That's true!* sets the stage for information to be gathered, knowledge to be reviewed, statistics to be compiled, and opinion to be formed. The installation appears to house the aftermath of a rally: an environment all too often blurred between the forces of religion and the passion of the individual. This rally is no different. The potential for real change is suggested. At once everything is presented as self evident, while nevertheless the "truth" remains obscure. This obscurity means that the installation is haunted by deathly absence, and a sense that although a site has been prepared, the speakers installed, the speeches written, and the voting booths unfolded, it is all too late.

The installation is held together by glue made of human labor. Layers upon layers of tape stretch across texts that seem to have only recently slipped off the spirit duplicator (the Banda machine once common in schools and churches). Hands have made this work. They have stuck, they have screwed, they have bolted, and time has been invested. The space is filled with grubby sweat marks and the sticky residue of gaffer tape. Rusting pillars mark out spatial dimensions, scraps of wire hang from the exposed ceiling. The walls are a neutral (battleship) grey. The installation space is full of entreaties. In a desire for clarity, grim statistics abound; framed and bolted to billboards they layer, overwhelm. This is information overload. Above us, fast-speaking voices that read through loud speakers are unintelligible over the clash and clatter within the confined (voting) booths that line one wall. Someone is giving a lesson in architecture. Elsewhere a small box reminds us to "vote only once." Spaces are demarcated and a silhouetted pointer shows us the way through the labyrinth. The voice from above reminds via well-worn Marxist sound bites, that we have the potential to extract ourselves from this picture.

Throughout the installation are thousands of stacked newspapers. "Maintenance of Social Solidarity no.4: *The New Zealand Altruism Review*" is published by the same hand as the *social solidarity* tabloid piled on the table in *maintenance of social solidarity-version 5* but when opened *The Altruism Review* tells a different story. In this newspaper are spreads of cartoon-like drawings interspersed with truisms and blacked out memos. Truisms include: "Reason is the greatest enemy that faith faces," "Believe in faith not reason, no pain," "No one can be sure 'it' will go on the same way" alongside blacked out correspondence about political torture. These mimic the indecipherable texts heard through the sonic boom of the loudspeakers. Truisms are never altogether altruistic, but generic, clichéd, promises of hope. They are information devoid of content and context. It is as difficult to receive the written text as the spoken; both have been subjected to redundancy, noise as external interference, and noise introduced in the communications channel.

These narratives echo the muffled responses to the kind of information retrieved through online news search engines everyday. *That's obvious! That's right! That's true!* begins to occupy a kind of forever future past. In the center of the installation is a large viewing platform. It is degree zero, a stage, a platform for vocalization or a space for occupation. Maybe getting above the information means that sense can be made from above, a kind of escape from noise. However, noise is the very material through which this information travels. Et al. suggest we compensate for our anxiety by paying attention to the architects of disaster. And in so doing they suggest that attention to noise is the escape mechanism itself. Despite material, social, and economic fluctuations, the presence of noise is a consistent guarantee or measure of stability. Someone and/or something will make a mistake, errors will be transmitted, the wrong button will be pressed, or the wrong answer recorded. Noise is essential for any system to function, including those of art and the gallery.

Media art histories require description of the experiences of being within the artwork, but also demand a form of embedded analysis that engages not only the materials of the artwork but also the histories that potentially inform the work. The control-room aesthetics of et al.'s gallery installations recall the social, political, cultural, and economic environment being marked out as Shannon and Weaver imagined systems that would have much greater effects than simply those of information transmission. Et al.'s grey surfaces that house multiple technologies, both real and virtual, digital and analog, absorb the viewer within a complex mix of history and fiction. These are not installations that easily devolve meaning. Instead they suggest that viewers invest in noise as the matter of information. Materials experienced as noise become resonant within the installation. However there is something that remains absent from this discussion: the violence that information transfers on a daily basis. By continually attempting to control the very surfaces through which we speak, those who employ informational models seek to rid communication of its nuances and eccentricities. The noise is not always heard. Et al. demonstrate that the violent imposition of order generates spaces of rupture through which materials can resonate. These noisy spaces, which are contradictory, mediated, and somewhat anxious, are what art installations such as those by et al. contribute to our understanding of the world.

Notes

1. Michel Serres, *The Parasite*, trans. Lawrence R. Schehr (Minneapolis: University of Minnesota Press, 2007), 3.
2. Claude Shannon, "A Mathematical Theory of Communication," reprinted with corrections from *The Bell System Technical Journal* 27 (1948): 379–423, 623–656, http://cm.bell-labs.com/cm/ms/what/shannonday/shannon1948.pdf, 48.
3. et al. in Gregory Burke and Natasha Conland, eds. *et al. The Fundamental Practice* (Wellington: Creative New Zealand, 2005), 110.
4. Jonathan Sterne, *The Audible Past: Cultural Origins of Sound Reproduction* (Durham: Duke University Press, 2003), 12.

5   Michael Hobart and Zachary Schiffman, *Information Ages: Literacy, Numeracy, and the Computer Revolution* (Baltimore, MD: Johns Hopkins University Press, 1998), 208.

6   Gordon Pask, *An Approach to Cybernetics* (London: Hutchinson, 1961), 12. See also Norbert Wiener, *Cybernetics or Control and Communication in the Animal and the Machine* (Cambridge, MA: MIT Press, 1961), 97.

7   Ibid., 15.

8   Claude E. Shannon and Warren Weaver, *The Mathematical Theory of Communication* (Urbana: University of Chicago Press, 1949).

9   N. Katherine Hayles, *How We Became Posthuman* (Chicago: University of Chicago Press, 1999), 54. See also Umberto Eco, *The Open Work*, trans. Anna Cancogni (Cambridge, MA: Harvard University Press, 1989), 63–66.

10 Weaver, "Recent Contributions to the Mathematical Theory of Communication," *The Mathematical Theory of Communication*, 7–8. See also Mischa Schwartz, *Information Transmission, Modulation, and Noise: A Unified Approach to Communication Systems*, 3rd edn. New York: McGraw-Hill, 1980.

11 Alternative definitions of information were offered by Gabor and Mackay but were not widely adopted within the American canon. See Pask, 123.

12 N. Katherine Hayles, *Chaos Bound: Orderly Disorder in Contemporary Literature and Science* (Ithaca: Cornell University, 1990), 52.

13 Jim Barr and Mary Barr, "L. Budd et al.," in *Toi Toi Toi: Three Generations of Artists from New Zealand*, ed. Rene Block (Kassel: Museum Fridericianum, 1999), 123.

14 See "Complaints under Section 8(1)(a) of the Broadcasting Act 1989," *Broadcasting Standards Authority*, Wellington, 2004, http://www.bsa.govt.nz/decisions/2004/2004–153.htm. See also: Jeremy Hansen, "et al. Field Story," *Frontseat*, Television New Zealand, July 4, 2004, http://frontseat.co.nz/exclusive/transcripts/transcript.php?id=9 accessed July 4, 2004; John Daly-Peoples, "Urban Myths and the et al. Legend," *National Business Review*, August 20, 2004.

15 Brian O'Doherty, *Inside the White Cube: The Ideology of the Gallery Space*. Expanded Edn. (Berkeley, Los Angeles, London: University of California Press, 1999).

16 et al. *maintenance of social solidarity–instance 5* (2006), installation, Google earth feed, newspapers, sound. Exhibited in Scape Biennal of Art in Public Space Christchurch Art Gallery, Christchurch, September 30–November 12, 2006. Including et al. *social solidarity* 4 page tabloid, 42gsm newsprint, 4 pages color, 10,000 copies.

17 Sylvère Lotringer and Paul Virilio, *The Accident of Art* (Cambridge, MA: MIT Press, 2005).

18 Paul Virilio, "Continental Drift," in *The Virilio Reader*, ed. James Der Derian (Oxford: Blackwell, 1998), 183–184.

19 Paul Virilio, *The Vision Machine*, trans. Julie Rose (Bloomington: Indiana University Press/British Film Institute, 1994), 64.

20 Ibid., 63.

21 Shannon, 2.

22 Weaver, 19.

23 Ibid., 19–20.

[24] Shannon, 1.

[25] Hobart and Schiffman, 203.

[26] See the discussion in Hans Christian von Baeyer, *Information: The New Language of Science* (London: Phoenix, 2004), 28–34 and 215–221.

[27] et al. *the fundamental practice* (2005), Installation. Exhibited as New Zealand's National Pavilion, 51st Biennale di Venezia Palazzo Gritti, Calle della Pieta, Venice, Italy, June 12–November 6, 2005.

[28] Gregory Burke, "Interference: et al. and Technics," in *Arguments for Immortality*, eds. Jim Barr, Mary Barr, and Gregory Burke (New Plymouth: Govett-Brewster Art Gallery, 2003), 28.

[29] John von Neumann, "The General and Logical Theory of Automata," in *Collected Works*, ed. A. H. Taub (Oxford: Pergamon Press, 1963), 294.

[30] p. mule in conversation with Allan Smith. Allan Smith, "Allan Smith Talks to p. mule for et al.," in *Nine Lives. The 2003 Chartwell Exhibition* Exh. Cat., 5 (Auckland: Auckland Art Gallery Toi o Tamaki, 2003).

[31] et al. *that's obvious! that's right! that's true!* Christchurch Art Gallery, Te Puna o Waiwhetu. July 23–22 November 2009. Includes "Maintenance of Social Solidarity no.4." Winter 2008. 4 page tabloid, 42gsm newsprint, 16 pages color, 10,000 copies. Printed in association with the Adam Art Gallery Te Pataka Toi.

Chapter 4

Add-Art and *Your Neighbors' Biz:* A Tactical Manipulation of Noise

xtine burrough

California State University, Fullerton

In 1948 Claude Shannon established that since communication occurs despite noise in the system, "exact transmission is impossible."[1] According to Shannon, noise must be contended with in the process of transmitting information. Shannon's essay is significant for information theorists because it provides a diagram for an information system used to determine how to store and compress data. However, it is also meaningful for artists because it names and positions noise, an abstract and erroneous element in the communication system that is normally disregarded or sidestepped. In this chapter, examples will be provided to demonstrate how this position is loaded with potential for poetic and tactical manipulation or intervention. Frustratingly, Shannon devoted only a few words to his definition of noise, when he wrote that it "is considered to be a chance variable."[2] Noise sits in opposition to the intended communication, often appearing as an erroneous element the communicator avoids. In contemporary communication practices, spam, or unwanted email, is an example of such a variable. For information theorists, noise is categorized as error and its definition is insignificant because, as Justice Potter Stewart said of hard-core pornography in the famous U.S. Supreme Court case,[3] you'll know it (noise) when you see (or hear) it.

Noise may be understood as an unintentionally disruptive counterpart to a transmission of an intended signal to a receiver. However, the babbling also acts as an opening or a position where the other's voice can be inserted into the stream of data comprising everyday life. Linguistically, noise can have a positive influence that results in multiple readings or double-entendres by the receiver. Semantically, noise also has the power to communicate alternative meanings. A visual representation of noise could be as simple as static on a television screen. With such a common stimulus, the automatic human response for the viewer is to change the channel. In a more complex example of a televised narrative, the static visual could signify a loss of communication between two characters. Noise in one situation becomes a signifier of noise in another, where the original noisy signal is used to represent disconnection. In the narrative, the static is no longer noise. Instead, it is a vital part of the narrative, just like the actions, scenery, and dialog. Noise is used to represent miscommunication and the error it causes as an active territory that can be intentional and manipulated. In *The Open Work*, Umberto Eco writes of the potential of noise for poetic intervention, to the extent that art and poetry "depend on deliberately provoking incomplete experiences— that is, art deliberately frustrates our expectations in order to arouse our natural craving for completion."[4] Thus, Eco clarifies the relationship between Shannon's model and acts of poetic manipulations. Eco finds Shannon's work to be relevant for information theorists, though it is not a sufficient model for human communication. He writes, "Once the signals are received by a human being, information theory has nothing else to add and gives way to either semiology or semantics, since the question henceforth becomes one of signification—the kind of signification that is the subject of semantics and that is quite different from the banal signification that is the subject of information."[5] Information exchange between machines is strictly guided by syntax. Human communication is more complicated because it relies on syntax (such as the rules of grammar), historical positioning, and semantics. Since all information follows a syntactical order, its formal organization can be tactically usurped or appropriated in art and poetry, thereby creating a new formal organization. According to Eco, "All deviation from the most banal linguistic order entails a new kind of organization, *which can be considered as disorder in relation to the previous organization, and as order in relation to the parameters of the new discourse.*"[6]

Shannon names noise and Eco delivers an argument for the creative pursuit of noisy territories. The transmission of information relies on probability (either the message is delivered or it is not). The creative message requires "a departure from the linguistic system of probability, which serves to convey established meanings, in order to increase the signifying potential of the message."[7] If information delivery relies on order, the creative message embraces entropy because it positions itself in a space where expectations are unmet. Noise is found in a space assumed to contain error, so the territory where noise dwells is ignored in everyday practices. It is possible that noise is always present and human awareness of noise is the subjective factor.

Since the point at which noise is revealed as noisy or disruptive is unde-
termined, the existence of noise alludes to its own virtual space. Poets,
artists, and activists have positioned their voices within this virtual space,
manipulating the part of the communication system considered to be
erroneous or unintentional such that it becomes a significant channel
for meaningful communication. This group of meaning makers and
shape shifters values noise, rather than ignoring it, and utilizes it as a
significant part of the communication system. "Let us call the world
around us noise-machines,"[8] Hélène Cixous so aptly points out, as "there
can be no transgression without law."[9] Cixous envisions a system that
opposes Shannon's model, where the assumed, intended signal, full of
information, rules, and laws of syntax are in fact the noisy elements of
artistic expression. "What the law says frightens but its noise is pretty."[10]
As Cixous reminds us in one line (and as Eco theorizes throughout
The Open Work): to engage noise as a medium is to transgress the order
of the communication system, making way for a co-opted or appropriated message.

Dada and Schwitters's Merz Works

Visually, Dada artists have demonstrated this transgression as early as
the beginning of the twentieth century. More than twenty-five years
before Shannon's essay, Dada artists embraced noise as antimatter, to
be celebrated and codified. Tristan Tzara wrote that noise, in opposition
to logic, embraces one of the concepts of Dada. "Logic is a complication.
Logic is always wrong."[11] Tzara reviews concepts of Dada, from "a hobby
horse" to "a relapse into a dry and noisy, noisy and monotonous
primitivism."[12] At the start of his "Dada Manifesto of 1918," Tzara rejects
rules and manifestos because "To impose your ABC is a natural thing—
hence deplorable."[13] Tzara rejects "natural things" and everyday life,
and favors "noisy primitivism" in opposition to the actions and attitudes
of bourgeois culture. The Dadaists were a disconnected group. From a
historical perspective the movement can only be defined as a collection
of moments, as Robert Motherwell suggests in his introduction to *The
Dada Painters and Poets* anthology. Dada arrived in Zurich, Switzerland
with Hugo Ball's Dada theater at Cabaret Voltaire, and can also be attributed to "young intellectuals from all over warring Europe, to the visit to
New York of Cravan, Duchamp, Picabia and others, to the meeting in
Nantes of Jacques Vaché and André Breton—all events taking place in
1915 or 1916, bloody years in the first world war."[14] The Dadaists were

against anything that seemed to perpetuate the bourgeois culture and its war efforts. As artists, they were interested in new methods and representations, in search of a truth that did not rely on realism or idealism.[15] As Motherwell states, "Dada was indeed the first systematic attempt to use the means of *l'art moderne* in relation to political issues."[16] By opposing the war and the bourgeois culture it supported, the Dadaists were against the systems governing everyday life. In their collages, theater, and poetry, they used the noise of their culture to express their disdain.

In his 1920 essay, "En Avant Dada: A History of Dadaism," Richard Huelsenbeck claims to have founded Dada at the Cabaret Voltaire in 1916 with Hugo Ball, Tristan Tzara, Hans Arp, and Marcel Janco. The group was "agreed that the war had been contrived by the various governments for the most autocratic, sordid and materialistic reasons."[17] Although the movement was international, works created by the German Dadaists were more overtly political than those made by their Swiss, French, and American counterparts. Huelsenbeck explained, "While in Zurich people lived as in a health resort, chasing after the ladies and longing for nightfall that would bring pleasure barges, magic lanterns and music by Verdi, in Berlin you never knew where the next meal was coming from. Fear was in everybody's bones, everybody had a feeling that the big deal launched by Hindenburg & Co. was going to turn out very badly."[18]

Kurt Schwitters, a German contemporary of the Romanian Tristan Tzara, is remembered as a prolific Dadaist who embraced noise (or as some might label it, distracting junk) in works he created under the neologism, "merz." If high quality materials, organized compositions, expressionistic, realistic, or staged renderings connoted "high art," Schwitters, following Tzara's manifesto and Dadaist inclinations from France to New York, focused on the opposite. Schwitters was born in Hanover, where he lived for 40 years until he fled to escape the Nazis in 1937. In the 1920s, he met with Huelsenbeck and Grosz in Berlin, as he was interested in joining the Dada movement there. However, he was mistrusted and judged negatively because his art did not exude (enough) radical opposition to the German government. When Schwitters returned from Berlin he decided that since his art could not be officially categorized as "Dada," he had to create a name of his own. Assemblages, paintings, sculptures, and Schwitters's creative process all came to be defined as "merz." The name was a selection of letters from an advertisement for *Kommerzbank* that appeared in one of his assemblages. Merz was a wordplay on the words "commerce" and "annihilation."[19] Merz, as a concept for artistic expression, attempted to eradicate bourgeois culture, most notably as expressed through commerce or other everyday practices. In his introduction to *Lucky Hans and Other Fairy Tales*, Jack Zipes explains that merz "did not align itself with a particular political party or theory, but it clearly scrambled forms and laws to challenge viewers to shape their own ideas of what art was supposed to mean in light of the social and political conditions of his times."[20] Schwitters combined found objects, everyday materials, and junk in his merz assemblages because those materials alluded to commonly understood life. Gestalt theory states that the whole is greater than the sum of its parts. Consequently, a literal pile of junk is transformed through the craftsmanship of Schwitters's assemblage into "art." Blurring the boundary between art and life, and recognizing an artistic practice as one that is politically motivated, Schwitters recognized the poetic or artistic value of visual noise or distraction. As Eco suggests that art frustrates (viewer) expectations with the question of signification, Zipes summarizes Schwitters's practice as abiding by "one governing principle, the spontaneous use and appreciation of all art forms and objects to free the imagination from the senseless and banal legislation of life."[21]

Perhaps noise, as is said of pornography and beauty, is in the eye of the beholder? Eco's understanding of the poetic potential of noise clarifies Schwitters's merz experiments as a conceptual shift in meaning from a literal reading of "junk" to an absurd and codified form of assemblage art.[22] Noise was transformed into a deliberate recording of everyday life selected and made visible in order to reveal political agendas as banal as the "legislation" of everyday life and as significant as the impact of the government's militaristic strategies on contemporary society.

Everyday practices are messy as they embed in daily performances the craft of the mundane, often disguising the legislation of everyday life. Tzara's essay and Schwitters's merz assemblages are celebrations of the resistance of mundane habits and attitudes of the 1920s in war-torn Eastern Europe. More than half a century later, Michel de Certeau would follow their expressive and poetic articulations with research that demonstrates the depth of organization embedded within structures beneath the surfaces of everyday practices.

Tactical Media: From Walking to Surfing

In de Certeau's *The Practice of Everyday Life*, patterns of dwelling, walking, living, and consumption were interpreted as individual moments of acceptance and resistance of the governing rules imposed on users of urban development (citizens) by the government and corporate planners of the shared space. In his seventh chapter, "Walking in the City," de Certeau demonstrates the differences between strategies and tactics by examining how cities are created and how people (users) walk through them. Although the city is conceived as a structured environment where citizens are directed to move in certain patterns by the architecture, government-made maps, and business locations, people who walk in the city create their own interpretation of the space by taking short cuts and leisurely strolls. The intentional layout of the city is a strategic outline provided by the government and businesses. While city dwellers employ tactical operations, using the rules and organization set in place by the government in order to perform individual movements that deviate from the implied structure of the city. Risto Linturi, a Finnish mobile technologies innovator, told Howard Rheingold in *Smart Mobs*, "If citizens have the freedom to set up ad-hoc wireless networks . . . they will create digital pathways on their own, the way people create pathways between buildings."[23] Linturi understands how tactical path making by walking in the city is translated to a deviant set of paths in a networked society.

For de Certeau the identity of the place is *the City*, "a shuffling of the pretences of the proper, a universe of rented spaces haunted by a nowhere or by dreamed of places."[24] This haunting is represented online in the early part of the twenty-first century through rented spaces such as online advertising, and dreamed of places in the form of "empty" databases common to Web 2.0 sites such as Facebook, MySpace, LinkedIn, YouTube, and so on.

Web 2.0 is defined by seven principles outlined by Tim O'Reilly in his 2005 article, "What is Web 2.0?"[25] Two of the principles, *harnessing collective intelligence* by relying on user-generated content, and creating *lightweight programming models* designed for hackability and remixability, allude to the potential of the web for fostering "dreamed of places." Imagine Facebook on the first day the web site launched. It was a white background with a blue logo in the top left corner, with the potential for containing status reports, photos, videos, and electronic messages. In its primary condition, the web site was an empty database, or a virtual space waiting for human interaction.

While Facebook.com was awaiting human interaction, the site's owners were hoping to seduce users into using the platform to gain page views and click-throughs on online advertising positioned on each Facebook page. Marketing is one of many instruments of control and the Internet is a popular platform for contemporary commercialism. Consumptive habits common to Internet surfing, online advertising, and marketing paradigms are controlled elements in the online form of the classic de Certeauian "City," that is, the commercial web. If "To walk is to lack a place,"[26] then to surf[27] the web is to lack a place online. The Internet provides a platform for the current immaterial era of online social networking that enables users to surf nomadically and create a habitual online practice as mundane as walking in the city during the machine age.

A Control Society

Unlike the machine age, where the sites of production could be sabotaged by literally inserting a wrench in a wheel, resistors to a computer-mediated society must rely on immaterial tactics that can be thought of as manipulations of noise. Cleverly, commercial web sites present noise within the infrastructure of computer-mediated society disguised as a part of everyday life. As physical or material noise is transformed in merz works, immaterial noise (data or image files) is transformed by way of tactical resistance on the Internet. Alexander Galloway has suggested, even within the distributed network that has come to be known as the Internet, "it is *through* protocol that one must guide one's efforts, not against it."[28]

In *Discipline and Punish*, Michel Foucault details how eighteenth century society transitioned from a sovereign state in the classical era to the early twentieth-century modern disciplinary period. The transition from one era to the next has social implications regarding power and ideals, control, and resistance. In light of Foucault's periodization, Critical Art Ensemble distinguished the placelessness inherent in the mediascape of the late twentieth-century as governments, businesses, and cultural practices transitioned from the Fordist empires of the machine age to the information era associated with digital production. "Before computerized information management, the heart of institutional command and control was easy to locate. In fact, the conspicuous appearance of the halls of power was used by regimes to maintain their hegemony."[29] As Deleuze noted in 1990, "We're moving towards control societies that no longer operate by confining people but through continuous control and instant communication."[30] A "control society" is one where "Individuals become dividuals, and masses become samples, data, markets, or 'banks.' "[31] In the Deleuzian control society "marketing is [now] the instrument of control."[32] Twenty years later, ours is a control society where cloud computing, social networking, and user-generated content constitute everyday new media practices. As Galloway notes, Deleuze posits a historical association between computers (the technology of the digital era) and a control society (the social phase of this era).[33]

In *Empire*, Michael Hardt and Antonio Negri argue that the Internet is a decentralized network: "[a]n indeterminate and potentially unlimited number of interconnected nodes [that] communicate with no central point of control."[34] Galloway disagrees—the Internet, a distributed network, "is a structural form without center that resembles a web or meshwork."[35] For Galloway a decentralized network is "a multiplication of the centralized network [where] instead of one hub there are many hubs, each with its own array of dependent nodes."[36] Galloway argues that while TCP/IP (Transmission Control Protocol and Internet Protocol) is decentralized, DNS (Domain Name System) is hierarchical. A central point of control is utilized in specific situations in which users are controlled at the root level. All named web sites use DNS to translate a series of digits into a name, so that *Facebook.com* is a translation of a group of numbers, such as 66.220.159.255. As an example of how DNS can be used to control a mediated networked culture, Galloway reports a 1999 lawsuit against the Swiss art group Etoy. The case was eventually dropped, but at the outset the Federal Bureau of Investigation shut down Etoy's New York-based Internet service provider, The Thing. "After some of this material was deemed politically questionable by the Federal Bureau of Investigation, the whole server was yanked off the Internet by the telecommunications company who happened to be immediately upstream from the provider. The Thing had no recourse but to comply with this hierarchical system of control."[37] Galloway suggests that whole countries could be removed from the Internet in a similar fashion within 24 hours.

Hegemonic Culture Online

Antonio Gramsci revitalized the spirit of Lenin's Bolshevik Revolution in tailoring socialism to the Western world through a recombinant approach to Marxism. As opposed to Lenin's model, "the war of movements," Gramsci's "war of position" relied on his most significant contribution to Marxists ideology, the theory of hegemony. Lenin proposed gaining control over society by violently conquering the State. Gramsci's method was less direct and more pervasive, using hegemony, or cultural domination, for cultural positioning.

A hegemonic society is made possible by ideological leadership. A ruling class may use force and coercion in addition to ideological positioning, but the dominant paradigm is maintained by normalizing culture. Luciano Pellicani writes that hegemony ". . . involves a capacity for determined leadership on the one hand and more or less spontaneous acceptance of this leadership on the other."[38] A collective understanding of societal rules is maintained by the cultural values of the hegemonic class.

Although the web seems endless,[39] a collection of corporately owned, well-advertised sites are frequently visited. Alexa.com, *The Web Information Company*, collects global user-viewing habits through its toolbar. In October 2009, the first five on the top 100 sites in the United States accessed through Alexa's toolbar were Google.com, Facebook.com, Yahoo.com, Youtube.com, and (Windows) Live.com.[40] The first pornography site appeared on the list in 36th position, LinkedIn placed 50th, and not too surprisingly, Google owned the 100th site listed. Although Alexa's "Top One Hundred List" is anything but exhaustive, user data supporting the popularity of these sites in a global community contribute to the creation of a mainstream online media. All one hundred sites listed were commercial entities. The commercial web, or sites identified by Alexa's toolbar, provides determined online usage and ideological leadership while millions of web users click their way to spontaneous acceptance.

If hegemony is the result of an ideological hold on culture by normative rules and values, anti-hegemony is the directive of the counterculture. John Fiske begins his article, "Global, National, Local? Some Problems of Culture in a Postmodern World," in response to Scott Sassa's (president of the Turner Entertainment Group in 1995) statement that "The 'middle class' of movie is going to be non-existent."[41] Fiske provides evidence that due to globalization this trend in movies extends to all of culture itself. In contrast to the global flow of a small but highly profitable range of cultural products, Fiske notes that local cultures are both fluid and hybrid, as "The hybridity of local culture carries always its own genealogy of that which it used to be, and whatever it is at the moment is always a point in a field of possibilities of that which it might become."[42] Global hegemony erodes national culture. This erosion may create socio-economic opportunities for such microscopic entities as local cultures on a transnational scale. For Fiske, a hybrid local culture is one that is always on the move, continuously deterritorializing and reterritorializing, and providing optimism for the creation of a "middle class" of culture—one that opposes subordination by the transnational mainstream. Local communities create nebulous cultures that form across borders. Fiske suggests "such a society will organize its key sociocultural relations not under the homogenizing concept of 'nation' but over a shifting and fluid terrain of interlocality, a terrain where local identities and cultures meet and intermingle."[43]

Transnational, corporate businesses create hegemonic consumption practices from entertainment industries to chain stores like Wal-Mart and cafés such as Starbucks or supersized hamburgers at McDonald's. Shopping locally is the consumer's resistance to corporate domination. If local or interlocal communities create anti-hegemonic practices by meeting and intermingling in the analog world, then this activity can also be practiced on the web.

Virtual communities take a variety of forms. Older communities include newsgroups, bulletin board systems, and multi-user dungeons. Contemporary virtual communities include massive multiplayer games such as Ultima Online and World of Warcraft, avatar-based virtual chat rooms such as Second Life and Webkinz, media sharing sites such as Flickr and YouTube, and social networking sites such as Facebook, LinkedIn, and MySpace. Howard Rheingold wrote that a virtual community is established "when people carry on public discussions long enough, with sufficient human feeling, to form webs of personal relationships."[44] While driving traffic to web sites that are not in Alexa's top one hundred list might prove difficult without a public relations team or advertising budget, artists and cultural producers create interlocal articulations on the web, using popular web sites as locations of deterritorialization. These actions take the form of tactical movements, akin to de Certeau's explanation that walking in the city is a unique interpretation of the system, and that the path one makes while walking is not counted, remembered, or charted. *Your Neighbors' Biz* and The Anti-Advertising Agency's *Add-Art* serve as case studies of online projects that tactically manipulate noise to support anti-hegemonic practices.

Your Neighbors' Biz

Facebook[45] is currently one of the largest social networking web sites in terms of unique visitors registered through online traffic. It trailed

MySpace prior to 2008, and has steadily gained visitors in advance of MySpace since approximately December 2008.[46] In November 2009, there were more than 350,000 active applications on the Facebook platform, more than 250 of them with over one million monthly active users.[47] The most engaging Facebook applications, measured in February 2009, were gambling and word games.[48] Facebook users tend to use the social networking site for online communication and entertainment purposes. A hegemonic Facebook practice might be a passive form of surfing through the top games and applications on the web site, whereas an oppositional practice could employ the noise associated with the site to support interaction among local communities, organized as defined by Fiske, "over a shifting and fluid terrain of interlocality."

Your Neighbors' Biz is a Facebook application that enables a user to buy or sell off-the-grid products and services within her social network. The application places the user and her location, not the product, at the center of the experience. This application differs from Facebook's Marketplace, Craigslist, Angie's List, and Ebay because it focuses on unusually local products and services, such as dog washing and babysitting, and artisanal products like soap, candles, and homemade food. *Your Neighbors' Biz* identifies "shopping locally" as an interlocal pursuit between friends of friends. Situating the application inside the Facebook community, and using the tools common to its platform, *Your Neighbors' Biz* transforms the entertainment-based web site into a place where consumers can avoid the hegemonic rule of corporate industries and shop locally while avoiding a recording of the transaction. The application is marginally used, encouraging users to rely on their human neighbors in the analog world, rather than their corporate neighbors online. Due to O'Reilly's principle of supporting lightweight programming models, *Your Neighbors' Biz* is constructed from the same digital application-programming interface (API) used to create entertainment-based applications on Facebook, thereby manipulating the Facebook infrastructure for political purposes. As Fiske notes, the existence of local cultures alludes to the optimistic possibility that global hegemony has left a few holes in the system. *Your Neighbors' Biz* harnesses collective intelligence by facilitating users who engage in the promotion and progress of local cultures while surfing in the Facebook digital community by collecting and maintaining user-generated content. For users of *Your Neighbors' Biz*, Facebook applications such as "Poke," "Gift," or lottery games are noisy digital elements, similar to spam. *Your Neighbors' Biz* reenvisions the Facebook application as one that can be used to transform consumer habits by co-opting the noisy structure of the Facebook application in order to empower the public to form a local consumer culture in the Facebook community.

Add-Art

Add-Art is a Firefox plug-in that inserts digital art into spaces reserved for online advertising. Firefox, a Mozilla web browser, allows the user to transform her web experience by installing third-party applications. Also relying on O'Reilly's lightweight programming model, "the Mozilla project is a global community of people who believe that openness, innovation, and opportunity are key to the continued health of the Internet."[49] The Add-Art plug-in works in concert with *Adblock Plus*, blocking advertising in the Firefox browser.[50] *Add-Art* literally replaces online advertising, noise of the online consumer culture, with rotating art exhibitions, such as *A story should have a beginning, middle, and an end . . . but not necessarily in that order (After Goddard)* curated by Ivan Tanzer, *Fragments of Defragmentation* curated by Daniel Caleb Thompson, and *Drawings and Photographs* from Tucker Nichol.

Paradoxically, for some users, rotating art exhibitions may look like noise. However, installing *Add-Art* (as well as *Adblock Plus*) is a choice. Unlike Schwitters's merz assemblages where noise, rendered as a pile of junk, is transformed into a unified work of art, *Add-Art's* "pile of junk" (advertisements) is replaced with artifacts from a different "pile." The unified work of art is the conceptual shift that takes place when the original web site (for instance, the plug-in is demonstrated on the *New York Times* and *Miami Herald* web sites[51]) is transformed from one that is loaded with information (text or images the user intends to view) and advertising (text or images the user avoids) to a web site only containing information (text the user wants to read and art the user is interested in viewing). As part of the Anti-Advertising Agency's portfolio, the *Add-Art* browser plug-in is a visual manifestation of the group's mission:

> The Anti-Advertising Agency co-opts the tools and structures used by the advertising and public relations industries. Our work calls into question the purpose and effects of advertising in public space. Through constructive parody and gentle humor our Agency's campaigns will ask passers by to critically consider the role and strategies of today's marketing media as well as alternatives for the public arena.[52]

The co-option of tools and structures used by mainstream media to oppose or resist the message, structure, or beliefs of the mainstream is the basis of the tactic Umberto Eco clarified in *The Open Work*. Following Cixous's description of the world as a noise-machine, the media created by the advertising and public relations industries is noise in a consumer culture. The Anti-Advertising Agency co-opts noise to resist the essence of the mainstream message in order to "increase awareness of the public's power to contribute to a more democratically-based outdoor environment."[53]

Hackers of Material and Virtual Worlds

Although Schwitters's merz assemblages were made long before users surfed the Internet, the scrambling of laws and viewer expectations links merz with *Your Neighbors' Biz* and *Add-Art*. All three projects rely on what McKenzie Wark refers to as "the potential of potential" to "express the possibility of new worlds, beyond necessity."[54] Applying the "hacker" label to Kurt Schwitters may seem contentious, but by Wark's manifesto "to hack is to abstract."[55] The term "hacker" emerged in computer circles, but since "the virtual is the true domain of the hacker"[56] any creative action that "touches the virtual—and transforms the actual"[57] can be defined as a hack.

The new world envisioned by Schwitters and other Dada artists is one where the noise produced by a society is believed to reveal its true nature, and is therefore celebrated as art. The abstraction is more than the physical transformation of materials used in the process of creating merz assemblages. For Schwitters, the abstraction—that is, the hack—is the linguistic shift that occurs when such materials are positioned in the political sign carried by "art." When Schwitters named his materials and the art that resulted from collecting and juxtaposing them "merz," he hacked the virtual process by which artists had historically created art. As Dada and merz assemblages were created for gallery exhibitions and publications, contemporary Internet projects meet consumers and users in mainstream online venues.[58]

Your Neighbors' Biz and *Add-Art* are web-based tactical manipulations that hack corporately funded online territories. For *Your Neighbors' Biz*, the noisy voice of the consumer culture is located in Facebook applications such as slots and lottery, poke, and silly quizzes, contributing to a passive culture, where entertainment takes the form of pushing buttons that have no effect on analog communities. In the case of *Add-Art*, online banner, pop-up, and interstitial advertisements are containers for noisy consumer culture. Both online projects create an abstract form of the web application or browser in order to transform the actual user experience in the consumer domain, at once resisting the consumer culture and providing an alternative to its hegemonic structure. Unlike merz assemblages, in which the newly envisioned materials created the abstraction central to understanding his high art hack, the web projects use the same material found on the web to express new possibilities. In the case of *Add-Art*, pixels are literally replaced with other pixels. The abstraction of these web projects relies on the virtuality of the medium. As with merz assemblages, these web projects challenge normative viewer expectations, and offer something else in its place.

Hacks are dreamed-of places, like de Certeau's cities, where abstraction is the directive and the outcome is superfluous, or "beyond necessity." In the case of merz assemblages, the new world is one where any scrap of material has the potential to become integrated into a work of art. The web projects create a new reality online that a user must choose to engage. Neither project is necessary to the operation of any web site; furthermore, users have to install the application or the plug-in to engage with each project. The history of art would be altered, but artists would continue to create exhibitions without the invention of merz; and web sites would continue to serve applications and ads to online communities in lieu of *Your Neighbors' Biz* and *Add-Art*. By envisioning a new world beyond necessity, the merz assemblages and the web projects create an opening for a new way of thinking about art materials, Facebook applications, and browser plug-ins. That is, they reveal "the potential of potential" embedded in systems of order and control from analog artistic media of the 1920s to web communities in the 2000s.

Eco summarizes contemporary art as one that "constantly challenges the initial order by means of an extremely 'improbable' form of organization."[59] The unlikely form of organization that these web projects employ demonstrates contemporary developments in opposition to consumer culture, or what Schwitters and the Dadaists would consider the bourgeois wheel of cultural production. The Dadaists transformed noise into visual art and poetry while contemporary artists tactically manipulate noise on the Internet to demonstrate that noise has evolved from an erroneous part of a communication system into a central channel for revisionist practices. While the definition and manifestations of noise remain elusive, the power it embodies comes from the extent to which it can be appropriated, co-opted, manipulated, and hacked.

Notes

[1] Claude E. Shannon, "A Mathematical Theory of Communication," *The Bell System Technical Journal*, 27 (1948): 379–23, 623–656, http://cm.bell-labs.com/cm/ms/what/shannonday/shannon1948.pdf, 48.

[2] Ibid., 19.

[3] *Jacobellis v. Ohio* 378 U.S. 184, 1964.

[4] Umberto Eco, *The Open Work*, trans. Anna Cancogni (Cambridge, MA: Harvard University Press, 1989), 74.

[5] Ibid., 67.

[6] Ibid., 60, his emphasis.

[7] Ibid., 58.

[8] Hélène Cixous, *Readings*, ed. and trans. Verena Andermatt Conley (Minneapolis: University of Minnesota Press, 1991), 111.

[9] Ibid., 7.

[10] Ibid., 9.

[11] Tristan Tzara, "Dada Manifesto 1918," In *The Dada Painters and Poets 2nd Edition*, ed. Robert Motherwell (Cambridge, MA: The Belknap Press of Harvard University Press, 1981), 80.

[12] Ibid., 77.

[13] Ibid., 76.

[14] Robert Motherwell, Introduction to *The Dada Painters and Poets 2nd Edition*, ed. Robert Motherwell (Cambridge, MA: The Belknap Press of Harvard University Press, 1981), xxiii.

[15] Richard Huelsenbeck, "En Avant Dada: A History of Dadaism (1920)," in *The Dada Painters and Poets 2nd Edition*, Ed. Robert Motherwell (Cambridge, MA: The Belknap Press of Harvard University Press, 1981), 24.

[16] The Italian Futurists, led by F. T. Marinetti from 1910–1914, predated the Dada movement with a manifesto in opposition to old traditions. Of particular significance is Luigi Russolo's 1913 essay, "The Art of Noises," in which the author claimed that noise was born with the invention of the machine, and it "reigns supreme over the sensibility of men." Russolo described noise as an irregular sound accompanying everyday life, "reaching us in a confused and irregular way from the irregular confusion of our life, never entirely reveal[ing] itself to us, [keeping] innumerable surprises in reserve." This interpretation of noise-sound resonates with the Dadaist's understanding of visual noise. See http://www.unknown.nu/futurism/noises.html.

[17] Huelsenbeck, 23.

[18] Ibid., 39.

[19] Jack Zipes, Introduction to *Lucky Hans and Other Fairy Tales*, Kurt Schwitters, trans. Jack Zipes (Princeton: Princeton University Press, 2009), 9.

[20] Ibid., 11.

[21] Ibid., 16.

[22] Artists such as Grosz, Huelsenbeck, and Schwitters, who's merz work exemplified methodologies of appropriation, were by no means the only artists working in this manner in the 1910s. Marcel Duchamp's *Fountain* is the ultimate example of early appropriation in Dada art works. Also see Dawn Ades, *Photomontage* (London: Thames & Hudson, 1986).

[23] Howard Rheingold, *Smart Mobs* (Cambridge: Basic Books, 2002), 14.

[24] Michel de Certeau, *The Practice of Everyday Life*, trans. Steven Rendall (Berkeley: University of California Press, 1984), 103.

[25] Tim O'Reilly, "What is web 2.0?" http://oreilly.com/pub/a/web2/archive/what-is-web-20.html?page-1.

[26] de Certeau, 103.

[27] I use the term *surf* to imply an entertaining, nomadic usage of the web, similar to the way the verb (channel surf) is used in relation to television.

[28] Alexander R. Galloway, *Protocol: How Control Exists after Decentralization* (Cambridge, MA: MIT Press, 2004), 17.

[29] Critical Art Ensemble, *Electronic Civil Disobedience and Other Unpopular Ideas* (New York: Autonomedia, 1996), 7–8.

[30] Gilles Deleuze, *Negotiations 1972–1990*, trans. Martin Joughin (New York: Columbia University Press, 1995), 174.

[31] Ibid., 180.

[32] Ibid., 181.

[33] Galloway, 22.

[34] Michael Hardt and Antonio Negri, *Empire* (Cambridge, MA: Harvard University Press, 2000), 299.

[35] Galloway, 3.

[36] Ibid., 31.

[37] Ibid., 10.

[38] Luciano Pellicani, *Gramsci: An Alternative to Communism*, trans. Mimi Manfrini-Watts (Stanford: Hoover Institution Press: 1981), 3.

[39] As of October 2009, Netcraft Web Server found over two hundred million (230,443,449) distinct web sites online. However, the survey was not perfect as the software used may not have found all of the sites, and the fact that sites were registered did not mean that any information was posted. http://news.netcraft.com/archives/2009/10/17/october_2009_web_server_survey.html.

[40] It should be noted that several million users have installed the toolbar. http://www.alexa.com/site/ds/top_sites?cc=US&ts_mode=country or http://www.alexa.com/topsites.

[41] John Fiske, "Global, National, Local? Some Problems of Culture in a Postmodern World," in *The Velvet Light Trap*, no. 40 (Austin, TX: The University of Texas Press, Fall 1997), 63.

[42] Ibid., 63.

[43] Ibid., 64.

[44] Howard Rheingold, *The Virtual Community* (Reading: Addison Wesley, 1993). http://www.rheingold.com/vc/book/intro.html.

[45] Launched in 2004 by Mark Zuckerberg while a sophomore at Harvard University, the site was initially open to users at one of the 30,000-plus recognized schools and companies in the United States, Canada, and other English-speaking

nations. In 2006 the web site launched open registration, so anyone could create a user account to post status reports, photos, and other user-generated
content. See Giselle Melanson, "New World Order: Thy Name is Facebook,"
Popjournalism, May 12, 2007. http://www.popjournalism.ca/pop/news/2007/
00288facebook.shtml.

46 "Site comparison of Facebook.com (rank #3), myspace.com (#15)," *Compete*,
http://siteanalytics.compete.com/facebook.com+myspace.com/?metric=uv&
months=12http://siteanalytics.compete.com/facebook.com+myspace.com/
?metric=uv&months=12.

47 "Facebook Statistics," *Facebook*. http://www.facebook.com/facebook?ref=pf#/
press/info.php?statistics.

48 Nick O'Neill, "The Top 20 Most Engaging Facebook Applications," *All
Facebook*, February 12, 2009, 1:53pm. http://www.allfacebook.com/2009/02/
engaging-facebook-applications.

49 "About Mozilla," *Mozilla*, http://www.mozilla.com/en-US/about/.

50 "Adblock Plus: Save your time and traffic," *Adblock Plus*, http://adblockplus.
org/en/ (accessed on November 30, 2009).

51 Anti-Advertising Agency, "Add-Art Intro Video," *Anti-Advertising Agency*,
http://add-art.org/.

52 Anti-Advertising Agency, "Our Mission," *Anti-Advertising Agency*, http://
antiadvertisingagency.com/our-mission/.

53 Ibid.

54 McKenzie Wark, *A Hacker Manifesto* (Cambridge, MA: Harvard University
Press, 2004), 14.

55 Ibid., 83.

56 Ibid., 74.

57 Ibid., 71.

58 Both online projects are freely available to web users, alongside other Facebook applications and Firefox plug-ins.

59 Eco, 60.

Chapter 5

Stock Imagery, Filler Content, Semantic Ambiguity*

Christopher Grant Ward

A central concern of media studies has often been to describe how transactions of meaning are established between the encoders and decoders of
media messages: senders and receivers, authors and audiences, producers
and consumers. More precisely, this discipline has aimed to describe
the semantic disconnects that occur when organizations, governments,
businesses, and people communicate and interact across media in order
to understand the causes of these miscommunications, and their social
and cultural implications. As increasingly multimodal messages are broadcast to more diverse, niche audiences, it is no surprise that error seems to
occur more (and not less) frequently, forcing difficult questions of mass
communication's value and efficacy.

By the middle of the twentieth century, many theoretical models of
communication error were described through the metaphor of a "signal"
transmitted from sender to receiver. Though such research varied in
approach, these theoretical models tended to characterize error as a dys-
function of a pure transmission process: interference that prevented the
otherwise successful relay of meaning from a sender to receiver. Berlo
saw error as a disruption of the "static" codes encoded by senders and
decoded by audiences.[1] Schramm's communication model described two
individuals sharing a "field of experience." Error and misunderstanding
occurred to the extent that these fields do not overlap.[2] Perhaps most
well known is Shannon and Weaver's information theory of communication, where miscommunication was described explicitly as "semantic
noise": distortions of meaning that resulted in the message received
being different from what was being transmitted.[3]

* For Stephen Todd Jordan, who encouraged me to navigate graduate school
and keep my own voice. Rest in peace, Stephen. We will miss you.

It is important to note that Shannon and Weaver were not humanities scholars; they were information theorists working for Bell Telephone Labs. Their mathematical model was originally aimed at measuring and maximizing communication capacity (think dial-up speeds, bits per second) across large telephone and radio networks. As such, their model broke down communication into five technical elements: (1) The "source," which creates a message, (2) the "sender," which changes the message into a signal, (3) the "channel," which transmits the signal, (4) the "receiver," which changes the signal back to a message, and (5) the "destination," where the message ends up. Beyond these elements, they also discuss the element of noise (literally *static*) that may be present in the channel, which may result in "variance" between the signal sent and the one received.

In particular, Shannon and Weaver's transmission metaphor became one of the most influential models for human communication studies of the twentieth century because it made information (and thus messages) measurable. Their technical approach to describing social systems, paired with the ambitious (if not misconstrued) goal of maximizing efficiency and effectiveness made for a very attractive theory that was soon referenced and appropriated by communication studies and humanities scholars throughout the world.

Certainly no serious communication theorist today would accept this as an accurate description of how human communication works. Specifically, we don't buy the implicit assumption that the comprehension of meaning is passive; that the receiver's only job is to convert or "decode" a transmitted signal back into the original message. Indeed, during the same time information theory was positing a streamlined understanding of communication, post-structural thinkers were questioning if intention (sending) and interpretation (receiving) of messages could be anything but distortion.

In particular, Jacques Derrida challenged the stability of any sender's intention or receiver's interpretation. Rather, he described communication as inherently "iterable . . . able to break with every given context, and engender infinitely new contexts in an absolutely unsaturable fashion."[4] While Derrida agreed that consistency and repetition of language might help to limit such iterability, he saw all meaning as ambiguous and never final. For Derrida, all signs (words, images, banner ads, emails, and so on) can and will forever signify a multitude of things to different individuals in different situations, at different points in time. Further, he believed that any perceived signification (and thus, meaning) is produced finally not by the sender, but by the receiver. To communicate, then, is to perpetually negotiate semantic ambiguity, not to overcome it, constrain it, or push it aside.

The presence of "uncertain value" in communication was not lost on Shannon, who described the effects of communication "entropy": a measure of the semantic uncertainty associated with an unknown communication variable.[5] Both Shannon and cyberneticist Norbert Wiener measured the entropy of random variables as a way to estimate the possibility for "error" in any communication system. Shannon quite literally measured an entropy rate to find the probability of potential error within a message or system of messages. Wiener saw entropy as a malfunction, the extent to which a system is disorganized because it lacked the necessary information to maintain order and structure.[6] Both approaches sought to limit error in communication by controlling the extent of entropy.

The critical difference between "entropy" and "iterability" is that for Shannon, Weaver, and Wiener, noise is described as an error that enters an otherwise pure information system. For Derrida, the noise of communication is not outside the system, but inherent in the undecidability of meaning. Derrida saw communication not as a linear, closed loop system of transmission, but as a perpetual, open-ended system of meaning making. However, and perhaps most importantly, Derrida and Shannon et al. would agree on a few key points: (1) there is a degree of instability (iterability, entropy) in all acts of communication, (2) this ambiguity creates noise and variability in a receiver's understanding, and (3) this noise is inherent in all acts of communication.

Considering modern media with these thoughts in mind, it becomes readily apparent that semantic noise is not only increasingly inherent in media communication, but it has sometimes become functional—even crucial—to a communicator's goals. Such circumstances are what Mark Nunes describes in the introduction to this collection as "cultures of noise": a term I attribute to organized media practices that seem to negotiate, function, and thrive by communicating ambiguously (or at the very least, by resisting the urge to signify explicitly). Cultures of noise are important to the study of media precisely for the ways they call into question our existing paradigms of what it means to communicate to a media audience. By suggesting that aberrant interpretations of meaning are not necessarily dysfunctions of what would be an otherwise efficient system, cultures of noise reveal how certain asignifying poetics might be productive and generative for certain communication goals.

I will expand upon this theory of how cultures of noise function by exploring one case study: the pervasive use of commercial stock images throughout mass media. I will describe how the semantic ambiguity embedded into stock images is productive, both to the stock photography industry and to advertisers and marketers as they communicate brand and corporate identity. I will first discuss the stock image's dependence upon semantic ambiguity and the productive function this ambiguity serves in supporting the success of the stock photography industry. I will then describe how this ambiguity comes to be employed by corporations and advertisers as "filler content," enabling these producers to elide the accountability and risk that is involved with more explicit communication.

Ambiguous Raw Material: The Stock Image Industry as a Culture of Noise

The photographic image has been a staple of corporate identity for as long as a visual identity has been a concern of corporations. Yet, it is estimated that more than 70 percent of the photographic images used in today's corporate marketing and advertising have been acquired from a select group of stock image firms and photography stock houses, which manage archives containing millions of images.[7] In fact, since its inception in the 1970s, increasing global dependence on stock imagery has grown the practice of commercial stock photography into a billion dollar a year industry.[8]

Commercial stock images are somewhat peculiar. Unlike other non-fiction genres of stock photography (e.g., editorial and journalistic) commercial stock images present explicitly fictive, constructed scenes. Indeed, many of the images of business workers, doctors, and soccer moms that one might find through a Google Image search are actually actors hired to stage a scene. In this way, commercial stock images share much more in common with the images produced for advertising campaigns, in that they are designed to support implicit branding and corporate identity messages. However, unlike traditional advertising images, which are staged to deliver a certain message for a quite specific application (think "Tide stain test" or "posh woman in the Lexus"), commercial stock images have been purposely constructed with no particular application in mind.

On the contrary, stock images must be designed to anticipate *many* diverse needs of cultural intermediaries—design firms, advertising agencies, and corporate marketing teams—who will ultimately purchase the majority of these images.[9] To achieve these goals, every commercial stock image is designed to be open-ended, in order to offer up a field of potential meanings. And yet, based on sales figures, these images also seem to anticipate the applications of use that will likely appeal to the discourses of corporate marketing and advertising. In this way, the commercial stock image might best be understood as undefined raw material, as a set of likely potentialities still lacking a final determination—what Derrida describes as "undecided" meaning:

> I want to recall that undecidability is always a determinate oscillation between possibilities (for example, of meaning, but also of acts). These possibilities are highly determined in strictly concerned situations . . . they are pragmatically determined. The analyses that I have devoted to undecidability concern just these determinations and these definitions, not at all some vague "indeterminacy." I say "undecidability" rather than "indeterminacy" because I am interested more in relations of force, in everything that allows, precisely, determinations in given situations to be stabilised through a decision . . .[10]

A stock image's ambiguity is the result of an intentional design process whereby the stock photography industry presents the maximum range of possible meanings, and yet, falls artfully short of deciding any of them. Rather, it is the advertisers, designers, and marketers who ultimately make these decisions by finding utility for the image's uncertain value for a certain context. The more customers that can find a use for a certain image, the more this image will be purchased, and the more valuable that particular stock image becomes. So not only do stock images seem have a bias toward iterability, but it also seems stock images thrive on a high degree of entropy. Where Shannon and Wiener sought to reduce the noise in the system, the producers of stock images are intentionally increasing entropy to allow a greater selection of possible meanings.

More to the point, the success of the stock photography industry quite literally *depends* upon the aberrant and unpredictable interpretations of buyers. It is now quite explicitly the "receiver," and not the "sender," who controls meaning by imbuing the image with meaning for a specific context and specific need. This is how the stock photography industry functions as a culture of noise, and raises questions of a sender-to-receiver model in modern media communication. Cultures of noise not only embrace semantic ambiguity; they rely upon ambiguity for their success.

Once a stock image is purchased, the potentialities of meaning within a stock image can become somewhat determined by placement within a certain context of circulation, such as its use for a banking advertisement or healthcare brochure. In many cases, the meaning of a given stock image is also specified by the text with which it is paired. (Figure 5.1) Using text to control the meaning of an image is what Roland Barthes describes as "anchorage . . . the creator's (and hence society's) right of inspection over the image; anchorage is a control, bearing a responsibility in the face of the projective power of pictures—for the use of the message."[11] Textual anchorage helps to control the ways an image should be interpreted. Now, the form, the subjects, and composition of a stock image can work to complement the textual message in a clear and defined way.

Such uses of stock photography reflect the classic corporate efficiency model of communication, aiming to keep noise down and ensuring that the message sent is the message received. Indeed, Shannon and Wiener would see anchorage as an attempt to organize the communication system by limiting the range of possible meanings, while Derrida might very well describe anchorage as a method to "pragmatically determine" the meaning of an image.

FIGURE 5.1   Barthes's textual anchorage: The subject, form, and composition of a stock image are made specific by the text with which the image is paired. Reproduced with permission of Washington Mutual

Filler Content: Advertising and Marketing as a Culture of Noise

In other marketing and advertising messages, stock images can be paired with text and yet used in quite a different way, as *filler content*: open-ended material that takes the place of more explicit, message-oriented elements. Operating within a culture of noise, filler content opposes the traditional goal of generating a clear and specific message. Rather, *the goal of filler content is to present an ambiguous message to consumers.*

Consider the image (Figure 5.2) used in a certain marketing design. Compared with Figure 5.1, this design makes little attempt to specify the meaning of the image through textual anchorage.

In this example, there is much less concern for making sure the message is decoded "correctly." In fact, when stock images are used as filler content, they are placed into advertising and marketing messages with virtually the same degree of ambiguity as when the image was originally constructed. As such, it stages the same ambiguous potential for final consumers as it did for the advertiser who originally purchased the image from the stock image house. Such images may receive only vague specificity from textual anchorage and little effort is made by the message producer to explicitly decide a message's meaning.

FIGURE 5.2 Filler content: What meaning(s) does the image have for you? Love? Happiness? Leisure? Freedom? The Outdoors? Perhaps you rode your bike today? Does it matter? Reproduced with permission of Visa.com

Consider how this strategy differs from a more classic understanding of mass communication, where the goal is maximizing the clarity and efficiency of the sender's intended message. The advertiser's main intention here is to *increase* entropy of the message. Filler content amplifies the semantic noise of ambiguous elements (such as images, but also text) by making little attempt to constrain these undecided elements to likely or particular contexts, and by placing much more of the responsibility for meaning-making in the hands of audience.

Yet stock images don't just rely upon audiences to fill in the blanks. They also leverage the cultural reinforcement of other, similar images. Consider the way that the image of "a woman with a headset" has come to signify customer service—a quick search of "Customer Service" images on Google will yield highly consistent results. All filler content depends upon the iconic status of certain stock photography clichés, categories and familiar scenes. The image itself doesn't represent this meaning on its own, but it works as part of a larger discourse, what Paul Frosh describes as an "image repertoire."[12] By bombarding us incessantly with a repetition of similar images, mass media can bolster the iconic value that stock images possess. The "woman with the headset" has become an icon for "Customer Service," simply because we are exposed to a repetition of images in the same context that repeatedly stage the same or similar scene of this idea. As Frosh suggests, " . . . this is the essence of the concept-based stock image: it constitutes a pre-formed, generically familiar visual symbol that calls forth relevant connotations from the social experience of viewers . . . "[13]

In her 2004 article, "Communication beyond Meaning: On the Cultural Politics of Information," Tiziana Terranova recognizes the role of iconography as an essential control to "manage" the potentialities of meaning present in modern media.[14] Specifically, the "iconic power" of logos lies in their innate ability to streamline communication, "shortcutting their way to the receiver by using the shortest possible route in the shortest possible time."[15] Advertisers and marketers have long relied on brands and logos to quickly, reliably, and effectively connect with their audiences. And perhaps more importantly for media is recognizing that what is easily processed can also be easily replicated. This "endurance" of icons allows them to be reused again and again, in different contexts, and yet, still reliably deliver more or less the same information. In fact, the more they are used, the more effective they are. As Terranova suggests,

Communication management today increasingly involves the reduction of all meaning to replicable information—that is, to a redundancy or frequency that can be successfully copied across varied communication milieus with minimum alterations. Whether it is about the Nike swoosh or war propaganda, what matters is the endurance of the information to be communicated, its power to survive . . . all possible corruption by noise.[16]

The stock image shares much in common with the logo and the brand. It is portable and replicable, it becomes more effective through repeated use, and it signifies in very broad ways. We should not be surprised, then, that professional communication managers have come to handle stock photography in much the same way as their logos. Amidst the array of stock photographic material at their disposal, might advertisers wish to align certain easily digestible, iconic photographic poses, compositions, and photographic subjects alongside the Nike Swoosh? Without question. So while advertisers rely on semantic ambiguity of stock photography, they also rely on these images as implicit logos that streamline the representation of specific ideologies, which can be repeated again and again in a highly visual, mass-media culture, and which are easily digested and understood by their audiences.

As a culture of noise, stock images function as filler content in two ways: (1) meaning is left undecided by the advertiser who intends for customers to create their own interpretations of an advertisement; (2) meaning is generated by the ideological constructs of an image repertoire that is itself promulgated by the repetition of these images in use.

The potentiality of meanings that was initially embedded into stock images in order to make them more attractive to cultural intermediaries is also being passed on to the final audiences by these same advertisers and marketers. The same noisy signification that supported the sale of stock photos from the stock industry to advertisers now also seems to support the sale of messages that advertisers pass on to their audiences. Just like in the stock photography industry, practices of filler content in advertising also create a culture of noise by relying upon ambiguous messages that end customers are now forced to both produce and consume.

The Corporate Imperative: Avoiding Identity Dissonance

Filler content signals a shift in the goals of modern advertising and marketing, where corporate messages are designed to be increasingly ambiguous, and meaning seems to be decided more than ever by the final audience. As marketing psychologists Kim and Kahle suggest,

> Advertising strategy . . . may need to be changed. Instead of providing the "correct" consumption episodes, marketers could give . . . an open-ended status, thereby allowing consumers to create the image on their own and to decide the appropriateness of the product for a given need or situation.[17]

Ambiguous communication is not, by itself, necessarily egregious. On the contrary, many media designers, including myself, believe that creating a space for thoughtful, open-ended discovery is one of the best ways to create a connection with end users. Media designer Phil Van Allen describes such approaches to ambiguous design as "productive interaction . . . an open mode of communication where people can form their own outcomes and meanings . . . sharing insights, dilemmas and questions, and creating new opportunities for synthesis."[18] In the twenty-first century, it is crystal clear that audiences are no longer content to be cast in the role of a passive receiver. On the web, in particular, virtually every media consumer has also become a media producer. There seem to be almost as many blogs as there are readers. News articles, videos, posts, and images are commented on, reposted, shared, combined, and repurposed almost as soon as they are published—each time appropriated by audiences with new contextual meaning and signification.

It is this shift toward a participatory digital culture (yes, Web 2.0) that has allowed filler content to emerge so ubiquitously and yet so often slip by unnoticed. As audiences become more comfortable producing their own meanings collaboratively, concern over properly decoding a corporation's primary message seems far less important (or even passé). Admittedly, "filler content" looks very much like productive open-ended design. However, it lacks the two-way street of sharing and synthesis that creates meaningful connections. An audience's connection to a corporate brand is not the same as their connection to one another on social media networks, because advertising and marketing messages are still largely promulgated to the masses, regardless of the progressive social media channels in which they operate.

It seems a critical definition of filler content must go beyond an understanding of how it operates in media. It must also describe what this material "fills in" for and the value it brings to the communication professionals who use it. Though ambiguous design might help to create deeper connections, in the context of marketing and advertising, it also allows corporations to obviate the production of explicit messages for their customers. Marketing and organizational communication research[19] suggests that as corporations manage their identities to increasingly disparate and diverse media audiences, misunderstandings are more likely to bring about "identity dissonance": disconnections between the identity projected by the organization and the identity attributed to an organization by its customer-public.[20] To grapple with identity dissonance, corporate identity researcher Samia Chreim believes that that top managers often choose to engage in the practice of *dissonance avoidance*: the use of ambiguous messages to provide flexibility in the interpretations of how customers can define a brand or organization:

> Dissonance avoidance can be achieved through the use of ambiguous terms . . . organizations use ambiguity to unite stakeholders under one corporate banner and to stretch the interpretation of how the organization, or a product or a message can be defined.[21]

Corporate brand identity is a tricky business, even more so with the expanding access of a digital marketplace. As audiences tap a virtually endless stream of media messages, grabbing mindshare has become increasingly difficult. Corporations have found value in an undecided, pliable identity that can be, at the same time, immediately recognizable, easily reproduced, and culturally reinforced.

In 2000, Jean Baudrillard adumbrated this potential for media images to be used for "deterrence . . . for the elusion of communication . . . for absolving face-to-face relations and social responsibilities. They don't really lead to action, they substitute for it most of the time."[22] Filler content has served this corporate imperative quite well, allowing companies to forgo the myriad disconnects and pitfalls of a noisy mass communication system by never crafting an explicit message, holding a position, or expressing a belief that their customers could demur or discount. In such instances, it appears at first glance that filler content services a shrewd corporate strategy of message deterrence.

A Hollow Core: The Implications for Reliance on Ambiguity

Yet Baudrillard also believed strongly that any strategy of deterrence incorrectly assumed "the naiveté and stupidity of the masses."[23]

Baudrillard believed that ambiguity would not serve media in the long-term because a media public will eventually learn to respond " . . . with ambivalence, to deterrence they respond with disaffection, or with an always enigmatic belief."[24] Indeed, as advertising and marketing shape the visual ground of our culture, Samia Chreim also warns corporations of the dangers of dissonance avoidance:

> What is gained in avoiding [identity dissonance] can be lost in the ability to create meaning for stakeholders. Over-reliance on abstract terms may well leave the organization with a hollow core, one that cannot be appropriated by [customers] in their quest for meaning and identification with the organization.[25]

While there may be room for ethical debate on a corporation's responsibility to message and the effect of filler content on a mass public, current research suggests that in the face of increasingly savvy customers, it is corporations themselves who are ultimately being harmed by vague and generic communication. In 2008, Forrester Research, Inc. surveyed the characteristics of the most successful and unsuccessful online brands. Their research suggested that pervasive, generic online experiences have caused over 21 percent of users to "trust marketers and corporations less."[26] Beyond trust, it seems audiences have also learned to "tune out" irrelevant, generic content, favoring "personalization" and "targeted function, content, and images." On the contrary, the most successful sites emphasize visual experiences that are both "informative and engaging." Specifically of stock imagery used in several leading travel sites, the research suggests that

> Building brand with imagery requires more than filling a site with happy faces. Firms need to strike a balance with imagery that helps users understand products and services while evoking specific emotions tied to a brand.[27]

This research suggests that dissonance avoidance may not be a functional long-term strategy in the face of an increasingly astute media public. While filler content allows visual/verbal messages to be communicated with the speed, simplicity, and semantic ambiguity of brands, Forrester's research suggests that such generic material fails in the most fundamental ways to "create empathy for users . . . to create products and services for real people with real, human wants and needs." It would appear old techniques of brand communication have given audiences marketing fatigue. Audiences are tired of consuming brands and now are responding to such deterrence offered by filler content with what Baudrillard rightly described as "disaffection."[28]

Semantic noise can be productive, and indeed, often vital, to certain tactics of media communication. Whether used as a way to create productive interactions with audiences, or as a hack for avoiding commitment to message, it is important to understand the role of semantic noise in media, especially as the lines between audience and sender are incredibly vague, and as audiences themselves thrive upon productive aspects of ambiguous communication. A close look at filler content also suggests that media messages should be critically evaluated in a slightly different way. Often, analyses of media communication root out the subversive and manipulative factors that reside "deep down" in culture. Cultures of noise suggest that it is also important to consider the effect of media's noisy, diluted, and facile surface, and to challenge the semantic depth of our visual-verbal culture, especially as it appears a reliance on ambiguity can sometimes cripple our ability to say anything meaningful at all.

Notes

[1] David K. Berlo, *The Process of Communication* (New York: Holt, Rinehart and Winston, Inc., 1960), 14.

[2] Wilbur Schramm, "How Communication Works," in *The Process and Effects of Mass Communication*, ed. Wilbur Schramm (Urbana, IL: The University of Illinois Press, 1961), 5–6.

[3] Claude F. Shannon and Warren Weaver, *The Mathematical Theory of Communication.* (Urbana, IL: The University of Illinois Press, 1964), 26.

[4] Jacques Derrida, "Signature, Event. Context," *Margins of Philosophy*, trans. Alan Bass (Chicago: University of Chicago Press, 1982), 320.

[5] Claude E. Shannon, "Prediction and Entropy of Printed English," *The Bell System Technical Journal 30*, no. 1 (1951): 50–64.

[6] Norbert Wiener, *Cybernetics in History* (Boston, MA: Houghton Mifflin, 1954), 5–27.

[7] Paul Frosh, *The Image Factory* (New York: Berg, 2003), 5.

[8] Gettyimages.com. October, 20 2007. http://gettyimages.mediaroom.com.

[9] Frosh, 57.

[10] Jacques Derrida, *Limited Inc.* (Evanston, IL: Northwestern University Press, 1988) 62.

[11] Roland Barthes, "The Rhetoric of the Image," *The Rhetoric of the Image*, trans. Richard Howard (Berkeley: University of California Press, 1977), 156.

[12] Frosh, 91.

[13]　Ibid., 79.

[14]　Tiziana Terranova, "Communication beyond Meaning: On the Cultural Politics of Information," *Social Text* 22, no.3 (2004), 55–56.

[15]　Ibid., 58.

[16]　Ibid.

[17]　Lynn R. Kahle and Chung-Hyun Kim, eds., *Creating Images and the Psychology of Marketing Communication* (Mahwah, New Jersey: Lawrence Erlbaum Associates, 2006), 63.

[18]　Philip Van Allen, "Models," *The New Ecology of Things* (Pasadena: Media Design Program, Art Center College of Design, 2007), 56.

[19]　See, for example, Samia Chreim, "Reducing Dissonance: Closing the Gap Between Projected and Attributed Identity," *Corporate and Organizational Identities: Integrating Strategy, Marketing, Communication and Organizational Perspectives*. ed. Moingeon, B. and G. Soenen (New York: Routledge, 2002); Kimberly D. Elsbach and Roderick M Kramer, "Members' Responses to Organizational Identity Threats: Encountering and Countering the Business Week rankings," *Administrative Science Quarterly* 41 (1996): 442–476; and George Cheney, *Rhetoric in an Organizational Society: Managing Multiple Identities* (Columbia: University of South Carolina Press, 1991).

[20]　Kahle and Kim, 63.

[21]　Chreim, 76.

[22]　Jean Baudrillard, "Aesthetic Illusion and Virtual Reality," *Reading Images* Ed. Julia Thomas (Houndmills: Macmillan, 2000), 203.

[23]　Jean Baudrillard, "The Implosion of Meaning in the Media," *Simulacra and Simulation*, trans. Sheila Faria Glaser (Ann Arbor: University of Michigan Press, 1994), 81.

[24]　Ibid.

[25]　Chreim, 88.

[26]　Ron Rogowski, "Web Site Imagery That Builds Brands: Brand Building Sites Succeed with Imagery That's Helpful And Engaging," Forrester's Research, Inc., August 2008, http://www.forrester.com/rb/Research/web_site_imagery_that_builds_brands/q/id/46945/t/2 .

[27]　Ibid.

[28]　Baudrillard, "The Implosion of Meaning in the Media," 81.

Game

Chapter 6

Gaming the Glitch: Room for Error

Peter Krapp
University of California, Irvine

Ever tried. Ever failed. No matter.
Try again. Fail again. Fail better.

(*Samuel Beckett,* Worstward Ho)[1]

Historic contingency and the concept are the more mercilessly antagonistic the
more solidly they are entwined. Chance is the historic fate of the individual—a
meaningless fate because the historic process itself usurped all meaning.

(*Theodor Adorno,* Negative Dialectics)[2]

Computing culture is predicated on communication and control to an
extent that can obscure the domain of the error. Between a hermetically
rule-bound realm of programmed necessity and efficient management of
the totality of the possible, this chapter situates a realm of contingency—
distortions in the strictest signal-to-noise ratio, glitches and accidents, or
moments where the social hierarchy of computer knowledge condescend-
ingly ascribes to "user error" what does not compute. Whether one sees
the graphic user interface (GUI) as a Taylorist discipline teaching ergo-
nomic interaction, or as yielding to reductions of interaction to resemble
what users had already learned—the GUI is pivotal for our culture.[3]
And the discursive formation of computer games runs parallel to that of
the GUI: interaction revolves around perception, hand-eye coordination,
and discerning errors. But while a typical GUI—from the desktop metaphors of Vannevar Bush's Memex to the Xerox Star or Apple Lisa, and
beyond—will aim for visibility and patient acceptance (or even anticipation) of user error, games probe the twitchy limits of reaction times and
punish user error with loss of symbolic energy. Audiovisual cues pull
users into feedback loops that a game might exhibit proudly, while other,
perhaps less playful user interfaces will tend to hide them in redundant

metaphor. As our digital culture oscillates between the sovereign omnip-
otence of computing systems and the despairing agency panic of the
user, glitches become aestheticized, recuperating mistakes and accidents
under the conditions of signal processing: "glitches can be claimed to be
a manifestation of genuine software aesthetics."[4]

If one postulates that computer games are an adaptive response to the
omnipresence of computing devices—and we do notice that gadgets
often come with a game preinstalled so as to teach us how to handle
them—then the fact that games afford users significant room for error
is an important deviation from the common assumptions about the stric-
tures of human-computer interfaces (HCI). Some observers have even
joked that if HCI practitioners were to try game design, the resulting
game would have a big red button with the label "Press here to win!"
But rather than discuss the tendency among game critics and designers
to resist or embrace HCI, this chapter will focus on a more theoretical
meditation on games as an enculturating force.

Games provide room for error; that room is typically understood to be
the construction of "the user." To the extent that this room for error, this
playful sense for potential deviations and alterations, is an essential part
of games, it pivots on the opposite or absence of complete necessity. As
Cogburn and Silcox argue, "our play-through of the game instantiates a
property that could not be instantiated in a computer program, and
hence that is not 'already there' in the lines of code that make up a finished game from the point of view of its designers."[5] Whether conceived
as choices, as rare constellations, or as mere accidents, in opening that
space where something is possible otherwise, those crevices in the continuity of experiential space are "revealing the folded-in dimensions of
contingency, which included those of experience and of its description,
very much in the noncausal and non-linear way in which the autopoiesis
of systems takes place in the descriptions currently given of them."[6] In
treating the (near) future in terms of possibilities, we distinguish between
the formal possibility of the imaginable, and the objective possibility of
what we can already anticipate. Gaming only unfolds as we play, and
although gaming videos and machinima are becoming more popular
now, archival records of actual gaming experiences are still relatively
uncommon. But by the same token, under highly technologized conditions, presence exists only insofar as past and future exist: the present is
the form of the unnecessary past and the unrealized future.[7] Without
such an opening to contingency and error, programs might seem to
close off that room for play.

Computer game studies seek to describe how protocols, networks, codes, and algorithms structure gaming actions. Recent criticism has begun to consider the implications of the allegories of control inscribed in games and simulations.[8] The difference between playing a game and playing *with* a game is crucial to gaming culture: while the former teaches one the game through navigating the game's commands and controls, the latter opens up to critical and self-aware exploration. Like learning to swim, gaming means learning to learn and initiates us into training systems; but like learning to read, it does not stop at a literal obedience to the letter. Still, computer games tend to obey a certain set of design rules, chief among them efficiency, expediency, and mastery of interfaces and interactions—since we value in such games that they "work." Indeed we can trace the interface design of computer gaming back to Gilbreth's applied motion studies that sought to transform the conditions of labor into the foundations of scientific management with the aid of chrono-cyclography and film.[9] This is to reduce neither gaming to ergonomics, nor critical game studies to an echo of early film studies—merely to point out that gaming still depends on controls and displays that are expected to be efficient while allowing for a lot of ludic latitude.

The mapping of abstract operations to an intuitive environment is a difficult task. The graphic user interface with its audiovisual registers is a more complex case for media studies than the keyboard, which similarly trains and constrains regular users of computers.[10] One widespread example is the "desktop" interface, where files are held in "folders" which may be "opened" or "thrown away"—as Dennis Chao observes:

> Current applications often do not leverage the rich vocabulary of contemporary mass media and video games. Their user interfaces are usually austere, reflecting the limitations of the machines of two decades ago. Our rectilinear desktops feature a Machine Aesthetic, not a human one. There have been attempts to create friendlier interfaces based on familiar physical settings (e.g. General Magic's MagicCap and Microsoft Bob), but they have often been derided as condescending or confining.[11]

In glossing his playful critique of the graphic user interface, Chao advocates an adversarial operating system that would throw into stark relief what the user faces, rather than relying, as most commercial interfaces still do to this day, on preconceived notions of accommodation and familiarity. Industrial designers argue that "a user interface is intuitive insofar as it resembles or is identical to something the user has already learned," yet it remains stuck on emulating work environments that are far less familiar to most computer users than games are.[12] The design of the Xerox Star, inheriting notions from Bush's Memex desktop and accommodating the office environment of the late 1970s, aimed for redundancy: "an important design goal was to make the computer as invisible to users as possible."[13] Chao's critique, written as executable code, comes in the shape of using the computer game *Doom* as an interface for Unix process management. Under Unix, the "ps" command displays currently running processes, analogous to the MS Windows "tasklist" command. (A similar project, but for the Apple Mac, is Chris Pruett's "*Marathon* Process Manager," using the *Marathon* game engine by Bungie.[14]) The downside of *psDoom* as Chao describes it is that it is too easy to kill the wrong processes if monsters (processes) can kill each other—and of course, it is hard to tell whether users are working.

As Chao glosses his project, *Doom* was chosen because as a classic game it was widely known and had been released under the GNU General Public License. The former meant users found the interface intuitive (one can quickly assess machine load by seeing how crowded a room is; while command line methods to slow down and kill processes are different, psDoom unifies them), while the latter fostered the spread of this project in the sysadmin community. Aware of objections against violence in games, Chao surmises that turning the single-player game environment of *psDoom* into an online world might reduce the level of violence by increasing a sense of community.

An example of a violent computer game turned to unexpected ends by exploitation of a glitch is Brody Condon's art work, "Adam Killer."[15] "Adam Killer" consists of a series of eight *Half-Life* mods. Multiple replicas of Adam, biblically pure in white slacks and a t-shirt, stand idle on a white plane, patiently inviting the player's interaction. Condon's art work is available as a playable, stand-alone mod of the game, or as a narrative-abstract machinima based on video documentation of his private performances of the work.[16] As Condon says, "White was an aesthetic decision, I felt it contrasted well with blood. As the characters were shot and bludgeoned with various weapons, an exploited glitch in the game's level editing software created a harsh trailing effect. This turned the environment into a chaotic mess of bloody, fractured textures."[17] In exploiting that software glitch, Condon's art installation juxtaposes a wry take on the rather bland advertising aesthetic of displays with the controversially popular transgression afforded by first-person shooters: a space where random acts of violence, up to and including needless killing, are possible over and over again.

If necessity and chance are complementary, then one might expect not only games of chance, but also games of necessity. Yet we can see that a fully rationalized game, like a state-operated lottery that allows for play, minimizes risks, and benefits the commons with profits, is simply a fiscal and moral calculation that recuperates contingency into a legally sanctioned form of gambling that serves to fill the state coffers.[18] Instead of opening iterative play chances, the contingency of a lottery holds out the totalized promise of one winner exiting the rat race, yet binds the collective more firmly into the game of labor and taxes. In short, the fully rationalized game is indistinguishable from work; and indeed many players of massively multiplayer online role-playing games have come to regret the "grind" required to level up to the advanced levels of exploration and play. If there was no intervening element of contingency, if play were to be totally rule-bound and programmable, then its complete rationalization would direct play as an unfettered pursuit into the mode of discovery and exploration. For instance, the endgame in chess is much more calculable and circumscribed, since most pieces have been removed and thus the possible combinations of moves are much less complex. Nonetheless, the endgame still involves play: namely in finding the strategy that may lead to a win condition and force an end-state. Certainly once chess becomes computerized (so as to be played between humans and machines), a formally represented tree of choices and consequences can be comprehended by a vast database of chess history and theory; but chess remains playable even against the computer, in the sense of not being entirely overdetermined by logical constraints. It would be premature, therefore, to assert that role-play and imitation are less directly reliant on the wager as a mode to comprehend and manage the hazardous and the contingent.

The paradoxical aim of good game design is to strike a balance between the protection of "play" through regulation and game design on the one hand (play as rule-bound), and the protection of "play" as free, unfettered, and improvisational activity on the other hand. Accordingly, game designers and critics alike have been calling for increased literacy about command and control structures that, like learning to swim or to read, enable us to move past fear and isolation into active engagement with the unknown and the incalculable. Put differently, key to what Galloway calls "counter-gaming" would be the quest to create alternative algorithms.[19] As he and others have diagnosed, many so-called "serious games" tend to be quite reactionary at the level of their controls, even if they cloak themselves in progressive political desires on the surface. Critical gaming would tend away from the infantilizing attraction of play as obedience to rules, and towards a "progressive algorithm"—if there can be such a thing. In a primary mode it would teach its players the lessons of the game's controls; in a secondary mode it would build upon the discovery and exploration of the game's regularities and irregularities and invite creative play with the game.

At this juncture, the discussion could branch out into how the structure of MMOs must carefully calibrate and balance labor with appeal, both in time-sinks and money-sinks, to regulate their virtual economy. A classic online game, Lucasfilm's *Habitat*, already provided a system of avatar currency (tokens) and stores where items could be purchased. When a glitch in the program produced a virtual money machine, a small group of *Habitat* avatars became virtual millionaires literally overnight. Against the wishes of the "poor," *Habitat's* programmers eventually decided to let the finders keep the virtual loot.[20] But instead of looking generally at self-regulation of games in terms of a risky, lawless world versus a safe, riskless world, let us stay with the systemic place of error, and particularly user error, in digital culture. Can simulation systems "think" their breakdowns if they are unable to predict them?[21] A recent example of how computer game technologies have come to be useful in precisely the kind of training and industrial production processes that made computer gaming possible in the first place is an adapation of the *Unreal Tournament* game engine for a training simulation of toxic spills in chip fabrication. The mod, as described by project lead Walt Scacchi, replaces weapons with clean-up tools and in-game opponents in the popular first-person shooter video game with critical situations that would disrupt the highly expensive and automated microchip production.[22] Similarly, engineering students at the University of Illinois, with a grant funded by the Nuclear Regulatory Commission, recently began modifying *Unreal Tournament* to simulate critical states of a reactor and train nuclear plant workers and emergency responders.[23] In such applications, contingency planning comes full circle: using game technology that only became possible—and cheap—with computer chips, to train groups interactively for the kind of potentially very costly scenario that is otherwise too rare to effectively train for. By the same token, even as the potential of computerized gaming technology for training is recognized and deployed in the military as well as in education, we return to the question of how much playful freedom can be harnessed by rules and controls before the game ceases to be a game.

Here we see the premodern opposition of luck and virtue flatten out into the tension between unfettered, improvisational play versus the grind as "productive" mode of play that nonetheless has become virtually indistinguishable from repetitive labor.[24] Yet gaming technology drew explicitly upon the measurement of accuracy, timing, and fatigue. Baer's patent application from 1968 for "an apparatus and method, in conjunction with monochrome and color television receivers, for the generation, display, manipulation, and use of symbols or geometric figures upon the screen" built upon time and motion analysis and coupled input devices with screens and users "for the purpose of training simulation, for playing games and for engaging in other activities by one or more participants."[25] At that juncture, some might argue about what differentiates simulations, as technically related to computer game mechanics, from interactive fiction. Indeed fiction emphasizes contingency and gives free range to the imagination of the writer. But arguably, narrative fiction in all its foregrounding of playful self-reflexivity may work with elements of chance and accident, and yet they are planned in their fictionalization.[26] Knights may go errant and experience adventures, but the adventures are planned to become meaningful. This remains true for interactive fiction as a transfer onto the realm of spelunking in the computerized database, where metonymic decision trees aggregate patterns into quests of overcoming obstacles to redeem the labors and errors of a fictional world as meaningful.[27]

Notably, the space of gaming, the conceptual and computational frame that provides us with room for error, is often modeled on the ancient concept of the labyrinth. It is no coincidence that the PacMan layout should resemble a maze, particularly the kind deployed by U.S. Army Mental Tests, decades earlier:

After touching both arrows, E[xaminer] traces through first maze with pointer and then motions the demonstrator to go ahead. Demonstrator traces path through first maze with crayon, slowly and hesitatingly. E[xaminer] then traces second maze and motions to demonstrator to go ahead. Demonstrator makes mistake by going into blind alley at upper left-hand corner of maze. E[xaminer] apparently does not notice what demonstrator is doing until he crosses line at end of alley; then E[xaminer] shakes his head vigorously, says "No — no," takes demonstrator's hand and traces rest of maze so as to indicate an attempt at haste, hesitating only at ambiguous points. E[xaminer] says "Good." Then, holding up blank, "Look here," and draws an imaginary line across the page from left to right for every maze on the page.

Then, "All right. Go ahead. Do it (pointing to men and then to books) Hurry up." The idea of working fast must be impressed on the men during the maze test. E[xaminer] and orderlies walk around the room, motioning to men who are not working, and saying, "Do it, do it, hurry up, quick." At the end of 2 minutes E[xaminer] says, "Stop!"[28]

Later, at MIT around 1960, one could play *Mouse in a Maze* on a TX-0 (the first programmable, general-purpose computer to dispense with vacuum tubes and rely on transistors, and the first to test the use of a large magnetic core memory), a game developed by Doug Ross and John Ward that had the user draw a labyrinth with a lightgun on a screen, then set a dot as the cheese that a stylized mouse then sought to reach. When the mouse made a correct turn, it drank a martini; along the way, it became increasingly inebriated.[29] One of the most interesting games played in the space of the Dynamic Modeling Group at MIT was the multiplayer *Maze*—"actually a legitimate test of a database system the group used for a research project"—later seen as *Mazewars*, a direct ancestor of *Doom*, *Quake*, and *Unreal*.[30] And Crowther said about his seminal text game *Adventure* or *Colossal Cave* that when he let his kids play, on his ASR33 teletype remotely connected to a PDP-10, it was meant as "a re-creation in fantasy of my caving."[31] Indeed, guides at Mammoth Cave National Park provide anecdotal evidence of first-time spelunkers finding their way without assistance simply due to their prior experience of the game. (Montfort adds that students at MIT also engaged in spelunking on campus, exploring off-limits sections and basements at night.[32]) Successor game *Zork* was judged to be less convincing, although it had better game mechanics; its underground setting was "based not on real caves but on Crowther's descriptions."[33] It should be noted here that Will Crowther was one of the original developers of routing protocols for the Interface Message Processor that gave rise to the ARPAnet—the very net that soon helped spread computer games. Even the creation of that most popular layer of computer-mediated communication, the World Wide Web, is said to have been inspired by interactive fiction and the adventure game: CERN physicist Tim Berners-Lee describes *Enquire*, the predecessor to his now famous proposal for a network interface that "allowed one to store snippets of information, and to link related pieces together" as a progress "via the links from one sheet to another, rather like in the old computer game *Adventure*."[34] Other witnesses are quoted as remembering how "Tim made bits of labyrinthine hyper-routes in *Enquire* that served no better purpose than to exploit the program's capacity for making them. 'I made mazes of twisty little passages all alike,' he explains, 'in honor of *Adventure.*' "[35] In short, between labors and errors, the labyrinth offers a powerful reduction of narrative and ideological complexity.

This allows us to read games as risk management. As an ancient cultural technique of coping with risks and accidents, with chance and fate, games are an exemplary resource for thinking about contingency. This may apply less directly to role-play and mimetic play than to the mode of scenario planning and anticipation that is characteristic for strategy games, from ancient chess to the most recent computer games. Neither would it apply strictly to games of skill—but it is evident that under the condition of the networked computer, all these game types increasingly rely on a skill-set that is determined by computer literacy and a certain twitchy familiarity with joysticks, keyboards, mice, and screens. Nonetheless, computerized play does not necessarily mean higher accuracy or less room for error in play: in fact, meeting the expectations of rule-bound play with others in computer-mediated communication still means the experience of the contingency of gamers' competence, and even the contingently variable experience of one's own skill in relation to the performance of the other players. Contributing factors include applause and positive reinforcement on the one hand, heckling and competitive pressure on the other—pride or shame. In short, one's own performance is experienced as contingent upon a host of contextual factors; what opens up the topology of play is a voluntary acceptance of the fictionalizing realm of contingency: the game. There, play is rule-bound as a way to circumscribe contingency, yet rules need not conform with moral or legal codes: they can model their own artificial worlds, where each move is just an actualization of one among manifold possibilities. Playing with dice not only requires one to accept the element of chance, it also allows one to recuperate it into rule-bound play actions that modulate randomness into games. In short, rules leave room for play, both for the calculations and anticipations of the players and for contingent turns.[36]

We call contingent whatever is neither impossible nor necessary—it might be, it could be otherwise, or it might not have been, it could have been otherwise.[37] The concept of contingency becomes interesting and fruitful where it diverges from the merely possible, the not-impossible, or imaginable.[38] But the concept also becomes difficult for productive thought where it ought to be differentiated from accident and chance. Systematic thought, whether as philosophy or theology, traditionally shied away from contingency: for religion, chance was blasphemy against sovereign divine guidance; for rigorously systematic thought, contingency is anathema insofar as it would undermine or even undo the logical stringency of necessity. Indeterminacy can figure in such systems only as failed necessity: and inversely, if there is chaos, accident, and chance, then it is unclear what real importance order and necessity have. Transfer of this setup to computer culture is fairly straight-forward: input devices and operating routines must be tightly controlled, or they appear faulty. If my clicking or pecking can yield unpredictable results or create the (correct or erroneous) impression of a variable and perhaps even uncontrollable situation, then a fundamental principle of interaction is at stake: a peculiar blasphemy against widely held beliefs about designing human-computer interaction. Ted Nelson was not the only one to warn against the perils of installing a computer priesthood, for precisely these reasons. Inversely, if my user interface were chaotic and irregular (regardless of whether it is a browser-window or game—or, for that matter, a buggy beta version or an intentionally artsy deconstruction of a browser or a game), then I would be highly unlikely to accord any value to such accidental elements—totalized contingency would indeed be the absence of meaning.

However, we tend to try and recuperate such moments. What that looks like can be seen, for instance, in the work of European net.art duo Jodi, who programmed a browser, the quintessential web interface, but taking the realm of HTTP errors as a starting point—for instance the notorious "error 404" that signifies that a file was not found on the World Wide Web; this protocol becomes the basis for a bulletin board that aestheticizes browser errors and network errors.[39] As Schultz comments, "Error sets free the irrational potential and works out the fundamental concepts and forces that bind people and machines."[40] The list of examples might also need to include the web site http://glitchbrowser.com, described there as a

> deliberate attempt to subvert the usual course of conformity and signal perfection. Information packets which are communicated with integrity are intentionally lost in transit or otherwise misplaced and rearranged. The consequences of such subversion are seen in the surprisingly beautiful readymade visual glitches provoked by the glitch browser and displayed through our forgiving and unsuspecting web browsers

—or indeed Net artist Heath Bunting's performance in Berlin of an actually non-existent "web art project" (supposedly hosted on a server in Cuba) that only returns error messages.[41]

The play of aesthetics shows itself in these examples as a reduction of contingency through form, harnessing adventure through aleatoric or stochastic management of the event and the surprise. Another example is Jodi's game mod *SOD*, a modification of the classic game *Castle Wolfenstein* that retains the sound track but scrambles the visuals, so as to render the screen a black-and-white abstract pattern of glitchy, non-representational patterns, appearing to make the user's computer run amok.[42] Yet another, related example is the artgame *ROM CHECK FAIL*, a recent "independent" game predicated on the glitch as an aesthetic tool, shuffling around the elements of a number of classic 2D computer games.[43] Apparently random switching to alternate rules complicates the game mechanics and allows *ROM CHECK FAIL* to recycle an oddball assortment of arcade and console classics, to be navigated with arrow and space keys. Of course, this kind of 1980s retro game-generator cashes in on nostalgia and aestheticizes the glitch. Going one step further in that direction is the commercial release *Mega Man 9*.[44] Though developed for recent consoles including the Nintendo Wii, *Mega Man 9* uses graphics and sounds harking back to the 8-bit era of the original NES. A "legacy mode" emulates the much lower frame rates of the game's ancestors, only partially renders sprites, and causes them to flicker when they crowd the screen; this degradation feature is achieved not via an emulator but using a dedicated engine that simulates the aesthetics of an outmoded technology.

The Aesthetics of Recuperation, or EBKAC

Aesthetic experience has long been the traditional refuge of chance and accident from their philosophical and theological exclusion; and art remains the refuge of chance and accident in the age of new media technologies, whether instantiated with deconstructed browser windows, reprogrammed games, or distorted sounds. Here, we encounter a shift in the historical understanding of creativity. Premodern thought had opposed sovereignty and powerlessness: a divine power creates, and the created do not have any choice about being constituted otherwise, or not at all. Wishing to be otherwise would have been blasphemous. This rejection of contingency in premodern thought is readily illustrated in the Christian legends of the saints, precluding experience as a learning process and reducing saintliness to two modes: an undeterred allegiance to God even under the greatest duress, or a turn away from God that is corrected by one redeeming turn back to God—in short, the martyr or the converted sinner. That hagiographic stricture was a reaction to the polytheistic personification of Fortune. Here we will instead prize multi-cursal labyrinths—precisely for the reason that they allow more than one turn. Moreover, accidental determination is also ascribed to things out of human control, including illness and death, acts of nature or "bad luck." This notion of contingency is what Kierkegaard derided as despairing of necessity in the absence of possibilities; it can also be described as one of the sources of metaphysical belief systems, in that they provide a means of coping with fate, and a reduction of contingency.[45] On the other hand, if contingency means something could always have been otherwise, this presents the dilemma Kierkegaard saw in an aesthetics despairing of possibilities in the absence of necessity; wholesale rejection of universal principles threatened to suspend art in a nauseating state of "anything goes." Arguably, modern and contemporary art practices exit from this dismal binary in games and adventures, as rendezvous with contingency. Modernity ascribed both sovereignty and impotence, both creativity and overdetermination to us humans. And now we find these ancient categories returning in new media technologies, where again it is not at all unusual to see placeless power ascribed to media, and the subject located in a powerless place.[46] One of the most striking illustrations of this setup is found in computer games, where everything depends on a user interface that multiplies incentives for repetition and replay.

Reduction of complexity is a highly successful strategy in the experimental laboratory, in systems theory, and thus in many academic disciplines: but in aesthetics, it may run into an endemic problem, one that also comes to the fore in this early age of digital transcoding of audiovisual information. How much image can be omitted before an image is no longer an image? How much music can be compressed in lossy sampling before it ceases to be music? Inversely, however, this argument is too often hijacked by the kind of cultural conservativism that denies aesthetic dignity to any product of computing culture. Regardless of whether the reduction at issue is an aesthetic strategy or a technical constraint, each system will react to its own instabilities with a justification. As Foucault writes, the very first risk he took and his initial hypothesis was to ask: what if

the very possibility of recording facts, of allowing oneself to be convinced by them, of distorting them in traditions or of making purely speculative use of them, if even this was not at the mercy of chance?

If errors (and truths), the practice of old beliefs, including not only genuine discoveries, but also the most naïve notions, obeyed, at a given moment, the laws of a certain code of knowledge? If, in short, the history of non-formal knowledge had itself a system?[47]

That kind of systemic closure—neither at the mercy of chance nor affected by error, other than by recuperating error into its order—would reduce the archeology of the human sciences to a code-set of epistemological regularity; in demanding necessity throughout, speculative idealism knows, yet subsumes, the concept of an absolute accident.[48] Yet where systematic philosophy sought to eliminate contingency, media archeology expects to uncover contingency as that which is governed neither by divine providence nor by absolute reason. Thus instead of emphasizing coherence, totality, and continuity, media studies after Foucault foregrounds breaks, conflicts, and discontinuities.

Of course it is evident that the glitch only becomes more palpable with the advent of higher expectations from audiovisual resolution. What is perceived as a lapse in frame rates in a current game would have been part and packet switch of earlier interactions on networked screens; what is heard as noisy clicking on a current soundfile might have gone unnoticed in the scratchy days before standardization of vinyl production quality and cartridge sensitivity. But what Manovich describes as an accelerating "aestheticization of information tools" covers up the very material fact that aesthetic choices in the visual and audiovisual arts never sought to suppress the roughness of a brushstroke or the hairy attack of bow on string: these were effects at one's disposal, not simply noise to be cancelled.[49] Arguably, the era of noise-canceling only truly takes off with the advent of digital technologies.

The productivity of distortion for media culture was first formalized as a cybernetic insight, whereby noise is the negation of communication, and its valorization allows for operational observation of communication systems. Shannon's diagrams show the information source passing a message from a transmitter as signal to a receiver destination, via the contingent yet systemic interference that is the noise source. However, this renders the distorting interference not only as a "parasite" but also as a second-order signal source, where it can act both as negation and as generation of received signals.[50] One logical consequence is that for computer-mediated communication, redundance can turn from an attribute of security to an attribute of insecurity. When recurring noise patterns become signal sources as their regularity renders them legible and receivable, the systemic function of distortion doubles over as deterioration of message quality and as enrichment of the communication process.

Information theory owes pivotal impulses to anticipating interception by an enemy cryptanalyst: "From the point of view of the cryptanalyst, a secrecy system is almost identical with a noise communication system. The message (transmitted signal) is operated on by a statistical element, the enciphering system, with its statistically chosen key. The result is the cryptogram (analogous to the perturbed signal) which is available for analysis."[51] Thus the synthesis of signal and key decrypts the noisy transmission into a received message for the destined receiver: but it is encoded so that the anticipated distortion of communication, namely the interception and decoding of the message by enemy cryptanalysts, might be avoided. Consequently, Shannon's communication theory of secrecy is always both noise (to foes) and message (to friends), and always has to anticipate both desirable and undesirable negation. What is noise to one may be message to another, and this radical relativity is pivotal for cybernetics. It seems that distortion is systemic, but it is not merely a matter of chance or accident whether there will be noise, nor is it simply a matter of fate whether one is being understood or intercepted. These situations are never completely impossible, and always somewhat likely; it may seem reasonable to anticipate them—but to do so requires a concept that differs, if only slightly, from indeterminacy or accident, chance or fate.[52] The problem is that erroneously introduced or accidental noise is neither impossible nor necessary.

Fiction, art, interpretation, and irony are forms of knowledge that can convey contingency: as a kind of knowledge that implies an awareness of its own contingency.[53] But while the executable codes of Oulipian poetics and prose generators, or more generally, the criticism and systematic study of possible worlds have tended to carve the space of imagination into analytic categories that can be harnessed in expressive form, criticism often proved unable to systematize that which is by definition unsystematic. In the age of cybernetics, it can seem as if human fallibility is what keeps systems from achieving their full potential—from systematic closure. Yet rather than our becoming abstractly "post-human" in information society, one might instead be tempted to argue that people, citizens, and individuals in fact become realized for each other and for themselves in unprecedented ways through networks of computer-mediated communication. For must we not recognize that new media necessarily involve the operation of human embodiment to close their feedback loops? Attending to the modes of embodiment of data in information machines, or focusing on technologies of inscription, allows for close analysis of innovation in cultural technologies.

Where we can thus allow ourselves to conceive of human-computer interaction as organized around the glitch, it is not to smuggle a covert humanism in through the back door of technological determinism: rather it is to emphasize what Galloway calls "the cultural or technical importance of any code that runs counter to the perceived mandates of machinic execution, such as the computer glitch or the software exploit, simply to highlight the fundamentally functional nature of all software (glitch and exploit included)."[54] One might conclude here, however provisionally, that gaming glitches are part of the art form in the same way that brushstrokes are part of oil painting. Game developers may be tempted to dismiss this as little consolation to a user who just had their program crash, or indeed to a programmer trying to debug the system. However, it may be a crack that can widen onto new vistas, and better mistakes. A plain pedagogical imperative to "learn from our mistakes" seems to suggest that the shipwrecked ought to make the best travel guides, as Ortega y Gasset joked.[55] But this sardonic recommendation might be all too easily appropriated in a culture of efficiency, especially after dot-com failures had to be reinterpreted in the resumes of a multitude of entrepreneurial types as a valuable life lesson.[56] Indeed ancient myth already indicates that only the failure of guidance allows for a happy end. In what Pias calls the most prominent narrative of the failure of cybernetics, the god Morpheus drips some water from the river Lethe into the eyes of Palinurus as he steers Aeneas' ship. Palinurus falls asleep and is washed off the deck into the sea. However, Aeneas later descends into Hades and meets his prematurely deceased helmsman there, only to learn that the attack on Palinurus was in fact not aimed at thwarting Aeneas from reaching his destiny. It was just that the ship's commander had to die to make a happy ending possible.[57] Thus perhaps the shipwrecked cybernetes does not make the best guide after all. Instead, what one needs to learn from mistakes is not their avoidance but something else altogether: namely to allow for them; to allow room for error.

To summarize this long story of glitchy computer-mediated communication, we may remind ourselves that in 1968, Licklider and Taylor could still assert that "men will be able to communicate more effectively through a machine than face to face."[58] However, it should also be remembered that their essay ends with the ambivalent prediction that "unemployment would disappear from the face of the earth forever, for consider the magnitude of the task of adapting the network's software to all the new generations of computer, coming closer and closer upon the heels of their predecessors until the entire population of the world is caught up in an infinite crescendo of on-line interactive debugging." The error remains the future.

Notes

[1] Samuel Beckett, "Worstword Ho," in *Nohow On: Company, Ill Seen Ill Said, Worstward Ho* (New York: Grove, 1996), 89.

[2] Theodor Adorno, *Negative Dialectics* (New York: Routledge, 1990), 359.

[3] For usability and technology, see Steve Woolgar, "Configuring the User: The Case of Usability Trials," in *A Sociology of Monsters. Essays on Power, Technology, and Domination*, ed. John Law (New York: Routledge, 1991), 57–99; as well as Thierry Bardini and August Horvath, "The Social Construction of the Personal Computer User," *Journal of Communication* 45, no. 3 (1995): 40–65. For theoretical issues raised by technologized society, see Carl Mitcham, *Thinking through Technology. The Path between Engineering and Philosophy* (Chicago: University of Chicago Press, 1994).

[4] Olga Goriunova and Alexei Shulgin, "Glitch," in *Software Studies: A Lexicon*, ed. Matthew Fuller (Cambridge, MA: MIT Press, 2008), 111.

[5] John Cogburn and Mark Silcox, *Philosophy through Videogames* (New York: Routledge, 2009), 107. They add that Neil Stephenson's novel *The Diamond Age* (New York: Spectra, 2000), 171 has a computer programmer code "a series of interactive games to teach his daughter about the basic limitation theorems of computability theory."

[6] Wlad Godzich, "Figuring Out What Matters," in H. U. Gumbrecht, *Making Sense in Life and Literature* (Minneapolis: University of Minnesota Press, 1992), xv.

[7] Indeed there is a danger that "we jeopardize the most important option offered by the materialities approach if we dream of a new stability for renewed concepts in a future age of theory," Gumbrecht warns: "This most important option might well be the possibility of seeing the world under a radical perspective of contingency." Hans Ulrich Gumbrecht, "A Farewell to Interpretation," in *Materialities of Communication*, ed. H. U. Gumbrecht and K. L. Pfeiffer (Stanford, CA: Stanford University Press, 1994), 402. Compare Karl Mannheim, *Ideologie und Utopie* (Frankfurt: Klostermann, 1952), n.38.

[8] See above all Alexander Galloway, *Gaming: Essays on Algorithmic Culture* (Minneapolis: University of Minnesota Press, 2006), and McKenzie Wark, *Gamer Theory* (Cambridge, MA: Harvard University Press, 2007).

[9] See Claus Pias on informatics versus thermodynamic pessimism, "Wie die Arbeit zum Spiel wird," in *Anthropologie der Arbeit*, ed. Eva Horn and Ulrich Bröckling (Tübingen: Narr, 2002), 209–229.

[10] We are locked by history into the QWERTY keyboard, designed for typewriters whose keys were too slow for our fingers. More than two-thirds of English

words can be produced with the letters DHIATENSOR, yet most of these are not in easily accessible positions (on home row to be struck by the first two fingers of each hand). QWERTY removed the vowels from the strongest striking positions, leaving only a third of the typing on home row. For a contemporary perspective, see *How Users Matter. The Co-Construction of Users and Technology*, ed. Nelly Outshoorn and Trevor Pinch (Cambridge, MA: MIT Press, 2003).

[11] Dennis Chao, "Doom as an Interface for Process Management," *SIGCHI 2001* 3:1, 152–157 (March 31–April 4, 2001: Seattle, WA); see his extensive project documentation at http://cs.unm.edu/~dlchao/flake/doom/chi/chi.html.

[12] Apple interface designer Jef Raskin, "Intuitive equals familiar," *Communications of the ACM* 37, no. 9 (September 1994): 17–18. Compare Brad Myers, "A Brief History of Human Computer Interaction Technology," *ACM Interactions* 5, no. 2 (1998): 44–54. As Chao jabs: "Children are now growing up on MTV and Nintendo and are therefore more literate in the languages of video games and mass media than in that of the traditional office milieu." Chao, 154.

[13] Compare Lawrence Miller and Jeff Johnson, "The Xerox Star: An Influential User Interface Design," in Marianne Rudisill, Clayton Lewis, Peter Polson, and Tim McKay ed. , *Human-Computer Interface Design: Success Stories, Emerging Methods, and Real-World Context* (San Francisco: 1996), 70–100; and William L. Bewley, Teresa L. Roberts, David Schroit, and William L. Verplank, "Human factors testing in the design of Xerox's 8010 Star office work-station," *Proceedings of the ACM Conference on Human Factors in Computing Systems* (1983), 72–77.

[14] For Chris Pruett's "Marathon Process Manager" for the Mac, see http://www.dreamdawn.com/chris/projects.html.

[15] Brody Condon, "Adam Killer" 1999–2001. http://www.tmpspace.com/ak_1.html.

[16] On machinima, see Peter Krapp, "Of Games and Gestures: Machinima and the Suspensions of Animation," in *Machinima Reader*, ed. Henry Lowood and Michael Nitsche (Cambridge, MA: MIT Press, 2010).

[17] Brody Condon, "Where Do Virtual Corpses Go?" http://www.cosignconference.org/cosign2002/papers/Condon.pdf.

[18] Hacker lore of course points to filibustering as an asocial way of winning the prize in such contests: tying up the phone lines with automatic redialers to guarantee the winning call, tying up computer processors with requests to force the desired breakpoint, etc. The conceptual model is already analyzed in J. C. R. Licklider and R. W. Taylor, "The Computer as a Communication Device," *Science and Technology* (April 1968): 21–41; republished online for *In Memoriam: J. C. R. Licklider 1915–1990*. Research Report 61, Digital Equipment Corporation Systems Research Center, August 1990. http://gatekeeper.dec.com/pub/DEC/ SRC/research-reports/abstracts/src-rr-061.html.

[19] Compare Alex Galloway, *Protocol: How Control Exists After Decentralization* (Cambridge, MA: MIT Press, 2004), and Eugene Thacker and Alex Galloway, *The Exploit: A Theory of Networks* (Minneapolis: University of Minnesota Press, 2007).

[20] Julian Dibbell, *My Tiny Life: Crime and Passion in a Virtual World* (New York: Henry Holt, 1998), 70–72.

21   Wulf Halbach, "Simulated Breakdowns," in *Materialities of Communication*, ed. H. U. Gumbrecht and K. L. Pfeiffer (Stanford, CA: Stanford University Press 1994), 343. Compare also Denis Kambouchner, "The Theory of Accidents," *Glyph* 7 (1980): 149–175.

22   Walt Scacchi, "Visualizing Socio-Technical Interaction Networks," DOI – http://www.isr.uci.edu/~wscacchi/.

23   Greg Kline, "Game Used to Simulate Disaster Scenarios," *The News-Gazette* Sunday, February 17, 2008, http://www.news-gazette.com/news/u_of_i/2008/02/17/game_used_to_simulate_disaster_scenarios.

24   We have perhaps all heard a colleague's brag: "the harder I work, the luckier I get"—once again reinforcing the common notion that labor and luck are in a dialectical pairing.

25   Here cited after Shaun Gegan, "Magnavox Odyssey FAQ," http://home.neo.lrun.com/skg/faq.html; compare W. K. English, D. C. Engelbart, M. L. Berman, "Display Selection Techniques for Text Manipulation," *IEEE Transactions on Human Factors in Electronics* 8, no. 1 (March 1967): 5–15.

26   A theoretical contribution to the literature on play is found in Hans-Georg Gadamer, *Truth and Method* (London: Continuum, 2004), 102–161 (original German published as *Wahrheit und Methode. Grundzüge einer philosophischen Hermeneutik* (Tuebingen: Siebeck, 1972), 107–108). For a historical sketch of the concept of contingency, see Hans Blumenberg, "Kontingenz," *Handwörterbuch für Theologie und Religionswissenschaft* (Tuebingen: Mohr, Paul Siebeck, 1959), 1793–1794.

27   A recent reconstruction of the history of interactive fiction is Nick Montfort, *Twisty Little Passages* (Cambridge, MA: MIT Press 2003).

28   Compare Clarence S. Yoakum and Robert M. Yerkes, *Army Mental Tests* (New York, 1920), 83, and Claude Shannon, "Presentation of A Maze-Solving Machine," *Transactions of the 8th Conference entitled 'Cybernetics', Eighth Conference on Cybernetics 1951*, ed. Heinz von Foerster (New York, 1954), 173–180. See also Ivan Sutherland, "A Method for Solving Arbitrary-Wall Mazes by Computer," *IEEE Transactions on Computers* vol. C-18, no. 12 (December 1969): 1192–1197, as well as Wayne H. Caplinger, "On Micromice and the first European Micromouse Competition," *AISB Quarterly* no. 39 (December 1980), http://davidbuckley.net/RS/mmouse/AISBQ39.htm.

29   http://www.bitsavers.org/pdf/mit/tx-0/memos/Ward_MOUSE_Jan59.txt—compare J. M. Graetz, "The origin of Spacewar", *Creative Computing* 1981, archived at http://www.wheels.org/spacewar/creative/Spacewar-Origin.html.

30   Compare David Lebling, Marc S. Blank, and Timothy A. Anderson, "Zork: A Computerized Fantasy Simulation Game," *IEEE Computer* 4 (1979): 51–59; as well as Tim Anderson and Stu Galley, "The History of Zork," *The New Zork Times*, nos. 1–3 (1985), the latter here cited after Montfort, 80.

31   Dale Peterson, *Genesis II: Creation and Recreation with Computers* (Reston, VA: Reston 1983), 188. See Rick Adams, "A History of *Adventure*," http://www.rickadams.com/adventure/a_history.html.

32   Montfort, 100–101.

[33] Graham Nelson, "The Craft of *Adventure:* Five Articles on the Design of Adventure Games," http://www.if-archive.org/if-archive/programming/general-discussion/Craft.Of.Adventure.txt.

[34] Tim Berners-Lee, "Information Management: A Proposal," http://www.w3.org/History/1989/proposal.html. In the web's pivotal precursor, Bush's concept of the MEMEX, the user "builds a trail of his interest through the maze of materials available to him. And his trails do not fade." Vannevar Bush, "As We May Think," *Atlantic Monthly* July 1945, 101–108.

[35] James Gillies and Robert Cailliau, *How the Web Was Born* (Oxford: Oxford University Press, 2000), 170. Inversely (and this connection will prove to be highly relevant for another part of this project), Barlow warns in discussing early computing culture that "cracking impulses seemed purely exploratory, and I've begun to wonder if we wouldn't also regard spelunkers as desperate criminals if AT&T owned all the caves"—John Perry Barlow, "Crime & Puzzlement" (1990), http://www.eff.org/Publications/JohnPerryBarlow, here cited after Wendy Grossman, *net.wars* (New York: New York University Press, 1997), 129.

[36] As Montfort notes, "*Labyrinth* was a hypertext catalog of the 1970 *Software* exhibition at the Jewish Museum"—and Ted Nelson, who had coined the term hypertext, called this catalog the first publicly accessible hypertext. Montfort, 73.

[37] On the complex etymology, history, and use of "contingency," one ought to footnote H. Schepers, "Möglichkeit und Kontingenz—Zur Geschichte der philosophischen Terminologie vor Leibniz," *Filosofia* (1963): 901–914; Richard Rorty, *Contingency, Irony, and Solidarity* (Cambridge: Cambridge University Press, 1989); Jacques Monod, *Chance and Necessity: An Essay on the Natural Philosophy of Modern Biology* (New York: Knopf, 1971); and Stephen Jay Gould, *Wonderful Life: The Burgess Shale and the Nature of History* (New York: Norton, 1989).

[38] Bloch makes important distinctions between the merely (passively) plausible and the creatively (actively) variable: Ernst Bloch, *Das Prinzip Hoffnung* (Frankfurt: Suhrkamp, 1969), vol. 1, 258–259. Compare also H.P. Bahrdt, "Plädoyer für eine Futurologie mittlerer Reichweite," *Menschen im Jahr 2000*, ed. Robert Jungk (Frankfurt: Umschau, 1969), 143–144., and Ossip K. Flechtheim, *Futurologie. Der Kampf um die Zukunft* (Frankfurt: Fischer, 1972), 154–155.

[39] See http://404.jodi.org.

[40] Pit Schultz, "Jodi as Software Culture," in *Install.exe* (Berlin: Merian, 2002).

[41] As recounted in Alexander Galloway, *Protocol*, 225–226.

[42] http://sod.jodi.org/—Jodi also produced a CD-ROM with 12 mods of the classic first-person shooter Quake and a list of cheats for the videogame *Max Payne:* see http://www.untitled-game.org or http://maxpaynecheatsonly.jodi.org.

[43] http://www.farbs.org/games.html—Also worth mentioning: Joan Leandre's R/C and NostalG—see http://runme.org/project/+SOFTSFRAGILE/ and http://www.retroyou.org.

[44] Megaman 9 (Capcom, 2008).

132　　　*Error: Glitch, Noise, and Jam in New Media Cultures*

45　Gregory R. Beabout, *Freedom and Its Misuses: Kierkegaard on Anxiety and Despair* (Milwaukee: Marquette University Press, 1996).

46　In this context it is interesting to note Ian Hunter's discussion of "the aesthetic as an ethical technology" incorporated into the governmental sphere, in his essay on "Aesthetics and Cultural Studies," in *Cultural Studies*, ed. L Grossberg, C. Nelson, P., and Treichler (New York: Routledge, 1992), 347–372.

47　Michel Foucault, *The Order of Things* (New York: Routledge, 2002), x.

48　Dieter Henrich, "Hegels Theorie über den Zufall," *Hegel im Kontext* (Frankfurt: Suhrkamp, 1971), 157–186, with reference to G. W. F. Hegel, *Die Vernunft in der Geschichte* (Hamburg: Felix Meiner, 1955), 29.

49　Lev Manovich, "Interaction as an aesthetic event," *receiver* 17 (2006), http://www.receiver.vodafone.com; see http://www.manovich.net. Compare also Gary McCarron, "Pixel Perfect: Towards a Political Economy of Digital Fidelity," *Canadian Journal of Communication* 24, No. 2 (1999), http://www.cjc-online.ca/viewarticle.php?id=520&layout=html.

50　Compare Claude Shannon, "Communication Theory of Secrecy Systems," *Bell Technical Journal* 28 (October 1949): 656–715, and Michel Serres, *The Parasite* (Baltimore, MD: Johns Hopkins University Press 1982).

51　Shannon, "Communication Theory of Secrecy Systems," 685. One recent invocation of Shannon and Weaver for game theory is Mia Consalvo's essay "Lag, Language, and Lingo: Theorizing Noise in Online Game Spaces," in *The Video Game Reader 2*, ed. Bernard Perron and J. P. Wolf (New York: Routledge, 2009), 295–312.

52　Erhard Schüttpelz illustrates the conceptualization of distortion as accidental to some, meaningful message to others with reference to the jump cut introduced by Meliés: a celluloid strip messed up, resulting in the apparent metamorphosis of a bus into a hearse. (See Georges Sadoul, *L'Invention du cinéma*, cited after Siegfried Kracauer, *Theorie des Films: Die Errettung der äußeren Wirklichkeit* (Frankfurt: Suhrkamp, 1975), 60). Erhard Schüttpelz, *Signale der Störung* (Munich: Wilhelm Fink 2003), 15–16.

53　Or as Gadamer put it, "the work of art cannot simply be isolated from the contingency of the chance conditions in which it appears", 115.

54　Alexander Galloway, "Language Wants to be Overlooked: On Software and Ideology," *Journal of Visual Culture* 5, no. 3 (2006): 326.

55　José Ortega y Gasset, *Obras Completas* vol 4 (Madrid: Editorial Taurus, 1983), 398. Here cited after Yvette Sanchez, "Der bessere Fehler – als Programm oder Fatum," *Fehler im System: Irrtum, Defizit und Katastrophe als Faktoren kultureller Produktivität*, ed. Y. Sanchez and F. P. Ingold (Göttingen: Wallstein Verlag, 2008), 22.

56　See the exemplary success of the popular business book by Paul Ormerod, *Why Most Things Fail: Evolution, Extinction and Economics* (London: Faber and Faber, 2005).

57　Claus Pias, "Der Auftrag. Kybernetik und Revolution in Chile," in *Politiken der Medien*, ed. D. Gethmann and M. Stauff (Zurich: diaphanes, 2004), 131–154.

58　J. C. R. Licklider and R. Taylor (see note 18).

Chapter 7

The Seven Million Dollar PowerPoint and Its Aftermath: What Happens When the House Intelligence Committee Sees "Terrorist Use of the Internet" in a *Battlefield 2* Fan Film

Elizabeth Losh
University of California, Irvine

The digital file that created so much consternation and alarm among the members of the U.S. House Permanent Select Committee on Intelligence on May 4, 2006 was first created months earlier and several time zones away in a peaceful neighborhood of planned housing in the Netherlands. The link to the video file that was posted to the *Planet Battlefield* forum was intended to celebrate the computer game *Battlefield 2* and was addressed to fellow enthusiasts of this kind of first-person shooter play. It also incorporated the sound track from the comedy puppet movie *Team America*, an audio clip from a press gaffe by George W. Bush, and music from the 1981 Anthony Quinn film *Lion of the Desert*.[1] The video file represented an imperfect rendering of a popular mash-up aesthetic that was created for a particular remix culture around global gaming, a heterogeneous assortment of media elements and rhetorical appeals that would prove difficult for federal investigators on the intelligence committee to rationalize and interpret: signal would be taken as noise by lawmakers, and noise would be received as signal.

The main message of online smack talk, virtuoso individual performance, and parodic reappropriation was overlooked, because it did not fit the narrative of serious terrorist evangelization. At the same time, amateur production that relied on footage captured accidentally and the defaults of the game engine was interpreted as though it conveyed carefully crafted political statements from a group of malevolent opponents who controlled every aspect of the software. Furthermore, this story about the clash between official narratives about terrorism and technology and vernacular fan production around game-based machinima video has continued since 2006. At least one of the parties involved has subsequently used digital authoring tools to record his own version of events in the cascading failure of replicated misunderstandings and to restage and reframe his own media-making efforts in the hope of avoiding future misinterpretation.

During the segment of the video shown at the hearing that would end up capturing the most attention of congressional representatives and the wire service reporters in attendance, eerie music plays as a character voiced by Trey Parker condemns the "infidel" and declares that he has "put a jihad" on them, as aerial shots move over flaming oil drilling compounds and mosques covered with geometric patterns. Suddenly a rocket is fired at a helicopter, and the screen is filled with a violent explosion. As the action accelerates, much of the game-play footage that follows takes place from the point of view of a first-person shooter, seen as if through the eyes of an armed insurgent. Occasionally, the footage is interspersed with third-person action in which the player appears as a running figure in a red-and-white checked kheffiyeh with a rocket launcher balanced on his shoulder. Among the other items in the player's hand-held arsenal is a detonator that triggers remote blasts. While jaunty music plays, helicopters, tanks, and armored vehicles burst into smoke and flame. Exclamations in Arabic are limited to simple phrases like "God is great" and "thank you." At the triumphant ending of the video, a green and white flag bearing a crescent is hoisted aloft into the sky, symbolizing a victory by the MEC or Middle Eastern Coalition in the final stages of the game.

By the time the *Battlefield 2* video reached the congressional hearing entitled "Terrorist Use of the Internet," it had already been misidentified as material for terrorist indoctrination at least twice: once by self-proclaimed experts from the military contractor SAIC or the Science Applications International Corporation and once more by Pentagon witnesses who had reviewed SAIC's reports about the risks to national security posed by anti-American digital content on the web. In what one congressional representative described as "a colorful and sobering presentation" and another called a "very compelling and sobering presentation," a PowerPoint slideshow starring this *Battlefield 2* fan film was used to dramatize the grave threat supposedly posed by games that glorified terrorism and to demonstrate the potential link between interactive media and a worldwide amorphous conspiracy to produce digital Islamist propaganda and distribute it to impressionable young people.

There were a number of mistakes that U.S. investigators made just because they could not imagine that in many American commercial video games playing from the position of a conventional "enemy" might be a desirable standard feature—or, in this case, part of a legitimately purchased *Special Forces* expansion pack. They did not account for the fact that the industry practice of allowing such role-reversal in the first-person shooter genre strengthens brand loyalty by extending play time and player engagement, and therefore it also solidifies the commitment of so-called "hardcore gamers" to a given company's products and guarantees profits in the most reliably lucrative segment of electronic entertainment's market share.[2] These mistakes by the military and its contractors were further compounded by assumptions that congressional witnesses made about other common digital practices, such as editing video files, claiming membership in fan communities, posting to forums, and even surfing web sites.

To pander to the public's seemingly endless fascination with lurid moral panics over the exposure of children to adult Internet content,[3] speakers repeatedly emphasized supposed age discrepancies between creators of digital media supporting the insurgency in Iraq and their presumed child-like audiences. For example, U.S. public diplomacy specialist Dan Devlin conjured up images of violated innocence with references to extremely young children in his appearance before Congress.

> They have a strong emotional appeal, and I can't stress how agile they are. The target audiences—recruitment age, seven to 25. Seven may be a little high. We've seen products that are aimed at ages even lower than seven.[4]

Witness Devlin argued that this vulnerable young population was made even more helpless by "Islamic clergy" and the detachment of "mothers and parents" from their normal duties of supervision. In her response to the witnesses' testimony, ranking minority member Democrat Jane Harman returned uncritically to several of Devlin's emotional appeals by asserting that "we're all parents here, and some of us are even new grandparents" and by agreeing that "the comment made that the target audience is 7 or perhaps younger than 7 is truly chilling."

Before the congressional committee, the witnesses from the Defense Department and SAIC also called attention to the purity of their roles as information intermediaries who were supposedly providing House representatives with direct access to original unedited—and even untranslated—sources from which elected representatives could draw their own conclusions. In the group's PowerPoint presentation Devlin introduced the web sites, songs, photos, and Internet videos that he would be citing as evidence by attesting to their digitally unaltered condition.

> The only thing I would add to the introduction to the briefing is that all of the images you're about to see, the sound bytes, the video, the multimedia products, were taken off adversary web sites. They have not been edited in any way except truncated to fit into the briefing.[5]

In contrast, the DoD/SAIC experts presented America's political and military adversaries as skilled computational media manipulators with cynical plans to run what would seem to be an expertly calculated Internet propaganda machine.

> The thing that I think I should emphasize to you is that al Qaeda advertised on the 'net for specialists in media skills . . . They advertised, they recruited and hired, and then they told us that the team was put together. It's a unit called the Global Islamic Media Unit. And most of the productions that we have seen, both in Arabic and in other languages, have some origin from this unit.[6]

Elsewhere, witnesses discussed how the jihadists "advertise" for "specific skill sets" in digital production and even offer "internships," much like corporate headhunters seeking to fill highly specialized positions of new media knowledge work.

Despite the jihadists' apparent corporate veneer, in the testimony before the House Intelligence committee, terrorists were also described as thieving "modders" and "hackers" who subvert those who properly author, possess, purchase, license, or regulate intellectual property. Undersecretary of Defense for Policy and International Affairs, Peter Rodman claimed that "any video game that comes out, as soon as the code is released, they will modify it and change the game for their needs."[7] To explain embarrassing breeches involving the digital files of the U.S. military, witness Ron Roughead said, "We're not even sure that they don't even hack into the kinds of spaces that hold photographs in order to get pictures that our forces have taken."[8] The jihadists' skill at editing digital video was explicitly tied to presumed acts of theft in the words of another witness who told how "footage was stolen from third anniversary coverage by U.S. networks" and then insurgents "us[ed] these difference pieces" to create a new narrative.[9]

Yet, as Rita Raley has pointed out, it was precisely the lack of aesthetic and technological skill in the *Battlefield 2* video that could have tipped authorities off to the fact of its non-terrorist source. Other gamers on the Planet Battlefield forum who had viewed the film the previous December had quickly reached a consensus, as Raley puts it, that the video was "unoriginal, too long, too repetitive, even perhaps too boring"[10] and that the video's creator, "SonicJihad," had appropriated too much of his game-play footage from another gamer, "JihadJoe" without enough artful remixing and remediation. Critical comments about the video and the sources of its creator were there for all to see, and as SonicJihad himself noted, they were only a Google search away.[11]

In an August 2009 interview for this collection, SonicJihad, whose real name is Samir and who works as a policy analyst for a hospital board, described getting a message from a fellow gamer with a line stating "you're famous" and a link to an article from Reuters that seemed to describe the video Samir had posted to the *Planet Battlefield* forum.[12] In the emailed article, reporter David Morgan opened with a grim assessment of the dangers posed by the products of unwitting American videogame companies to the American public as a whole.

> The makers of combat video games have unwittingly become part of a global propaganda campaign by Islamic militants to exhort Muslim youths to take up arms against the United States, officials said on Thursday.
>
> Tech-savvy militants from al-Qaida and other groups have modified video war games so that U.S. troops play the role of bad guys in running gunfights against heavily armed Islamic radical heroes, Defense Department official and contractors told Congress.
>
> The games appear on militant web sites, where youths as young as 7 can play at being troop-killing urban guerillas after registering with the site's sponsors.[13]

Once Morgan's story hit the wire services, it was republished in a number of national newspapers and reposted on Internet news sites with titles that included "Islamists Using US Video Games in Youth Appeal," "Islamic Militants Recruit Using U.S. Video Games," and "Experts: Islamic Militants Customizing Violent Video Games." Like a similar story from the Associated Press, the Reuters coverage highlighted the sampling and remixing of former American president George W. Bush's voice and his diplomatically insensitive use of the word "crusade."

> "I was just a boy when the infidels came to my village in Blackhawk helicopters," a narrator's voice said as the screen flashed between images of street-level gunfights, explosions and helicopter assaults.

Then came a recording of President George W. Bush's September 16, 2001, statement: "This crusade, this war on terrorism, is going to take a while." It was edited to repeat the word "crusade," which Muslims often define as an attack on Islam by Christianity.[14]

Morgan would subsequently become a butt of jokes on the Internet, because he failed to recognize the lines from *Team America* about supposed Blackhawk attacks and the voice of Trey Parker narrating them, despite a humorous mention of "goats screaming" in the clip played before Congress.

In a cascading chain of failures, mistakes made about the video by SAIC "experts" were not corrected by Pentagon officials, and then those uncorrected mistakes reached legislators in Congress, where somehow no one present recognized *Battlefield 2* or *Team America* from popular American culture, so those errors were then replicated by the media. Members of all four estates in this story—researchers from a think tank, military officials, elected representatives, and professional journalists— seemed to have totally failed to detect multiple errors. Almost all of the salient details in the Reuters story turned out to be incorrect. SonicJihad's film was not made for or by terrorists: it was not created for "Muslim youths" by "Islamic militants." The video game it depicted had not been modified by a "tech-savvy militant" with advanced programming skills. Even the type of software at issue had been misidentified: SAIC witnesses apparently believed that the video capture program for the PC advertised on the clip could be used for retasking the proprietary software of the *Battlefield 2* videogame engine itself.

Of course, the very framing of perspective in SonicJihad's film may have seemed disconcerting to the intelligence committee and to its invited guests. Alexander Galloway has observed that the first-person shooter genre is characterized by a merging of the point of view of camera and character, which in film usually signifies intoxication, alienation, or predatory interest, but in videogames is coded as phenomenologically normal and even socially appropriate.[15] Galloway also emphasizes the importance of the relationship of the iconic weapon to this first-person visual framing as an extension of the player's agency and his ability to control action on the screen. No one at the hearing had seemed to have had the basic digital literacy to challenge SAIC witnesses successfully, particularly when they claimed to manage a large team of sophisticated interpreters analyzing the digital data in its full cultural context.[16] There were no stakeholders familiar enough with videogames or machinima (computer-generated filmmaking that uses a game engine or other 3-D simulation technology) present to contradict the SAIC briefer directly when he claimed that their "analysts employ a method called full cultural context, where we team native speakers with fluent, culturally aware analysts, so that when they look at a product, they can capture as much of the nuances in the analysis as possible."[17] This assertion that SAIC experts were uniquely qualified to read subtle "nuances" and also understand the "full" frame of rhetorical and historical "context" seems laughable in retrospect, but it also shows how they imagined their position as being both inside and outside the field of discourse, fluently able to participate in cultural conversations with opponents and yet aware of the patterns that might be naturalized by the unconscious natives who would be unable to perceive how they were being manipulated by jihadist factions. If the reversal of signal and noise is understood as a strong possibility in the postmodern situation of reading, the hubris of SAIC that posits an impossibly perfect mechanism for reading transmedia texts only seems to make an aberrant reading more likely.

Nonetheless, even as the hearings were taking place, not all of the representatives accepted the authority of SAIC experts uncritically. Congressman Rush Holt took the hearing as an opportunity to object to the way that many domestically produced videogames presented the political enemies of the United States as dehumanized virtual objects to be destroyed. He also may have wished to undermine the holier-than-thou position of U.S. investigators looking at foreign-made digital media, because he was aware that several of the witnesses speaking before Congress had been involved with unsuccessful U.S. propaganda efforts in Iraq, including the ill-fated Iraq Media Network.

REP. HOLT:	Okay. I guess, you know, as I look at computer games that are out there, I don't think of them as our finest and proudest output. I'm wondering, are U.S. computer games regarded in some cases as anti-Muslim and supporting a crusading point of view? Is there commentary on U.S. computer games and that sort of thing?
MR. RODMAN:	I think I would have to go back and say that we're looking at computer games that are being used for development of materials on the web, and we don't actually spend a lot of time looking at—
REP. HOLT:	But if you're looking—if I may jump in here, you're looking at how the United States is portrayed.
MR. RODMAN:	Yes.

REP. HOLT: And one of the ways that I suspect they might present deleterious information is by saying, "Look, here in the United States the kids are playing games that are anti-Muslim or crusading." And I'm just wondering if you see that kind of presentation.[18]

Furthermore, Representative Holt points out that the repurposing of U.S. digital content in the Islamic world may actually involve appropriating the norms of already violent commercial videogames from the West. Holt may have even been aware that some of this violent digital content in first-person commercial games glorifying battlefield violence and celebrating allegiance to what Holt characterizes as a "crusading" politics was originally produced for military training games and simulations. As Gonzalo Frasca asserts, American videogames, such as the recruitment game *America's Army*, could even be taken as a form of political propaganda.[19] Unlike *Battlefield 2*, one can only play as an American in *America's Army*, and one can never switch to being a U.S. opponent, even in the online multiplayer version of the game. (In the interview for this collection, Samir also pointed to *America's Army* as an example of U.S. hypocrisy about military videogames aimed at the young.)[20]

Representative Alcee Hastings also raises substantive doubts about SAIC media expertise, despite the fact that he opens his remarks by saying, "I'll make it very clear that I don't disparage anything," and "I hope you're successful in your research." In particular, Hastings questions whether the outside experts from SAIC actually are bringing new information to the hearing, since members of the intelligence committee are already well aware of America's poor image abroad from traditional broadcast and print sources.

I'm not surprised by anything I saw. I also recognize that these are, put in proper context, moments in time that are spread on the Internet . . . And I can't even begin to tell you the number of other things that are out there. I have the good fortune of traveling a considerable amount. And when I am in Europe, particularly, I look at Arab television. I don't understand a single word, but I look for images, the same as, I'm sure, that you-all do. And quite frankly, I've seen stuff that I think is equal to your research, and that's just on television and not on the Internet.[21]

Perhaps inspired by SAIC's tendency to present materials to the committee that are already in English, Hastings also wonders aloud about the team's linguistic competence. At one point he asks if they "have a considerable number of Arab linguists working" on the project. Although assured that SAIC has "25 Arabic linguists," who have "all of the dialects of Arabic covered" so they can "tell the origin of the people that are speaking or writing in," Hastings continues to express skepticism. It is significant, however, that the congressman from California does not choose to expose the limitations of the interpretive activity itself, particularly within the national security framework, or the ways that mobility, bilingualism, and biliteracy of political subjects create zones of indeterminacy about definitive origin and authorship. What Derrida has called the "monolingualism of the Other"[22] and the impossibility of ever being either entirely inside or entirely outside the linguistic frame would seem to make absolutely scientific judgments about meaning impossible, especially when, as Megan Boler has pointed out, so much political content on the web uses tropes of satire and parody.[23]

When Samir heard that he had been labeled a terrorist by U.S. intelligence analysts, initially he "thought it was a joke." [24] He characterized himself as extremely taken aback by the misinterpretation of his creative efforts at machinima.

> The least they could do is some online research before publishing this. If they label me al-Qaeda just for making this silly video, that makes you think, what is this al-Qaeda? And is everything al-Qaeda?[25]

As he explained in the interview for this collection, rather than intending to foster "brainwashing" or "training our kids with violent video," he had simply used material from the inside-camera option of the game to respond to other gamers who had made similar videos starring U.S. or Chinese soldiers in *Battlefield 2*.[26] He had considered his video to be a form of friendly joshing on a moderated forum that rarely carried inflammatory or abusive comments, one where the "positive reviews" that he received on his fan film were "great," and the "less positive" ones were still "okay," because he considered the forum to be a generally supportive learning environment where constructive criticism came from all types of community members.[27]

Samir described his participation in the *Battlefield* online community of gamers as a cross-cultural enterprise rather than one devoted to the one-sided jihadist agenda that SAIC investigators had depicted. He portrayed gaming as a "great tool to meet people from different cultures" and a "great way to socialize with people that you would have never met otherwise."[28] Rather than express the uniformly anti-Western feelings attributed to creators of pro-Islamic digital media in the congressional hearing, he asserted that "English is very important to online life" and that "it is a positive thing to communicate . . . [in a] universal language that can bring people together and bridge different cultures and backgrounds," even though he saw these linguistic exchanges as the by-product of historical conditions in which "America is the last great empire," and in ten years the universal language may be Chinese.[29] He even offered some sympathy for overwhelmed U.S. investigators and indicated that he understood the similar "frustration" experienced by "other agencies" around the world faced with the fact that "there are no more borders online" and that this ability for large-scale collaboration without crossing geographical borders that can be policed and monitored is "the wave of the future."[30]

After the media controversy over the supposedly subversive character of "SonicJihad: A Day in the Life of a Resistance Fighter" generated more public interest in his work, Samir created another machinima "masterpiece" to follow up on his initial effort. This new machinima film also incorporated news footage about the misunderstandings of SAIC, the Pentagon, Congress, and Reuters. However, with the new "SonicJihad: The Last Stand" video he was careful to post a disclaimer making his intentions as a creator explicit.

> This video is NOT intended as a propaganda tool or a Jihadi recruiting video. This video is being exclusively published by me on the Planet-Battlfield site . . . Anybody posting this video on other sites will take his or her own responsibility! I repeat this is a Battlefield video for the community only.[31]

Like "A Day in the Life of a Resistance Fighter," "The Last Stand" begins the narrative chronologically from the perspective of boyhood: "As a child, I never imagined that I would one day lead my people in a battle for their freedom."[32] In addition to incorporating *Battlefield 2* footage and some of the audio clips used on the first video, the more ambitious score of the second video integrates music from rapper T-K.A.S.H.'s *Turf War Syndrome* album, from "Ave Bushani" by Paris, and from the instrumental number "La Caution" from Thé à la Menthe, which was also used in the popular American film *Ocean's Twelve*. "The Last Stand" also includes material from a Malaysian-made cartoon about the historical figure Saladin, an important military leader against the Crusades and political unifier of Islamic city-states in the Middle East.

This inclusion of animated footage could be read ironically against the thesis of Congressional investigators about SonicJihad/Samir as a maker of Islamic propaganda appealing to very young children. Like machinima, AMVs (anime music videos) or other "fan vids" that repurpose animated cartoons are often explicitly created by adults for adults and reference sophisticated systems of visual and auditory signification that younger viewers would not find meaningful.[33] He also incorporates a seemingly incriminating photograph of a group of men in Middle Eastern clothing whose faces have been digitally blurred out assembled around a computer, one that attentive *Battlefield* fans will see displays a sequence from the game rather than the jihadist materials that Bush officials might expect.

It is worth noting that much of "The Last Stand" is devoted to credits, as is the case with many machinima films. In these credits, SonicJihad lists himself as "Executive producer" and the creator of the "movie score," but he also thanks the "entire Battlefield community for sticking around" and expresses his appreciation for the web site for the Saladin television show as the source of his animated footage. Peter Krapp has argued that elaborate forms of citation are recognized conventions in machinima filmmaking and that references to source materials are often part of the genre's appeal to its audience.[34] Thus SonicJihad shows himself to be conscious of the expectations of playing the system within the norms of machinima filmmaking, as he also clarifies the video's provenance for future audiences to prevent mistakes of attribution in the future like those he has experienced in the past. As Lisa Nakamura notes in an essay on the emerging genre, machinima often points to the circumstances of its own production in other ways, because such digital media may be "more forms of practice than they are texts."[35]

In "Of Games and Gestures: Machinima and the Suspensions of Animation," Krapp claims that "gestures and their citability mark the performative space of theater or cinema that is cited by machinima," although "precisely calibrated in-game gestures remain particularly difficult."[36] Krapp argues that orchestrating such gestures in machinima is an important part of creating a virtuoso piece for the public and ensuring its recognition in the surrounding culture of reception that values playing with and around the rules, but the lack of granularity in expression available in game engines also draws attention to the limitations of the art form, much as silent film depended on stock movements and exaggerated gesticulations in the early days of the moving image.

Expertise about which game mechanics cause a particular gesture to be produced is often an important part of membership in machinima knowledge communities. For example, after the debut of the trailer for another SonicJihad video, "Warfare," one viewer asks, "How did you make the guy kiss his AK?"[37] In response, Sonic Jihad explains, "In special forces and regular BF2 the Arab anti-tank forces kiss their weapon if you stand still and do nothing for a while."[38] In other words, while most gestures in machinima are created by a form of button-pushing puppetry, SonicJihad explains that this section of the film is generated by a lack of interaction with game play. Although SonicJihad admits that he doesn't "know if the US / Chinees [*sic*] / Russian forces do the similar thing" when the game is in an inactive state, another commentator has the answer: "All the armys [*sic*] in BF2 and the Exs do different things with their weapons. The U.S. guys spin their pistols if you have it out. and the Russian guys in SF check their Aks when they have them out."[39] In addition to news segments about the SAIC fiasco, clippings about the Iraq war, and stock footage, the "Warfare" trailer uses material from a trailer for the game *Metal Gear Solid 4*. In his jump to experimenting with the genre of the trailer, SonicJihad/Samir indicates that he has developed further in exploring the rhetoric of the production cycle of a machinima film and in mixing conventions of cinema with conventions of games.

Although the actual "Warfare" video has received fewer views in comparison to the videos most directly related to the misguided Congressional hearing, it appears to be the longest and most technically ambitious of SonicJihad's machinima projects. It was posted in three parts on his YouTube channel two months after the trailer appeared. In introducing the video SonicJihad/Samir continues to encourage "creative suggestions" from his viewership. In addition to in-game footage, the full version of "Warfare" uses cinematic cut scenes and in-game advertising spots from Burger King.

In "Warfare" SonicJihad/Samir points to the structural agnosticism of videogame play, particularly in games that allow fighting for either side in combat scenes, and how such games encourage players to *game* systems and pursue exploits in ways that complicate master narratives of manifest destiny. Since a player's commitment to a given side produces certain handicaps and advantages independent of any ideological considerations associated with loyalty to that political, religious, ethnic, or national affiliation, the choice to illustrate part of the text of the sixth-century treatise *Art of War* seems particularly appropriate. After all, this classic by famed Chinese battlefield strategist Sun Tzu is taught in military academies throughout the world, regardless of cultural identification, and is standard reading at West Point and many other American officer training institutions. The video presents five separate units, which in

Samir's translation appear as segments on these five principles from Sun Tzu: "The Way," "Heaven," "Ground," "General," and "Law." Although the video opens with a creation story about the MEC (Middle Eastern Coalition) in *Battlefield 2* that compares the coalition to Saladin's army, many of the tactics shown could be used by any side in the conflict and have to do with situational advantage rather than moral rectitude.

Unlike the kinds of just war doctrines at work in other videogame discourses,[40] SonicJihad often applies an attitude of pragmatic opportunism that focuses on the constraints placed upon his enemy by the physics of falling objects, limited vision, adverse weather conditions, open ground, narrow passageways, or lower altitudes. A typical chapter in "Warfare" is introduced with advice directly out of Sun Tzu's text. For example, one section, which ends with music from the Disney movie *Pirates of the Caribbean*, includes a passage that reads: "Amour brings a unique combination of firepower, mobility and protection to the battle."[41] Other references to the soundscape of epic cinematic scores include snippets of music from *The Last Samurai* and from the pop classical string quartet Bond. In the "Warfare" video, credits include a trademark symbol associated with SonicJihad's content. He also marks his video as a transnational, multicultural product by presenting credits in both English and Arabic, although viewer moha13881 gripes about the "poor" quality of the Moroccan-born and European-reared Samir's Arabic.[42]

Following "Warfare," Samir had promised another "SonicJihad Trilogy" to be based on the games *Call of Duty 4*, *Far Cry*, and *Assassin's Creed*.[43] In August of 2008 he had even posted a trailer for the film, which opens with the prequel title "SonicJihad: The Story Begins."[44] By this time, some of his viewership was complaining that his machinima videos were not pro-Islamic enough to be satisfying to Muslim gamers. Since the game *Call of Duty 4* only allows players to assume significant parts in the roles of U.S. marines, British commandos, or other invading outside forces, his fans expressed their irritation that they could not "play like MUJAHIDEEN not a USA pig."[45] After viewing Samir/SonicJihad's latest trailer, moha13881 objects that any machinima based on *Call of Duty 4* inevitably resembles "a stereotypical Hollywood propaganda movie talking about the 'dirty sand niggers,'" and suggests that the game engine would need to be retasked in the way that SAIC experts had once feared possible: "There should be a mod where you play as an arab and toss away that american content. You start breaking free from Abu Ghraib and later becoming a commando."[46] In an email, Samir has said that he doesn't know "who put that statement up there and why" and that he only tolerates such "personal expression" on the understanding that YouTube comments have "nothing to do with me or my views."[47] Although the *Call of Duty 4* player does adopt the point of view of the president of a Middle Eastern country in the beginning of the game, this character is held hostage and is executed before becoming truly playable. Moreover, as a U.S. marine, the player can fire on other American soldiers in the chaotic "Charlie Don't Surf" episode, but engaging in friendly fire generally does not allow the player to level up.

Based on this prequel trailer, it is difficult to gauge how SonicJihad plans to use the material from *Assassin's Creed* that he had earlier promised to incorporate. It is significant that, using *Assassin's Creed* as a case study, game researcher Magy Seif El-Nasr has argued that many popular games are read differently by Western and non-Western audiences. In particular, El-Nasr argues, games that cast Middle Eastern countries as exotic locales may be received very differently by those who recognize landmarks of Muslim cultural practice or public space as part of what Salman Rushdie has called their own "imaginary homelands."[48] Thus, she describes how a player from Egypt or Saudi Arabia may react to these landscapes and built environments with feelings of nostalgia rather than with the fear and alienation that North Americans or Europeans may experience.[49] In this way, SAIC analysts may have thought that they had found the definitive and correct (as well as correcting) reading of the game play space, but El-Nasr reminds interpreters of a game's meaning that a supposedly standard reading may actually be another example of what Umberto Eco has called the practice of "aberrant decoding."[50]

After all, weeks after the scandal broke, the U.S. military was still defending its experts and attacking Samir/SonicJihad, as an email from Lieutenant Colonel Barry Venable to an Australian reporter covering the story indicates:

> I don't know, but it certainly appears from the various media inquiries I've received that the creator wants to claim, for reasons I don't care to judge, credit for creating the game modification. Have you asked this "young man" how he feels about having the product of his intellect hijacked by extremists who, if able, would seek to deny him the freedom to do so?[51]

Not only is Venable still insisting that the machinima film was a "game mod" long after this reading had been discredited, but he also wishes to make an argument about the relationship between freedom and intellectual property that argues both for and against remixing gaming content to save face.

In contrast, Samir now describes himself as an inactive gamer, who gave up *Battlefield 2* after playing it and other PC games intensely for "about a year or a year and a half." As he puts it, "It was a nice game; it was fun; that's it." Now he has an Xbox "at home" that he "maybe played ten times." After "two sequels," he says "it's done," and "let's get on with other things." He also distinguishes his personality type sharply from those who are immersed in MMORPGs (Massively Multiplayer Online Role-Playing Games), because "I'm not confused about what's real and what's not real," because "I'm living in the real world. I'm not playing *World of Warcraft* all day."[52]

When Reuters revisited their reporting and actually interviewed Samir in May of 2006, they found him to be a "clean-cut youth" who obviously didn't merit the title of terrorist, because the video created by this consumer of American pop culture and fast food was "just for fun, nothing political," and his footage was not a "serious game" intended for education or indoctrination, since it had "nothing to do with recruiting people or training people."[53] However, in the August 2009 interview for this collection, Samir looked back at the congressional incident and offered a more sustained critique of the media that once presented him as a terrorist. When asked about what he thought of "real Internet materials created by jihadists," he emphasized the importance of seeing both sides of the U.S./Iraq conflict. He argued that materials posted by jihadists on the Internet "let us see the other side" through an "important medium" and allowed citizens of many countries not to depend just on "footage that is approved by the Pentagon" that is produced by embedded reporters with the U.S. army. [54] He protested that sanitized media coverage did not present "the whole picture" to Western audiences unlike "the real uncut footage like Vietnam."[55] Although the U.S. military may be gaming the media, Samir argues that other political actors can also be media-creators that see the relationship of signal to noise differently.

This form of "seeing the other side" digitally with portable hand-held video cameras may be less palatable to those who champion computer-mediated communication and games in particular as a venue for U.S. public diplomacy that can persuade others to support American policy by bridging gaps between diametrically opposed political sides and encouraging new perspectives.[56] Nonetheless—in a world in which U.S. intervention in the affairs of Islamic countries is extremely unpopular with the planet's one-billion-plus Muslims—it is an attitude about even-handedness in media representation that is difficult to ignore. Samir asserts that he is not arguing for the legitimacy of "videos of beheading" or other "gruesome" content, but he does insist that "videos about statements made by both parties in the conflict" contribute to "good judgment" and a public sphere that is "fair and democratic."[57]

Lawrence Lessig has described remix culture as being at the heart of mainstream Western democratic institutions and part of a history of progressive politics that has stretched over centuries in the United States.[58] But Samir seems to see his own remix practices as part of a more complicated form of critique that is harder to account for in Lessig's patriotic and quintessentially American narrative, one that wants to see "western and non-western media" able to "fight" fairly with equal advantage in the game,[59] based on the very democratic principles that liberal democracies supposedly propagate. Samir certainly does not see his videos as terrorist propaganda, but neither does he want those who revisit the SonicJihad story to trivialize the perspective on war and conflict that gaming facilitates. To miss his legitimate concerns about how liberal democracies tolerate oppressive police powers in the West and the occupation of civilian populations in the Middle East is to replicate another kind of error, one which overlooks the lesson about "seeing the other side" that he offers.

Notes

[1] *SonicJihad: A Day in the Life of a Resistance . . . Part 1 of 2*, 2007, http://www.youtube.com/watch?v=ATSz9ulJflg&feature=youtube_gdata.

[2] For the more idealistic argument about the pedagogical benefits of such role-reversal, see James Paul Gee, *What Videogames Have to Teach Us about Learning and Literacy* (New York: Palgrave MacMillan, 2003), 141.

[3] See Sarah Banet-Weiser, "Surfin' the Net: Children, Parental Obsolescence, and Citizenship," *Technological Visions: The Hopes and Fears that Shape New Technologies* (Philadelphia, PA: Temple University Press, 2004), 270–292.

[4] "Panel I of the Hearing of the House Select Intelligence Committee, Subject: Terrorist Use of the Internet for Communications," *Federal News Service*, 4 May 2006, 5.

[5] Ibid., 4.

[6] Ibid., 9.

[7] Ibid., 21.

[8] Ibid., 9.

[9] Ibid., 7.

[10] Rita Raley, *Tactical Media* (Minneapolis: University Of Minnesota Press, 2009), 76.

[11] Jake Tapper and Audery Taylor, "Terrorist Video Game or Pentagon Snafu?" June 21, 2006, June 30, 2006 *ABC News Nightline*, http://abcnews.go.com/Nightline/Technology/story?id=2105128&page=1.

[12] Samir, interview by the author, August 30, 2009.

[13] David Morgan, "Islamic militants recruit using U.S. video games," *ZDNet*, http://news.zdnet.com/2100–9595_22–195896.html.

[14] David Morgan, "Islamists using U.S. video games in youth appeal," *Reuters* May 4, 2006, May 19, 2006, http://today.reuters.com/news/ArticleNews. aspx?type=topNews&storyID=2006–05-04T215543Z_01_N04305973_RTRUKOC_0_US-SECURITY-VIDEOGAMES.xml&pageNumber=0&imageid=&cap=&sz=13&WTModLoc=NewsArt-C1-ArticlePage2.

[15] Alexander R. Galloway, *Gaming: Essays On Algorithmic Culture* (Minneapolis: University of Minnesota Press, 2006).

[16] For more about how elected representatives focused on the digital practices of their constituents at home rather than the use of the Internet by terrorists abroad, see the first chapter of Elizabeth Losh, *Virtualpolitik: An Electronic History of Government Media-Making in a Time of War, Scandal, Disaster, Miscommunication, and Mistakes* (Cambridge, MA: The MIT Press, 2009).

[17] May 4 Hearing, 4.

[18] Ibid., 21.

[19] Gonzalo Frasca, "Extra! Reuters Using News Story to Manipulate Readers," online posting, May 5, 2006, *Water Cooler Games*, May 6, 2006, http://www.watercoolergames.org/archives/000526.shtml.

[20] Samir, interview.

[21] May 4 Hearing, 12.

[22] Jacques Derrida, *Monolingualism of the Other, or, The Prosthesis of Origin* (Stanford CA: Stanford University Press, 1998).

[23] Megan Boler, *Digital Media and Democracy: Tactics in Hard Times* (Cambridge MA: MIT Press, 2008).

[24] Samir, interview.

[25] Ibid.

[26] Ibid.

[27] Ibid.

[28] Ibid.

[29] Ibid.

[30] Ibid.

[31] *SonicJihad The Last Stand part 1of 2*, 2006, http://www.youtube.com/watch?v=w66bdYiXPbk&feature=youtube_gdata.

[32] Ibid.

[33] Dana Milstein, "Reframing Fan Videos," in *Music, Sound and Multimedia: From the Live to the Virtual*, ed. Jamie Sexton (Edinburgh: Edinburgh University Press, 2008).

[34] Peter Krapp, "Of Games and Gestures: Machinima and the Suspensions of Animation" (presented at the Center for Computer Games and Virtual Worlds Seminar Series, University of California, Irvine, November 12, 2009).

[35] Lisa Nakamura, "Digital Media in *Cinema Journal*, 1995–2008." *Cinema Journal* 49, no. 1 (2009): 154–160. http://muse.jhu.edu/ (accessed December 23, 2009).

[36] Krapp.

[37] *SonicJihad: Warfare Trailer*, 2006, http://www.youtube.com/watch?v=3d2A5wn43OY&feature=youtube_gdata.

38 Ibid.

39 Ibid.

40 See Elizabeth Losh, "Regulating Violence in Virtual Worlds: Theorizing Just War and Defining War Crimes in *World of Warcraft*," *Pacific Coast Philology* (forthcoming, 2010).

41 *SonicJihad: Warfare (the movie part 2 of 3)*, 2007, http://www.youtube.com/watch?v=bQ9yLNN_q0U&feature=youtube_gdata.

42 *SonicJihad: Warfare (the movie part 3 of 3)*, 2007, http://www.youtube.com/watch?v=IDPW7vqSXxg&feature=youtube_gdata.

43 Ibid.

44 *SonicJihad - Call of Duty (teaser)*, 2008, http://www.youtube.com/watch?v=XasCOdSOULk&feature=youtube_gdata.

45 Ibid.

46 Ibid.

47 Samir, email message to author, January 12, 2010.

48 Salman Rushdie, *Imaginary Homelands: Essays and Criticism 1981–1991* (New York: Penguin, 1992).

49 Magy Seif El-Nasr, Magy Seif El-Nasr, Maha Al-Saati, Simon Niedenthal, and David Milam, "Assassin's Creed—A Multi-cultural Read," *Loading* 2, no. 3, http://www.sfu.ca/~magy/publications.html.

50 Umberto Eco, *The Role of the Reader: Explorations in the Semiotics of Texts* (Bloomington: Indiana University Press, 1979).

51 "Email from Lt. Col. Barry Venable to Media Watch," http://www.abc.net.au/mediawatch/transcripts/ep15barry.pdf.

52 Samir, interview.

53 "Dutch Gamer's Clash with U.S. Government," *Reuters*, May 24, 2006, May 25, 2006, http://news.com.com/Dutch+gamers+clash+with+U.S.+government/2100-1043_3-6076255.html?tag=nefd.top.

54 Samir, interview.

55 Ibid.

56 See the work of the Center on Public Diplomacy at USC as an example of these initiatives and stories such as "Reinventing Public Diplomacy through Games Awards Ceremony | USC Center on Public Diplomacy | Events Detail," http://uscpublicdiplomacy.com/index.php/events/events_detail/1686/ and "Future Tense: Video Games as Cultural Exchange," http://www.publicradio.org/columns/futuretense/2006/05/video-games-as.html.

57 Samir, email message to author.

58 Lawrence Lessig, *Remix: Making Art and Commerce Thrive in the Hybrid Economy* (New York: Penguin, 2008), 27.

59 Samir, email message to author.

Chapter 8

Disrupting the Public Sphere: Mediated Noise and Oppositional Politics

Ted Gournelos
Rollins College

A number of the chapters in this volume address the concept of error through a related concept: noise. This association relies on an implicit acknowledgment of noise as a positive (if sometimes destructive) distortion that is significantly distinct from an "anti-aesthetic" proposed as the basis of postmodern aesthetics in the work of scholars such as Hal Foster[1] and Frederic Jameson.[2] Rather than a leveling, surface-level, cynical, or wholly self-referential exercise of artists and architects, the use of noise and error as a negotiation of the politics of late capitalism often encourages us to look at community rather than individual *ennui*, global politics rather than local aesthetics, and capitalism as a vehicle for opposition rather than an end game in and of itself. This chapter will attempt to emphasize these differences by connecting noise with an overtly political engagement with the "public sphere," an elusive concept often ignored in the embrace of fragmentation and alienation foregrounded in early, and particularly U.S.-based, versions of postmodernism.

In three case studies that focus on capitalism as a mediated institution (i.e., as a concept that is inextricable from interpersonal and technological mediation), I suggest that noise represents not only a useful challenge to "clear" pathways of capitalism (e.g., alienation → lack → desire → exploitation), but also the symbiotic evolution of *both* capitalism *and* the opposition to capitalism. In this way, "disrupting" the public sphere is not just an attempt to take control of, subvert, or oppose the interests of the Dominant, but is also a process of social and cultural change based in part on the use of noise to cause, express, and exploit error in the lines of cultural communication themselves. Improv Everywhere's staged "freeze" of volunteers in New York City's Grand Central Station, for instance, challenges a sense of apathy or banality in everyday life by fragmenting both space and "reasonable" activity. Although that activity was immediately co-opted by an advertising agency, it is far from clear that the co-optation removes the distortion of the original media event. Similarly, in the appearances of Stephen Colbert and Bill O'Reilly on each other's shows on January 18, 2007, disruption becomes a kind of mediated Socratic dialogue that challenges the coherence of both con-structed "sincere" identity and the parodic, "ironic" identity that stems from it. When The Yes Men replaced thousands of copies of *The New York Times* with their own, oppositional text, they also relied on noise to redefine not only the issues at hand, but also the claim to news "truth" as an unquestioned meta-narrative.

In each case, we see a mediated public sphere that is repeatedly engaged and altered through noise, not to destroy it but to accentuate existing (internal) counternarratives. In other words, they productively disrupt the system in order to accentuate its fallibility, gaming the system by playing it with enough devotion that they reveal the cynicism of treating politics as a closed system. Perhaps more importantly, they do this with enough skill that the Dominant's hegemony begins to break down simply because it appears to lose at its own game.

It is not surprising that theorists that focus on anti-aesthetic approaches to culture and politics tend to avoid discussions of the public sphere. The "public sphere" as a concept, most commonly associated with Jürgen Habermas's canonical *The Structural Transformation of the Public Sphere*,[3] is unabashedly modernist in its faith in the individual as a coher-ent, stable, and reasonable entity. Moreover, the concept often expands on that faith to consider a society or culture (and even the politics that define and are defined by that society or culture) to be an amalga-mation of a reasonable set of discussions between those individuals. It makes sense, then, that Hal Foster's concept of the anti-aesthetic, in which noise represents a reaction to stability, is not framed in terms of community or coherent political action. Similarly, in Frederic Jameson's discussion of postmodernism as a "cultural logic" that not only draws from but embraces late capitalism, we see a desire to focus on a fragmented and eternally self-reflexive individual that draws from a slick surface of cultural production to pacify any concept of cohesion or overt political challenge. In other words, signifiers are flattened in order to express general *ennui* or angst, rather than specific political engagements.

This is, thankfully, not the only way to discuss the public sphere, although this is not the place to discuss the ongoing battle over defini-tions of postmodernism or the public sphere in any depth. However, two fundamental critics of the Habermasian public sphere, Nancy Fraser and

Michael Warner, are important for an understanding of how noise might function to introduce productive error into this supposedly stable, modernist concept. In Nancy Fraser's *Justice Interruptus*,[4] she argues that Habermas neglects to mention that power-differentials were a key component to the "ideal" public spheres he advocates, and that gender, race, class, and other components of identity not only render those "ideals" in terms of conflict, but that such conflict was mediated, contained, and presided over by the Dominant. Michael Warner expands on this idea in *Publics and Counterpublics*,[5] in which he suggests that publics are built not only demographically, but also on the circulation of and participation in discourse. In both cases, we see Habermas's concept developed (rather than discredited) through a growing recognition that the public sphere is based in community, but with three significant modifiers: first, that community is formed by individuals that are themselves fragmented and conflicted entities; second, that community is inflected by (and inflects) multiple arenas of discourse stemming from the fragmentation and conflict of its members; and third, that communities are not isolated or insular, but instead overlap with other communities (i.e., other public spheres) that are often placed in direct opposition. In other words, by acknowledging politics and power-differentials as a central aspect of the public sphere, and by expanding the idea of the "public" to encompass "publics" (in the plural) that might be "counterpublics" (i.e., communities that are in some way oppositional to other communities), we understand the public sphere to be a shifting terrain of desires, interests, needs, oppression, power, and expression that is *always* and *a priori* in a state of conflict and change.

This broadened understanding of the public sphere is essential to a discussion of noise as an agent or manifestation of political change because it suggests that noise can (1) silence certain areas of discourse in order to emphasize one conflict over another, and (2) augment conflict in order to suggest the internal incoherence of a system itself. As Tiziana Terranova suggests in her brilliant *Network Culture*,[6] contra the dystopian visions of many postmodern critics:

> If there is an acceleration of history and an annihilation of distances with an informational milieu, it is a creative destruction, that is a *productive* movement that releases (rather than simply inhibits) social potentials for transformation. In this sense, a network culture is inseparable both from a kind of *network physics* (that is physical processes of differentiation and convergence, emergence and capture, openness and closure, and

coding and overcoding) and a *network politics* (implying the existence of an active engagement with the dynamics of information flows).[7]

This physics and politics of network culture is fundamentally about a shift in the processes of cultural politics as they are produced, exchanged, and exploited by late capital *in concert with* (rather than against) publics in a broader and more contentious media ecology.[8] More importantly, because this media ecology is heavily concerned with noise (e.g., establishing noise for propaganda, or removing it to sustain a brand identity), it becomes concerned not with the construction or maintenance of Identity as such (i.e., static or coherent identity), but rather with representation and community as levels of communication that "displace the question of linguistic representation and cultural identity from the centre of cultural struggle in favor of a problematic of mutations and movement."[9]

For the purposes of this chapter, it is most useful to understand noise as a symptom of "error" in the mediated culture of late capitalism. Particularly because noise is thus a symptom of both the creation of error in an otherwise "friction-free" system *and* a way for the system to mask and distract from already existing errors, this method avoids the binaries between alternative and mainstream culture, Dominant and oppositional politics, and organic and reified/valorized cultural production. It takes Bourdieu's suggestion that we see capital as a cultural and symbolic process as much as a "concrete" process of exchange seriously, but does so in order to problematize any claim to "authentic" culture separated from the market.[10] This does not mean that culture cannot be oppositional; instead, as I have argued elsewhere, it means that we should see oppositional culture as a *necessary but crucially overdetermined* aspect of late capitalism.[11] In other words, the ever-increasing rate of change, shift, and evolution required by the cultural and symbolic exchanges in the contemporary media climate both remove the chance to operate outside of the Dominant *and* allow production, distribution, and reception methods that threaten the system itself (which relies on the illusion of its stability and omnipresence). This is particularly the case in digital culture. As Terranova argues, "a piece of information spreading throughout the open space of the network is not only a vector in search of a target, it is also a potential transformation of the space crossed that always leaves something behind—a new idea, a new affect (even an annoyance), a modification of the overall topology."[12] This almost biological model of noise charts communication as simultaneously productive and destructive, yes, but also as unpredictable processes that might or might not lead to anything. They can remain static, as do so many would-be media viruses on YouTube or JibJab, but they can also become far more than ever expected or "veer off catastrophically in unexpected directions."[13] I will focus on actively (and intentionally affective) political interventions into the public sphere, rather than media viruses that become affective by chance. In each case, moreover, the intervention was either staged as or subsequently incorporated into a valorized media production.

The first intervention is not, on the surface, a mediated event. In fact, drawing from a history in the aesthetics of political groups like the Situationists International or the so-called "happenings" and other performance art events (by fluxus, etc.), it would often fall under the rubric of the "avant-garde" or High Art. However, it is not the event itself in which I am primarily interested, but rather the pseudo-documentary videos released on YouTube immediately afterwards. Significantly, without having named the event or film, many if not all readers of this chapter will have a few examples in mind that fit this description: videos of a "screaming" Howard Dean at a campus campaign rally, Susan Boyle's first appearance on *Britain's Got Talent*, or the hilarious answers given by Lauren Caitlin Upton at the Miss Teen USA pageant in 2007. This is not an accident; I suggest that many media viruses *became* media viruses precisely because they represent a challenge to the conventions of the public sphere. In this case, I will discuss the videos of a January 2008 "flashmob," led by Improv Everywhere, in which they "froze" New York City's Grand Central Station.

The idea was relatively simple: approximately 200 "agents" of Improv Everywhere (i.e., people in their social network reached via social media outlets) organized in Bryant Park for an undisclosed intervention. Before entering the large, extremely busy space, they synchronized their watches, and at precisely 2:30 p.m. they froze in whatever action they were performing at the time (of course, some of them chose obviously distinct "frozen" positions, e.g. kissing or reading maps). They remained in the positions for several minutes. Passersby were understandably fascinated by the event, and the various videos circulated of it have received millions of views. The "official" videos of the event foreground the meeting in Bryant Park, with instructions given via megaphone to a large crowd, and then transition through semi-professional cuts through "amateur" video sequences of individual moments in the event.

It is the fact that it was planned as a *mediated* event that relied on social media organizing and more crucially, on social and viral media *distribution*, that interests me here. This is not significant in and of itself, of course; documentary filmmakers, performance artists, and voyeurs have "documented" strange events since the birth of photography. However, the decentralized network distribution patterns of this particular filmed event allow it to become a pedagogical as well as aesthetic experience. In other words, as I argue elsewhere, disrupting the public sphere in this way is not as interesting for the initial disruption as it is for the compelling lesson in *how to disrupt*, and is thus indicative of a critical pedagogy in which the public sphere is conveyed as a contained space (i.e., a space of power relations in which one does not act out of bounds) that, *because it is contained*, becomes an aesthetically rich and viable medium. This is distinct from the "wisdom of crowds"[14] or "smart mobs"[15] concepts, in which "collective intelligence" or networked communication alter the Dominant in some way (often, significantly, in terms of the use of technology). Instead, it is a planned and guided intervention into a public space in order to call into question the use or sanctity of the space (or of the "public") itself.[16] This is limited in terms of political affect, of course. A momentary questioning of one's physical space, or even one's place within a power structure, is not necessarily much different from any self-reflexivity inspired by a wealthy person passing a homeless person on the street, an employee crossing a picket line, or a spectator watching a bar fight. However, by framing this event as both *fun* and *easy*, the media virus that resulted could potentially change the way people think about their place as producers as well as participants of a "public."

Significantly, this event and events like it have been immediately co-opted by capital, both in terms of YouTube's advertising revenue from the video and in terms of similarly staged events by advertisers or other corporate interests. Is this co-optation the same as the "original" or "authentic" staging? If there was any political space opened by this event and the videos of it, are they foreclosed by the co-optation, as critics like Thomas Frank might argue?[17]

In the case of the "frozen" Grand Central Station event, a television commercial for the Xbox 360 that directly draws from it was released within a year. In this advertisement, which became a media virus in its own right, dozens of people in a large public building that looks a great deal like Grand Central Station begin to act as if they are participants in a "first person shooter" game. In other words, what begins as a "normal" set of relationships, in which people are individual, atomized elements in a mass public, turns into a game reminiscent of childhood play. At first it seems halfway serious, as the first "gamers" point fake guns (made by mimicking pistols with their hands) at each other, acting out the tense pre-shootout scenes ubiquitous in contemporary action films. As the "virus" spreads and the entire station begins to play along, a happy sound track starts and the game spirals into the entire public, thus reconnecting the atomized individuals through a mediated community based on the shared rules of the game. Notably, this childish play is perfect; unlike children, these participants act fairly. In other words, if someone is "shot" she/he dutifully performs his or her "death," sometimes in spectacular fashion, as in when a worker tethered to a second floor balcony falls and is suspended in midair by his tether.

In one scene, in fact, a gamer "shoots" a man in a suit talking on a mobile phone in his car. The man, who seems to be on a business call, holds up his hand in a "wait a minute" signal. The gamer sighs in obvious frustration in a classic "alternative vs. Dominant" binary. However, after the man completes his call, he whips himself back in recoil from the "impact," falling face-forward onto his steering wheel and blowing the horn. This scene is paradigmatic of the appeal of this advertisement and its reliance on the legacy of flash mob activity like the Improv Everywhere event. It highlights not only a communal game in a "perfect" system, but also the potential *breakdown* of that game due to differences in status and power (in this case economic power). In other words, it suggests that the community formed by this staged flash mob transcends *without ignoring* power relations in the public sphere. Notably, this mimics Improv Everywhere's discussion of the event on their web site, in which they foreground participation by multiple classes and races, and reactions by a police officer.

I am not attempting to equate the staged advertisement with the staged intervention, of course. However, it is important to question whether or not there is a significant difference between the events, especially once they are mediated online. In the first, the event remains singular and attempts to intervene in the banality or alienation of everyday life. That intervention is particularly significant, as Terranova points out, in an informational culture "where meaningful experiences are under siege, continuously undermined by a proliferation of signs that have no reference, only statistical patterns of frequency, redundancy and resonance."[18] The advertisement is thus an attempt by a corporation to retain a coherent brand identity in a world obsessed and consumed by noise.[19] However, leaving aside issues of "authenticity" vs. "co-optation" or "valorization" for the moment, the mediated events do not seem significantly different. In other words, they both represent interventions into a public sphere that are relatively easy to accomplish, and frame them in such a way as to make them seem fun and accessible. Although the disruptive potential of these interventions internalize (and are immediately reincorporated into) what Guy Debord called the "society of the spectacle," that does not necessarily render either one useless in terms of political or social affect.[20] It might be more interesting to argue along with Terranova that "rather than capital 'incorporating' from the outside the authentic fruits of the collective imagination, it seems more reasonable to think of cultural flows originating within a field which is always and already capitalism."[21] That suggests not only that it is impossible to operate outside of the system, but also that operations performed both *within* and *by* the system become potential areas for opposition, particularly as they might escape original signification within the mediated world of incessantly circulating and mutating information on the Internet. Although the disruptive possibilities of noise as a game are always already systemic disruptions, the interest in the intervention transcends the interest in its reappropriation. In other words, in this case the Xbox 360 brand is overdetermined, relying on rather than containing the oppositional politics of community inherent in flash mob action.

Mediated noise takes a different direction when the "game" is less in terms of play (in the chaotic, unpredictable sense) and more in terms of *a play* (with scripted identities and interactions). However, as we will see in the next case study, in which Bill O'Reilly and Stephen Colbert became guests on each other's shows, the mediation of the events once again overdetermines the events themselves. The resulting clash of constructed identities and scripted performances, in which O'Reilly is forced to acknowledge his own constructed identity when he attempts to critique Colbert's parody of it, places them in a dialogue that produces signification from the disruption of previously mediated discourse. In other words, O'Reilly and Colbert function as relatively coherent discourses separately; however, in the mediated world in which they come into contact *on each other's shows*, those discourses break down to suggest something greater than the sum of their parts. This again expands how we should think about an oppositional "gaming" of the system. On the one hand, participation in this game of constructed identity plays into the standard idea of parody, in which the process of appropriation calls into question internal errors in the codes on which the original relies. On the other hand, the mediation foregrounds error as a *practice* in which viewers are not just "in on the joke," but actively participate in challenging the codes of "the news" itself.

On January 18, 2007, Stephen Colbert appeared as a guest on *The O'Reilly Factor* in a much-hyped "confrontation" between the hyper-conservative Bill O'Reilly and the self-reflexive parody performed in *The Colbert Report*. Although O'Reilly begins the interview in much the same way as he frames his other guests, the discourse quickly spirals out of his control. This is not because he is faced with an overtly oppositional figure, as has occurred in appearances by Jon Stewart, Al Franken, and others, but instead because he is faced with a parody of his own "authentic" identity and the absurdities of his hyper-nationalistic or hyper-conservative performance (e.g., "I catch the world in the headlights of my justice"). When he attempts to attack Colbert, for instance by using a "French" pronunciation of his name rather than the "American" pronunciation (i.e., with a hard "t" sound at the end), Colbert responds by amplifying his own "authentic" American-ness in connection to O'Reilly's, twisting it into an argument about authenticity within a capitalist system: "you've got to play the game that the media elites want you to do, okay? You and I have taken a lot of positions against the powers that be and we've paid a heavy price. We have TV shows, product lines, and books." Moreover, in repeated sexual innuendoes (none of which O'Reilly acknowledges), he removes any possibility for O'Reilly to truly play along without alienating his target audience. When O'Reilly attempts to attack Colbert's associates, specifically Jon Stewart, Colbert again amplifies the performance, agreeing with assertions of heavy drug use, etc. In both cases, O'Reilly is obviously stuck between agreeing with his own rhetoric, albeit in a "strengthened" state, or defending the ideologies he supposedly opposes. His last technique, which occasionally works with hostile guests, is to overwhelm them by being aggressive or yelling (and/or cutting off their microphones). However, Colbert meets his aggression with silence, pausing for an uncomfortable few seconds to amplify the absurdity of the rhetorical technique rather than falling into an argument he could not possibly win. Colbert then draws attention to that technique itself: "you know what I hate about people who criticize you? They criticize what you say but they never give you credit for how loud you say it." The interview ends in one of the only complete and utter rhetorical failures in an *O'Reilly Factor* interview.

It does not end there. Later in the day, O'Reilly appeared as a guest on *The Colbert Report*, and rather than discussing politics as such, he was forced to address his incredible failure to contain Colbert's parodic identity or to match it with his own "authenticity." They begin, in fact, as O'Reilly responds to a catcall from the audience by saying "that's Jon Stewart!" Colbert responds with "I have a restraining order" on him, and calls Stewart a "sexual predator" (an allusion to O'Reilly's own alleged sexual harassment of a female employee in 2004). Quickly afterwards O'Reilly even admits that the co-appearances were "a huge mistake," and seems to give up on beating Colbert at his own game. When he tries, he either ends up falling into a rhetorical trap, as when he answers that NBC news is more dangerous than "activist judges, illegal immigrants, [and] gay marriage" because it incorporates the others on the show, or makes a gaffe, as when he says that NBC anchor Brian Williams shouldn't be trusted because "he's Venezuelan." Significantly, because *The Colbert Report* and *The O'Reilly Factor* do not share ratings demographics, most viewers would not watch the "live" versions of the performances. However, the subsequent joining of the interviews on *The Colbert Report*'s web site and on sites such as YouTube and Google Video foreground the interviews as a duet, making the appearances easily seen and compared by both audiences.

This is not the place to discuss either O'Reilly or Colbert, and they have been discussed at length elsewhere in a series of excellent articles and book chapters.[22] However, it is useful to consider that Colbert inhabits an ironic identity in *The Colbert Report* that both asserts a (disjointed) political viewpoint drawn from pundits like O'Reilly and an overt attack on both the rhetoric and perspectives of those pundits. Because that ironic identity is itself a dialogue, when it comes into contact with the "sincere" and "authentic" identity proclaimed on *The O'Reilly Factor* it reclaims a contact with the "real" that transforms it from an in-group dialogue (to borrow Goffman's terminology) to a Socratic dialogue between in-group and out-groups.[23] More importantly, perhaps, because the *mediated* event on the Internet involves *two* such Socratic dialogues, each one initially placed in a different in-group, it avoids being seen as potentially preachy, arrogant, and self-righteous and becomes rather a disruption of coherent identities themselves. This is not self-evident in the events, of course; if O'Reilly did not attempt to match Colbert as a performed identity, or if Colbert did not respond to rhetorical attacks by either increasing his performance (in supposed agreement with O'Reilly) or becoming silent (thus amplifying the noise of O'Reilly's rhetorical sledgehammer), the event would have lost much of its political affect.

Its success was, I argue, in large part due to the characteristics of the mainstream U.S. political system itself, which has become increasingly focused on polls, image, so-called "spin," and the performance of distinct identities by "pundits" (i.e. paid actors, technocrats, amateur "commentators," and public relations operatives rather than journalists) rather than a more public sphere oriented approach to the political (rather than politics). As Terranova argues:

> In this context, the opponent becomes noise and the public becomes a target of communication: not a rational ensemble of free thinking individuals, endowed with reason, who must be persuaded, but a collective receiver to which a message can be sent only on condition that the channel is kept free of noise (competing politicians, but also the whole noisy communication environment to which such politicians relate, where, for example, more young people vote for reality TV shows than for general elections).[24]

This implies that Colbert's parody is not as important as the moments in which his parodic identity comes into contact with (and thus begins to discursively break down) the authenticity of other, supposedly coherent identities. Thus parody here becomes a living engagement with both shifting events and shifting personalities, acting as an *embodied* parody rather than the more traditional definition (especially when "parody" is confused with "satire") in which one text appropriates from another in a negative critique.

As I argue in *Popular Culture and the Future of Politics*, the form of parody associated with appropriation (what I call the "allusive" mode of political engagement) is often limited by the discursive trends of the original text. Even embodied parody (what I call the "responsive" mode) can only act with or against established discourse. This is the case with *The Colbert Report* in general, as its parody is decidedly an in-group critique on the "stupidity" of the out-group. Although it certainly relies on "error" within the system, particularly in terms of hypocrisy, hyperbolic performance, and illogical or self-contradictory arguments, it is only able to (1) generate noise to disrupt O'Reilly's existing rhetoric or (2) through silence expose error in O'Reilly's own noise. However, when the clips of *The O'Reilly Factor* and *The Colbert Report* appear and are circulated together, the critique of contemporary political discourse becomes not so much about the political concepts involved, but rather about the *discourse* by which those political concepts are established, maintained, and potentially disrupted. In other words, by refusing to acknowledge O'Reilly's established methods of silencing opposition through noise, and by inserting his own parodic noise into O'Reilly's favored rhetorical tools, thereby disrupting any claim to "authenticity" or "coherence," Colbert places O'Reilly *in dialogue with himself* as a constructed political figure rather than a "real" expression of policy, news, or even emotion. This suggests that "gaming" the system might be more active as a willing participant in a game that the players acknowledge is impossible to "win"; that is, the goal might be to sustain the game itself (an endless battle or state of play) rather than to gain supremacy or foreclose any form of discourse. As Terranova argues:

> It has become impossible to ignore the way in which much communication is not simply about access to information and public debate, but is also about manipulation by way of positive (spin, propaganda, hegemony) and negative tactics (censorship, exclusion, distortion, etc.). The impossible task of the public sphere thus becomes that of returning communication to an older, purer function by combatting [*sic*] the corrupting influence of manipulation, censorship, propaganda and spin. However, this activity would be pointless if this public sphere did not aim to represent and address an inherently interested and enlightened public to whose opinion governments are bound. If such reasonable public opinion was shown not to exist or to be ineffectual in influencing the actions of politicians, then the organs of public opinion would lose much of their power.[25]

This suggests that the power of these mediated events comes not so much from their establishment or disruption of a public, but rather from a *reinscription* of the importance of the public itself as a conflicted and contradictory set of interests. In other words, by showing how the dominant political mode in the contemporary United States is internally inconsistent and incoherent, it reasserts the existence of voices lost in the fight for control over those voices. As with the flash mob example, this is a somewhat limited approach, no matter how useful it might seem. Even in its ideal state, this sort of disruptive practice relies on a celebrity status that both increases its immediate affect and limits the sustainability of the opposition. In other words, fame allows for oppositional *moments*, but not for oppositional *movements*. In that way, the potential *immediate* affect is stronger than the first case study, but the *extended* affect is far weaker. Eventually, the performance of a Socratic dialogue becomes a mere performance, and thus ceases to be Socratic. Significantly, however, reinscribing the importance of the public to some extent is itself a practice by which the game might not be changed, but the rules for playing it might be.

In my third case study, the replacement of approximately 80,000 copies of *The New York Times* in Los Angeles and New York City on July 4, 2009, both the potential and downfall of this form of performance becomes even clearer. Rather than using noise as a game, or establishing noise as a dialogue, the strategy of the New York City-based artist/activist group The Yes Men relies heavily on noise as a prank. Pranking by its very nature implies that the Dominant is fallible/gullible. Moreover, its fallibility stems from the system's own hubris, myths of coherence, and attempts to retain control over all aspects of discourse, mediated or not, and thus these characteristics become the target. Pranks do not attack the people, events, organizations, etc. that are their most obvious victims; instead they attack the systems and processes of control endemic to the Dominant's assertion and maintenance of hegemony.

Like the first two case studies, The Yes Men's pranks are inherently based in the mimicry and enjoyment of the spectacle (and, of course, potentially subverting the spectacle). Also like the first two case studies, they are heavily mediated. Interestingly enough, The Yes Men rarely produce material that intervenes in established media outlets directly. This is not due to lack of vision or ability; in some cases, for instance in their infamous attack on Union Carbide and its parent company Dow Chemical for the humanitarian and ecological disaster in Bhopal (through a widely publicized interview on the BBC no less) they do so to great effect.[26] Instead, however, they prefer to "create scenes" that are not often recognized as pranks. These are often not covered by the mainstream media much at all, and only gain life on the blogosphere, on YouTube, or through The Yes Men's own publicity. Their creation of a fake web site for George W. Bush for the 2004 election, for instance, linked to areas of his history and policy not covered in the mainstream. Similar web sites for corporate giants led to events filmed for their first documentary, *The Yes Men* (2005), in which they document their interventions into corporate gatherings and conferences, often as they impersonate official representatives from established corporations (e.g., Exxon). Indeed, despite the film's hilarious content (e.g., claiming that the Iraq War was ended, that Congress passed National Health Care legislation, or that Bush indicted himself for war crimes) and massive scale, few news outlets covered the *New York Times* prank at all (including the *New York Times* itself). Indeed, most coverage of the event is either in the group's second film, *The Yes Men Fix the World* (2009) or in reviews of the film.

As we saw with the first two case studies, coverage of an event or context for an event is not completely reliant on mainstream media outlets any longer. The Yes Men's intervention could be seen as a "performative surprise" that implicitly critiques the dominance of mainstream media itself.[27] In fact, as Amber Day argues, the use of spectacular visuals and incredible situations elicits a pleasure that is "a key concern in their design, a pleasure often conceptualized in opposition to the potential displeasure of the straightforwardly didactic."[28] The Yes Men's form of "identity correction," as they call it, is thus a *pleasurable* and *marketable* product; it is, in other words, a spectacle. However, what is so important about The Yes Men's mediated spectacle is that in most cases it is *not* part of the larger Spectacle itself until they themselves produce and distribute the media covering it. This lack of coverage is transgressive not because of the events themselves, but rather that the outrageousness of the events rendered them *too spectacular for the Spectacle*, at least at the time. These forms of pranks, therefore, are revealed to be interventions into a *controlled* public sphere rather than the "ideal" public sphere Habermas (and many politicians and journalists) argue we either have or desire. In many cases this means that the pranks are largely silenced, but because they are so spectacular the silencing *itself* overdetermines them. Because mainstream media outlets either do not cover the pranks and so reveal themselves to be disturbed or vulnerable, or they cover the pranks and so reveal corporate (and corporate media) systems to be flawed (as happened in the Bhopal incident, but also in their more recent faux press conference for the Chamber of Commerce, in which they "revealed" that the Chamber had changed its policy to be in favor of green legislation), The Yes Men's prankster approach reveals that the Dominant is not only limited in its ability to eliminate noise and error, but also that its *attempts* to limit discourse often target "real" or "true" information. In other words, it reveals that the "trusted" elements of the mediated public sphere classify some signals as "noise" or "error" if they deviate from established or hegemonic norms, regardless of those signals' claims to truth or importance. This does not only intervene in media for its own sake, but rather in the concept of the public sphere constructed by the climate of media conglomeration and hegemony:

> Thus, the public sphere of the welfare state and mass democracy is described by Habermas in terms that are markedly different from those of the bourgeois public sphere. While the bourgeois public sphere comprises individuals engaged in public discussion, within mass democracy, the state and private interests are directly involved, as the pressure on journalists and the aggressively televisual nature of politics demonstrates. The current public sphere is not a sphere of mediation between state and civil society, but the site of a permanent conflict, informed by strategies of media warfare. Communication is not a space of reason that mediates between the state and society, but is now a site

of direct struggle between the state and different organizations representing the private interests of organized groups of individuals.[29]

As Terranova continues, however, the difficulty in this understanding of the public sphere as a problematically mediated entity lies in the fact that it is overwhelmingly controlled by mainstream media and the powers that influence/own them. What events like The Yes Men's pranks illustrate, however, is that performances do not need to operate outside of the mainstream media to have an effect, but can rely on the noise created by their interactions with the media as they are either amplified or silenced by it.

In each of these case studies we see different strategies by which oppositional politics can disrupt the illusion of an ideal, power-free, open, and informed public sphere. As Terranova has argued, it

> does not really involve a metaphysics of being and appearance so much as a kind of information ecology which also includes a dimension of warfare—a warfare to determine the differential power and dominance of some types of images over others. It is no longer a matter of deception, but of the tactical and strategic deployment of the *power of affection* of images as such.[30]

Political affect is no small thing; it might not change contemporary politics (i.e, have a demonstrable political effect), but it might change the way in which people think about and interact with their everyday lives. The age of "convergence culture" is an age where power does not disappear, but rather becomes more and more apparent even as its processes become increasingly subtle. This might seem like a paradox, if one chooses to see opposition as apart from the Dominant. However, I believe that to do so is problematic at best and counterproductive at worst. If we are raised and educated in a system of advanced capital, with all of the media production and distribution tricks of contemporary entertainment, journalism, and advertising, then shouldn't oppositional politics follow the same media pathways, only contrapuntally? Doesn't it make sense for opposition to be strongest in the areas where capital is weakest: in controlling the richly profitable but immensely risky world of conflict?

The creation of noise in the system is endemic to late capitalism, even more than other systems of control, because of the constant need for new symbolic areas to valorize and exploit. However, by amplifying or exposing that noise into the error of a flawed or exploitative system, noise also has the potential to generate a transformative affect. As Henry Jenkins has argued, "if we think about contemporary politics in the age of YouTube, one of the things we've seen again and again is the public exploiting openings into the system provided it by trying to negate the rules of the system."[31] Exploiting error and noise, therefore, is not about challenging the Dominant itself as much as it is about challenging the symbols and codes by which the Dominant retains control. This reflects an incredibly powerful system, yes, but it also suggests the existence and increasing skill of "a political and cultural milieu that can no longer be subsumed (if it ever was) under a majority, led by a class/avant-garde/ idea or even made to form a consensus that is not inherently fractured and always explosively unstable."[32] If we want to locate and encourage the production and use of the kinds of noise that fragment that illusion of consensus, aggravate the instability, and generate opposition, then we have no farther to look than our everyday lives. It is not the generation and reification of the public sphere as an arena of compromise and com- munication that sets us free; it is the disruption of that public sphere, the noise of our own dissatisfaction, and the sound of insane laughter or ironic snicker that gives us hope.

Notes

[1] Hal Foster, ed., *The Anti-Aesthetic* (Port Townsend: Bay Press, 1983).
[2] Frederic Jameson, *Postmodernism, or, The Cultural Logic of Late Capitalism* (Durham: Duke University Press, 1991).
[3] Jürgen Habermas, *The Structural Transformation of the Public Sphere* (Cambridge, MA: MIT Press, 1991).
[4] Nancy Fraser, *Justice Interruptus: Critical Reflections on the "Postsocialist" Condition* (New York: Routledge, 1997).
[5] Michael Warner, *Publics and Counterpublics* (Cambridge: Zone Books, 2005).
[6] Tiziana Terranova, *Network Culture: Politics for the Information Age* (Ann Arbor, MI: Pluto Press, 2004).
[7] Ibid., 2–3.
[8] Ibid., 7–8.
[9] Ibid., 10.
[10] Pierre Bourdieu, *The Field of Cultural Production*, Trans. Randal Johnson. (New York: Columbia University Press, 1993).
[11] Ted Gournelos, *Popular Culture and the Future of Politics: Cultural Studies and the Tao of South Park* (Lanham, MD: Lexington Books, 2009).
[12] Terranova, 51.
[13] Ibid., 105.

14 James Surowiecki, *The Wisdom of Crowds* (New York: Anchor, 2005).

15 Howard Rheingold, *Smart Mobs: The Next Social Revolution* (New York: Basic Books, 2002).

16 Judith A. Nicholson, "Flash! Mobs in the Age of Mobile Connectivity," *Fibreculture*, no. 6 (2005), http://journal.fibreculture.org/issue6/issue6_nicholson.html.

17 Thomas Frank, "Why Johnny Can't Dissent," in *Commodify Your Dissent*, ed. Thomas Frank and Matt Weiland (New York: W.W. Norton and Company, Inc, 1997).

18 Terranova, 14.

19 Ibid., 16.

20 Guy Debord, *Society of the Spectacle* (New York: Zone Books, 1995).

21 Terranova, 80.

22 Megan Boler and Catherine Burwell, "Calling on the Colbert Nation: Fandom, Politics and Parody in an Age of Media Convergence." *The Electronic Journal of Communication* 18, no.2 (2008); Jeffrey Jones, *Entertaining Politics* (Lanham, MD: Rowman and LIttlefield Publishers, 2004); Jeffrey Jones, *Entertaining Politics*, 2nd Edn. (Lanham, MD: Rowman and LIttlefield Publishers, 2009); Amber Day, "And Now . . . the News? Mimesis and the Real in the Daily Show," in *Satire TV: Comedy and Politics in a Post-Network Era*, ed. Jonathan Gray, Ethan Thompson and Jeffrey Jones (New York: New York University Press, 2009), 85–104; Geoffrey Baym, "Representation and Politics of Play: Stephen Colbert's *Better Know a District*," *Political Communication* 24, no. 4 (2007): 359–76.

23 Erving Goffman, *Strategic Interaction* (Philadelphia: University of Philadelphia Press, 1969).

24 Terranova, 16–17.

25 Ibid., 133.

26 Ted Gournelos, "Hacking the News," Paper presented at the National Communications Association Annual Conference (Chicago, IL, November 12, 2009).

27 Kate Kenny, "'the Performative Surprise': Parody, Documentary, and Critique," *Culture and Organization* 15, no. 2 (2009): 221–35.

28 Amber Day, "Are They for Real? Activism and Ironic Identities," *The Electronic Journal of Communication* 18, no. 2 (2008).

29 Terranova, 134.

30 Ibid., 142.

31 Henry Jenkins, "Politics in the Age of YouTube: A Transcript of a Public Conversation 2/10/09 at Otis College of Art and Design," *The Electronic Journal of Communication/La Revue Electronique de Communication* 18 (2008), http://www.cios.org/www/ejc/EJCPUBLIC/018/2/01848.html.

32 Terranova, 154.

Chapter 9

Wikipedia, Error, and Fear of the Bad Actor

Mark Nunes
Southern Polytechnic State University

For a growing number of individuals, Wikipedia has become the source of choice for encyclopedic knowledge. More often than not, Wikipedia entries appear at the top of search engine results, and according to Alexa, Wikipedia consistently tracks in the top ten most frequented sites.[1] This popularity, however, belies the fact that Wikipedia is presented so often in the popular press as an unreliable source. In newspapers and magazines, on television and blogs, the fact that Wikipedia entries are subject to editing by any individual serves as grounds for anxiety over the degeneration of scholarship, and more broadly, the collective body of knowledge. This anxiety reproduces itself in a concern that finds Wikipedia critics accused of elitism and Wikipedia supporters charged with a populist anti-intellectualism. For example, in his response to Susan Jacoby's essay on "The Dumbing of America,"[2] Wikipedia founder Jimmy Wales champions the contributions of amateurs within a collective knowledge base and praises Wikipedia for offering up "a place where today's youth, in phenomenal numbers, are helping professors and graduate students to build a repository of living knowledge."[3] But the strength that Wales cites is, in effect, Wikipedia's fundamental flaw from the perspective of institutional authority. Collective intelligence is all well and fine when we are discussing *Survivor* or *The Matrix* trilogy,[4] but what will become of "true" scholarship in this leveled world of peer-to-peer collective endeavor?

We are well within our second decade of an ongoing debate over the redefinition of scholarly labor by way of distributed networks. The terms of the debate have shifted somewhat, but the underlying question remains the same: to what degree do the leveling effects of these networks provide a "liberation" of knowledge from institutional fields of power? It is in this context that Wikipedia has emerged as the poster child for institutional anxiety over Web 2.0 knowledge communities.

Consider, for example, Michael Gorman's critique of Wikipedia as an attack on knowledge itself, posted, appropriately enough, as part of Encyclopedia Britannica's "Web 2.0 Forum." He writes:

> The aggregation of the opinions of the informed and the uninformed (ranging from the ignorant to the crazy) is decidedly and emphatically not "what is known about any given topic." It is a mixture of the known (emanating from the knowledgeable and the expert) and erroneous or partial information (emanating from the uninformed and the inexpert).[5]

At issue here is a question of the definition of scholarly labor itself, who performs that labor, and the means by which he or she gains access to a scholarly community. After all, much of the leveling presented as a democratization of knowledge runs contrary to the institutional structuring that grants scholarship its social capital. In contrasting collective intelligence with what Peter Walsh calls the "expert paradigm," Henry Jenkins argues that in place of a knowledge system that differentiates expert from novice through various processes of inclusion and exclusion, and which defines knowledge fields by these same processes of bounding and differentiation, collective intelligence operates across disciplinary boundaries in an ad hoc fashion with individuals participating in granular ways that allow knowledge structures to emerge and dissipate to meet a dynamic, interactive communal need.[6] These "knowledge communities" result from what Jenkins calls "voluntary, temporary, and tactical affiliation, reaffirmed through common intellectual enterprises and emotional investments."[7] Wikipedia celebrates the democratizing impulse of the wiki, working to establish non-hierarchical strengths that outweigh the light hierarchy of editorial access. In contrast, hierarchy is reinforced within traditional knowledge institutions as a mode of meritocracy and is therefore championed at a site like Britannica.com. In discussing its change in policy to "make it easy for our readers to suggest edits, revisions, updates, amplifications, and corrections," the Britannica blog is quick to add: "The operative word . . . is *suggest*. Britannica users don't have the ability or authority to *publish* the edits they propose; only Britannica editors can do that, and that's the way it will stay."[8] Likewise, while Wikipedia enthusiasts highlight the potential for collaborative, dynamic activity within emergent knowledge communities, Gorman explicitly critiques the very idea of *collective* intelligence, insisting that *authentic* scholarship is predicated upon not only authority, but the authorial: scholarly labor is an *individuated* activity—shared within a community of scholars, but never in itself a collective expression.[9]

Gorman and the Britannica editors are by no means alone in this anxiety over competing ideologies of scholarship and the labor associated with the production of knowledge. While one can point to numerous instances reported in the popular press of Wikipedia's failure to live up to its own expectations of accuracy and reliability (a number of which I will mention below), this anxiety over who can and cannot labor within a knowledge community was perhaps best captured and exploited to comic ends in the July 31, 2006 broadcast of *The Colbert Report*, in which Stephen Colbert defined "wikiality" as the ability to "create a reality that we all agree on—the reality we just agreed on," and then instructed his viewers to edit the Wikipedia entry on "Elephants" to reflect a tripling in population in the past six months. The humor arises, of course, from what is presumed to be Wikipedia's greatest strength—that anybody can add, delete, or edit content—and that anyone includes individuals acting in bad faith. Colbert's spoof calls attention to our assumptions about the relationship between knowledge production and a fact-based stable reality. He does not simply introduce an error into the collective body of knowledge; he does so with rhetorical intent, namely to "create a reality" that fits his political and social viewpoint (or more accurately, the viewpoint of his on-air persona). Error, in this instance an intentional error, does not merely undermine the reliability of a collective body of knowledge; it threatens "reality" itself with a rhetorical construct—"wikiality."

While "bad facts" are at issue here, I would argue that Colbert's spoof plays off of a more primal anxiety connected with Wikipedia: namely, a fear of the bad-faith actor. The presumed good that arises from the collaborative endeavor of collective intelligence assumes that participants are united in "common intellectual enterprises and emotional investments." Colbert's wikiality, in contrast, foregrounds the actions of individuals operating at cross-purposes to a collective aim—or at very least who are operating by a different set of rules than the rest of the community. Certainly the "intellectual commons" model for Wikipedia has a well-established precedent in the free and open source software movement and in the principles of Linus's Law, that given enough eyeballs, all bugs are shallow. We can fully accept the notion of well-intentioned individuals laboring in common toward a collective goal, but how do we reconcile the role of individuals who are not acting in good faith? As Frank Ahrens notes in a 2006 piece in *The Washington Post* (in language quite similar to that of Gorman), for all of our faith in the good will and good work of a knowledge community, we cannot escape from the gnawing anxiety that as a collective endeavor, Wikipedia articles "may be written by experts or insane crazy people. Or worse, insane crazy people with an agenda. And Internet access."[10]

These "insane crazy people" with an agenda and Internet access interest me greatly—not so much for the damage (sometimes clever, sometimes sophomoric) they bring to the information commons, but rather, for the role they play in signaling contested ideologies within networked social space. And there is an undeniably long list of Wikipedian bad actors.[11] During the 2004 presidential campaign, vandals[12] regularly attacked pages for John Kerry and George W. Bush, leading Wikipedia to lock both pages from further edits. In December 2005, the press began to report the story of John Seigenthaler, a retired editor for *The Tennessean* newspaper who was surprised to find out from reading his Wikipedia biography that he was "thought to have been directly involved in the Kennedy assassinations."[13] In 2007, the story broke that the highly regarded Wikipedia contributor Essjay, cited in a 2006 *New Yorker* piece as "a tenured professor of religion at a private university,"[14] was in fact 24-year-old Ryan Jordan, who was neither tenured nor a professor, with no academic credentials beyond a bachelor's degree. Other high profile instances of Wikipedian bad actors include corporations editing entries to eliminate unfavorable commentary or add proper spin, universities making page edits that blur the line between recruitment brochure and encyclopedia entry,[15] and contributors assuming multiple IP "sockpuppet" accounts to support their own opinions and create the appearance of community consensus. As recently as 2009, Wikipedia was (once again) considering revisions to its editorial policy in response to vandalism to the entry for Edward Kennedy, which reported his death nearly a year in advance of his actual passing.[16] While the anxiety associated with the bad actor arises from the harm he or she can bring to a fact-based understanding of reality, the basis for this anxiety runs deeper, I would argue, stemming from the willingness of this agent to play by a different set of rules. In each instance, the bad actor operates within the collective commons with an ulterior motive. In doing so, the bad actor does more than merely corrupt the facts; he or she threatens to undermine the basic altruistic assumptions that underlie the collaborative project itself.

In contrast to this widespread anxiety, however, the reality seems to be that most acts of vandalism are quickly corrected, and overall Wikipedia remains a rather functional collective.[17] Perhaps the real challenge of the bad actor, then, is in fact summed up by a figure like Stephen Colbert, to the extent that it is the lack of *scholarly intent* that creates concern. In his case against Wikipedia, Gorman foregrounds traditional scholarship as a form of meritocracy, one that is "based on respect for expertise and learning, respect for individual achievement, respect for true research, respect for structures that confer authority and credentials, and respect for the authenticity of the human record."[18] Clearly vandal and comedian alike are operating under a different set of rules. Note, though, that in Gorman's articulation of knowledge, authority is, in effect, the social capital of academia that validates a peculiar form of labor that we call scholarship. Within Wikipedia and other collaborative, peer-to-peer knowledge communities, authority itself is transformed with and by changes in social capital. As Axel Bruns has noted in his recent book, *Blogs, Wikipedia, Second Life and Beyond*,[19] meritocracies can and do exist within what he describes as "produsage" communities, a term he uses to describe a mode of production after the product; these meritocracies are, however, outside of institutional hierarchy, defined ad hoc and on the fly by one's ability to help advance the collective usage needs of a given community.[20] This meritocracy is very much based upon an assumption of good faith action, precisely what the purposeful errors of the bad actor attempt to undermine. In his review of literature on accuracy, reliability, and credibility on Wikipedia, Ryan McGrady comments on the degree to which "authority" in Wikipedia translates into a form of respect, but a respect that has less to do with claims to mastery of subject matter than with the ability to advance the collective project itself.[21] McGrady concludes that this sense of "spirit" is a primary structuring ideal that trumps rhetorical and authority-based claims to knowledge, creating its own form of meritocracy. While much of this spirit is captured in the concept of "WikiLove,"[22] McGrady pays particular attention to the social norms expressed in the entry for "Wikipedia: Gaming the System," a practice that involves "using Wikipedia policies and guidelines in bad faith, to deliberately thwart the aims of Wikipedia and the process of communal editorship."[23] Wikipedia is guided by a spirit of good faith actors toward a collective knowledge project. This collaborative spirit—one's willingness to co-labor as a good faith participant within a communal endeavor—provides distinction (in Bourdieu's sense[24]) and social capital within a social field of power. McGrady goes as far as to assert: "it doesn't matter what you read; if you're not working towards the common good, you are in the wrong."

Such a meritocracy asks us to redefine standards for "quality work," and not without some concern. Certainly there is an ongoing critique, both in the academic and the popular arena, of the quality of some articles written by Wikipedia authors. As Danah Boyd notes, Linus's Law does not hold too strongly for all articles, since a large number of pages are composed by a single author and undergo little or no revision by a multitude of eyeballs.[25] Likewise, in his sample study of 100 Wikipedia entries, John Willinsky found a scant 168 source citations, only 2 percent of which came from open access peer reviewed journals.[26] What's more, Paul Duguid notes that our faith in Linus's Law might be somewhat ill-placed: in the world of software coding, there is always the foundation of the code itself that serves as final arbiter—how well the application runs.[27] The same cannot be said within other knowledge domains. It is somewhat curious, then, that however justified these concerns may be, they are far outnumbered in the media by concerns related to vandalism, sockpuppetry, and other shenanigans of Wikipedian bad actors. Perhaps it should be expected that the media would pay particular attention to these more scandalous controversies since scandal makes for better news. But as Elizabeth Byrd notes, "scandal" provides an important window into a cultural moment by making clear the ideological fault lines within a society.[28] Following on James Carey's cultural analysis of communication, which emphasizes the community-enhancing structure of communication, rather than the information it transmits,[29] Byrd notes that scandals present a context for debating, and often reinforcing, communal standards for behavior by revealing scenes of moral contestation within a culture or subculture.[30] What, then, in the errant practices of the bad actor, garners it the attention of scandal inside and outside of academia, and does so within a context that recognizes the increasing penetration and dominance of Wikipedia as an information source?

We may start to answer this question by noting that what captures headlines is more often than not a matter of error, not noise. While there has been some commentary on the "noteworthiness" of entries in Wikipedia, there seems to be little *scandal* associated with the increasingly large field of information encompassed by Wikipedia's 15 million and growing number of entries.[31] The only thing wrong with noise is that there's too much of it. The loss of authority in a sea of noise is apparently something that we are OK with, as long as the search tool returns the requested page, or a close enough approximation. Error, however, poses a threat of a different order, precisely because an "errant result" holds the potential of leading us unwittingly astray. Unlike noise, error suggests (with or without a human actor) a form of agency that "escapes" systematic closure, or worse still, that tricks us into unintended outcomes.

It's why Google's millions of hits are not newsworthy in and of themselves, whereas "Google bombing" and gaming the algorithm are. Error
retains if not authority per se then agency—the ability of an individual
act or a singular event to subvert the ruling order and disrupt systematic closure. But this offers only part of the answer, for when we look to
popular accounts of the failures of Wikipedia, it is clearly not the case
that all errors trigger the same degree of scandal. One finds instead a
disproportionate amount of attention paid to the intentional errors of
bad actors over what should perhaps be of equal concern, namely the
actions of the ill-informed or naively informed good faith amateur who
has muddied up the pool of collective knowledge with half-truths and
platitudes. As with the Colbert spoof, being wrong in fact (a tripling
in the elephant population) in and of itself is far less scandalous than
refusing to play by the rules—or worse yet, gaming the rules toward one's
own ends, rhetorical or otherwise ("if you're not working towards the
common good, you are in the wrong").

This distinction between error as "wrong fact" and the error of "being
in the wrong" parallels Wiener's distinction between the Augustinian evil
of entropy and the Manichaean evil of purposive opposition. Here it is
worth quoting on the subject somewhat at length from *The Human Use of
Human Beings.* Wiener writes:

> The Manichaean devil is an opponent, like any opponent, who is deter-
> mined on victory and will use any trick or craftiness or dissimulation to
> obtain victory . . . On the other hand, the Augustinian devil, which is
> not a power in itself, but the measure of our own weakness, may require
> our full resources to uncover, but when we have uncovered it, we have
> in a certain sense exorcised it, and it will not alter its policy on a matter
> already decided with the mere intention of confusing us further.
>
> Compared to this Manichaean being of refined malice, the Augus-
> tinian devil is stupid. He plays a difficult game, but he may be defeated
> by our intelligence as thoroughly as by a sprinkle of holy water.[32]

Control systems can correct errors of wrong fact—after all, feedback
systems depend upon "error handlers" to capture and contain these
sorts of events. Error as strategy or tactic—as "trick or craftiness or
dissimulation"—presents a challenge of an entirely different order. To
put it in terms of control systems: the Augustinian error, once captured
as feedback, will ultimately serve order's purpose; the Manichaean error,
in contrast, follows its own purpose. While Wiener is concerned with the

"difficult game" posed by Augustinian error, compared with the Manichean devil, who is "playing a game of poker against us and will resort readily to bluffing,"[33] the Augustinian devil's game is no *game* at all.

Wiener's mention of games in this context is by no means accidental. Gaming is inherent in the tricks and wiles of the bad actor, and von Neumann's game theory figures prominently throughout Wiener's discussion of purposive resistance. Error captured serves the purpose of a control system by providing feedback on the gap between intention and outcome. Error uncaptured marks a potential for a slippage toward nature's entropic tendencies. Error exploited, however, is *error agent.* Wiener returns to this subject at the very close of *The Human Use of Human Beings* in a chapter entitled "Language, Confusion, and Jam," again with explicit reference to von Neumann. Here Wiener is particularly interested in distinguishing between the scientist's attempt to wring order out of the passive resistance of nature and the communicative challenge involving two competing systems—one dependent upon a cooperation of participants (sender and receiver) and the other intent on jamming communication.[34] This chapter is haunted, so to speak, by the specter of the counteragent who acts at cross-purposes to a system of order. This is a challenging problem for Wiener to the degree that *purpose* is a defining characteristic for his understanding of order as a "temporary and local reversal" in entropy;[35] here, however, the counteragent presents the possibility of purpose-directed behavior in the service of failure and disorder. In part, this chapter serves as Wiener's attempt to dispel the Manichaean, ultimately by arguing that from the viewpoint of science, one must deny *as an act of faith* the place of purposive resistance in the quest for order. He writes:

> I have implied that Manichaeanism is a bad atmosphere for science. Curious as it may seem, this is because it is a bad atmosphere for faith. When we do not know whether a particular phenomenon we observe is the work of God or the work of Satan, the very roots of our faith are shaken.[36]

Faith provides a condition for science—not religious faith, Wiener insists, but rather a "faith that nature is subject to law."[37] The relationship of faith and science to "the law" is ultimately a faith in all actors abiding by the law, and more importantly, a faith that the order inherent in the law will reveal itself to all good faith actors.

Note, however, that for Wiener to dispel this Manichaean anxiety over the bad faith actor who would exploit the law toward his or her own ends, he must bracket off a wide range of rhetorical engagements in which operating at cross-purposes is in fact the norm. Cybernetics can handle the Augustinian "enemy" of *confusion* in what he terms "normal communicative discourse"; things get trickier, however, within "forensic discourse . . . such as we find in the law court in legislative debates and so on," which presents "a much more formidable opposition, whose conscious aim is to qualify and even to destroy its meaning."[38] Wiener is willing to acknowledge the role of game-playing in communication, but he insists on establishing two distinct language games: "one of which is intended primarily to convey information and the other primarily to impose a point of view against a willful opposition."[39] Anxiety arises, however, from the fact that one cannot easily bracket off the agonism inherent in forensic discourse from a presumed "normal" mode of communication "intended primarily to convey information." While Colbert may push the point to a parodic extreme, the challenge of his game is real: clearly the pursuit of knowledge and the sharing of that knowledge is rarely (if ever) free from strong rhetorical considerations. Wiener, rather than attempting to address this inherent agonism, suggests instead a means of extending "normal communicative discourse" and cybernetic control principles to a wider range of social arenas, ultimately arguing for an ideal legal system in which "the problems of law may be considered . . . problems of orderly and repeatable control of certain critical systems."[40] This "orderly and repeatable control" must capture any gap between intended and actual outcome in any system, legal or otherwise, for this gap marks an opportunity for the Manichaean exploit: "a no-man's land in which dishonest men prey on the differences in possible interpretation of the statutes."[41] Note that in Wiener's rendering of problem and solution, there is very little distinction between the malice of the bad actor and the agonistic positioning of the lawyer or the rhetoritician. In each instance, "the system" depends upon the collaboration of individuals united in cooperative purpose and is threatened by those practices that seek to exploit "the interstices of the law" toward other purposes and other outcomes.[42]

To some extent, then, the bad actor raises the same anxiety for Wiener as for critics of Wikipedia—by introducing the possibility of a purpose outside of purpose, and an agent willing and able to game the system. In the case of Wikipedia, advocates cannot as easily deny the existence of the Manichaean in the fashion of Wiener (the evidence is overwhelming that they edit among us). But as with Wiener's call to faith, documents such as "Wikipedia:WikiLove" and "Wikipedia:Gaming the System" strongly suggest that this community sustains itself through an act of faith in the collective process itself. In many instances, this is less a faith in human altruism than in system dynamics and the law of large numbers. As Bruns notes, collective intelligence projects are always playing a numbers game to the extent that one can anticipate a probabilistic increase in overall quality of knowledge, even when local problems exist.[13] Clay Shirky makes a similar argument in *Here Comes Everybody*: "A Wikipedia article is a process, not a product, and as a result, it is never finished . . . Wikipedia assumes that new errors will be introduced less frequently than existing ones will be corrected. This assumption has proven correct; despite occasional vandalism, Wikipedia articles get better, on average, over time."[44] In this regard, Wikipedia operates as a kind of community of inquirers as described by Charles Sanders Peirce (and as discussed in the Introduction to this collection) to the extent that "truth" is set at an infinite remove from a process of ongoing and collective approximation. Faith in Wikipedia as a collective project is ultimately a faith in a version of Peirce's Fallibilism, in that one must start from the assumption that whatever ideal statement of truth exists on a given topic, it exists *outside of* the inherently inaccurate approximation of the collective. Contrary to the truth claims implicit in an understanding of encyclopedic knowledge as a compendium of "what is known," such a perspective instead insists that Wikipedia need only provide an accurate and reliable representation of the state of knowledge *at a given moment*, not the "final word" on any subject. This concept is captured in founder Jimmy Wales's reference to Wikipedia as "a repository of living knowledge." Quoting Wikipedia's guideline for inclusion as "verifiability, not truth," Bruns likewise argues that in contrast to traditional models for encyclopedias, Wikipedia focuses on *representations* of knowledge, not presentations of a stable truth.[45] In Bruns's terms, it is a "produser's" project; without the focus on finished product, Wikipedia (and other flavors of user-based information commons) "always represents only a temporary artifact of the ongoing process, a snapshot in time which is likely to be different again the next minute, the next hour, or the next day."[46] It would seem, then, that what separates Gorman's critique of collective intelligence from Bruns's (and others') more supportive analysis is less a matter of how they view the damages posed by a bad actor than in how willing each is to assume a good faith belief not only in the *current* validity of a given representation of knowledge (how closely it fits in its approximation of truth), but also in the self-correcting workings of the system that represents that knowledge.

It may be too strong a claim to suggest that Wikipedia, as a community of inquirers, *advances* knowledge in Peirce's sense (given, for example, the emphasis Peirce places on scientific inquiry[47]). At the same time, Peirce's concept of Fallibilism provides meaningful insight into how Wikipedia *structures* knowledge within collective activity, and why the errant practices of the bad actor serve as a scandal and scene for a collective anxiety. Knowledge advances within the framework of Fallibilism when our belief in a systematic account of truth falls into doubt, which serves as an *irritation* (in Peirce's words) to our sense of certainty. Inquiry, Peirce maintains, is the means by which we reveal error (in both method and result) in order to reestablish belief in a truth and relieve the "irritation of doubt."[48] As a Web 2.0 parallel, we might think of Peirce's irritation of doubt as a motivation to participate within a collective endeavor—what Henry Jenkins describes as a "provocation" to its users.[49] In the case of Wikipedia, error—or more generally, doubt that a given entry presents no room for improvement—serves a good faith purpose by provoking participation. Within this scene of collective action, fallibility provides the basis for a systematic strength, but only to the extent that this community can sustain a collective belief in "the spirit of Wikipedia." This system can handle error as long as it can assume a shared purpose. Much as Wiener states, Manichaeism is apparently bad atmosphere for collective intelligence. The central role of fallibility in the production of a living repository of knowledge forces the community to fall back upon "the spirit of Wikipedia" not in spite of the possibility for error, but in good faith response to this possibility to the extent that these "failures" provoke participation within (and perpetuation of) a *cooperative* collective action.

What is one to do, then, given the presence of the bad actor, who instead of playing by the rules exploits opportunities to game the system? Can a systems-based, probabilistic view of purposive error overcome the challenge posed by the counteragent?

A similar problem faces Peirce's Fallibilism and his insistence on an ideal truth. As Joseph Margolis notes, if Peirce is to maintain a fallibility in all individual measures, and at the same time insist that the approximations of a community of inquirers provide a pattern of fit that transcends the failures of individual measures, where does one turn to find a measure outside of human fallibility that will insure a teleological movement toward increasingly better approximations of an ideal Truth?[50]

What is to insure, in other words, that the irritation of doubt will yield a *better* belief in truth rather than simply a *different* belief in truth? Margolis's concern, from an epistemological point of view, is that one is left with nothing more substantial than faith in that progress. In order to dispel the constructivist suggestions of his own theory of knowledge, Peirce falls back on a transcendental "agapism"—a hypothesis suggesting that "evolutionary love" drives all actions toward ideal expression.[51] Wikipedia seems trapped in a similar epistemological bind and offers as an escape a similar epistemology of faith in the spirit of Wikipedia—an epistemology that is actively undermined by the singular actions of the bad actor. Is WikiLove, then, merely Agapism 2.0?

As with Wiener's cybernetics, theories of collective intelligence seem well-positioned to handle Augustinian provocations. It is the Manichean provocation, however, that jams the works. And as with Wiener, the attempt to control for errant practice by citing some version of the law of large numbers ("vandals are hopelessly outnumbered by invested editors who want to see Wikipedia succeed"[52]) may be nothing more than an attempt to subsume purposive error within the passive, resistant drift of entropy. But clearly not all errant provocations serve the "creative evolution" of a collective intelligence. Users of the Internet have long experience with provocation in the form of "trolling," defined as off-topic, offensive, or otherwise disruptive contributions "with the primary intent of provoking other users into an emotional response."[53] As the oft-repeated advice within online communities suggests—*do not feed the trolls*—netiquette dictates that the only appropriate response to bad faith provocation is none at all. But again, such a perspective attempts to deny the possibility of an inherent agonism through supposedly clear distinctions between right and wrong action within a collective. Of course, not all flame wars start as a result of trolling—and in many instances, trolls and vandals are not born, but made.[54]

As might be expected, opportunities for bad faith action present themselves on Wikipedia most readily in instances where rhetorical stakes are highest—most recently, for example, on pages related to "climate change." At the start of 2010, in response to a range of actions by contributors that were reported as inappropriate, an editorial process involving Wikipedia community members resulted in these pages being placed on "article probation"; now, any editor perpetrating "disruptive edits, including, but not limited to, edit warring, personal attacks, incivility and assumptions of bad faith" will be sanctioned.[55] While reinforcing a "spirit of Wikipedia," such actions might also present, one might argue, provocations in their own right for further gaming of the system toward mischievous ends (You think *that* was disruptive? Well, watch *this*). Note also in the final principle of the terms for sanction, as with Wiener's dispelling of the specter of Manichaeanism, one is called upon not only to act in good faith, *but to assume good faith in the actions of others.* Such an approach is seen as a systematic strength that will allow differences of opinion to coexist. In doing so, however, it denies the *inherent* place of purposive resistance within a participatory culture, in effect taming all forms of agonistic engagement as expressions of rational disagreement—some of which unfortunately fall into the (Augustinean) error of straying too far from reason. Bruns notes, for example, that the "neutral point of view" policy at Wikipedia serves as an assurance that opposing perspectives sustain themselves in controversial issues, resulting in a "recasting of otherwise often bitter argument as a competition for discursive leadership without resorting to destructive means."[56] He goes on to note that while "destructive means" of engagement do occur, such events are short lived and ultimately counterproductive to the purposes of the offender. But note that Bruns and other commentators treat this bad actor and the errant practices that he or she engages in as outlying aberrations within the larger communal good, one that we are best off ignoring. Can we easily bracket off these practices in the name of championing collective intelligence? This becomes a particularly problematic move when we acknowledge just how widespread vandalism is as a provocation for participation, regardless of how short-lived the damage—not to mention the prevalence of a set of values in a range of online communities that valorizes hacks, exploits, and other opportunities for unintended "wins." Yes, one can play by the rules, but there is likewise a similar valorization of (and pleasure in) the ability to find and exploit a cheat, allowing one to operate outside of a system's protocological limits.

Gaming the system, for all the harm it presents to the collective endeavors of a project such as Wikipedia, likewise marks a potential in its own right and a locus for intentional/ rhetorical interventions beyond programmatic purpose. Shirky and others maintain a system-based view of Wikipedia, insisting on a probabilistic, long-term increase in quality. The scandal and anxiety over the actions of the bad actor, however, suggests that we are not entirely sure if this faith is warranted, *or even "good."* Perhaps, then, the bad actor offers a kind of dark wish fulfillment; if one individual can disrupt this networked collective, then perhaps there is still agency in the singular event.

Following on Fuller and Goffey's call for an "evil media studies,"[57] we might find in the errant practices of the bad actor a set of conditions that reveal a strategy of the object. Here, intention does indeed come into play, but now "bad" strategy takes the form of an action that plays with the rules themselves. Clearly the errant practice of the bad actor in and of itself offers nothing that is essentially external to the operations of a collective intelligence. Consider, for example, another wiki-based knowledge community: Encyclopedia Dramatica, in which a community of bad actors assemble in the spirit of "lulz": to produce humor without point (simply "for the lulz"), or to produce humor at another's expense. This is not to downplay the real harm at times inflicted on individuals by some participants within this community (as Mattathias Schwartz explains in a *New York Times* piece on trolling subcultures, "lulz is how trolls keep score"[58]), but it does call attention to a structuring of collective intelligence in which strategies of misdirection play a role in the *success* of a community. How, then, do we come to understand the functioning of this form of collective intelligence in which disruption is the rule, rather than exception? While we might find in "lulz" the sorts of "common intellectual enterprises and emotional investments" that definitions of collective intelligence assume, it would also seem that our theories of new media practices tend to assume a set of "good practices" with an eye toward (evolutionary) improvement, be that in politics, in social networking, or in grassroots scholarship. To the extent that an information commons must inevitably assume a spirit of "common good," it is structurally foreclosing those actions, engagements, and contributions that are inherently agonistic to a protocologically and ideologically determined understanding of *purpose*. If error teaches us something of the potential of potential, it does so in a fashion that does not permit a foreclosure of "good" and "bad" practices.

And here, perhaps, is where control systems fail to account for the productive possibility of an inherent agonism. Wikipedia would not be "better off" without its bad actors, no more than one can imagine Encyclopedia Dramatica a worse place for participants failing to abide by a spirit of lulz. Be it WikiLove or "doing it for the lulz," these collectives remain *vital* to the extent that participants are provoked into a collaborative process, *for purposes that are not always foreclosed by purpose*. Anxieties aside (be they Peirce's resistance to constructivism or Wiener's fear of cross-purposes), Fallibilism in new media cultures suggests something more potent than steady progress toward a final, stable state. Rather, it speaks to dynamic potential itself. While Wiener attempts to banish his own Manichaean dread, the errant practices of the bad actor suggest that networks articulate a mode of agency that cannot be captured by "the good" of order, or the "evil" of aberration that are implicit within the cybernetic imperative. Wikipedia policies may condemn gaming the system as a mode of "being in the wrong," but it would seem that the collective project succeeds only to the degree that it fails to exclude its errant practices—the tricks, wiles, and seductions that keep "the commons" from a fatal closure of program, protocol, and purpose.

Notes

[1] http://www.alexa.com/topsites. As of February, 2010, Wikipedia ranked #6.
[2] Susan Jacoby, "The Dumbing of America," *The Washington Post*, February 17, 2008, http://www.washingtonpost.com/wp-dyn/content/article/2008/02/15/AR2008021502901.html?sid=ST2008021801642.
[3] Jimmy Wales, "We're Smarter than You Think," *The Washington Post*, February 19, 2008, http://www.washingtonpost.com/wp-dyn/content/article/2008/02/18/AR2008021801248.html?hpid=opinionsbox1.
[4] To cite two examples of collective intelligence in a participatory culture, discussed at length in Henry Jenkins, *Convergence Culture* (New York: New York University Press, 2006).
[5] Michael Gorman, "Jabberwiki: The Educational Response, Part II," *Britannica Blog*, June 26, 2007, http://www.britannica.com/blogs/2007/06/jabberwiki-the-educational-response-part-ii/.
[6] Ibid. 51–54.
[7] Ibid. 27.
[8] "Is Britannica Going Wiki?" *Britannica Blog*, March 8, 2009, http://www.britannica.com/blogs/2009/03/is-britannica-going-wiki/.
[9] Michael Gorman, "Web 2.0: The Sleep of Reason, Part II," *Britannica Blog*, June 12, 2007, http://www.britannica.com/blogs/2007/06/web-20-the-sleep-of-reason-part-ii/.
[10] Frank Ahrens, "Death by Wikipedia: The Kenneth Lay Chronicles," *The Washington Post*, July. 9, 2006, F07, http://www.washingtonpost.com/wp-dyn/content/article/2006/07/08/AR2006070800135.html.
[11] For extended discussion of many of these highly publicized scandals, see (appropriately enough) the Wikipedia entry for "Criticism of Wikipedia." http://en.wikipedia.org/wiki/Criticism_of_Wikipedia.
[12] It has become common parlance both within Wikipedia and in the popular press to describe these errant practices as vandalism, be they on the order of deleting entire entries, inserting inaccurate or insulting material, or otherwise engaging in informatic mischief. The assumptions of information as property inherent in this usage of the term are worth further exploration but exceed the purposes of this chapter.
[13] John Seigenthaler, "A False Wikipedia 'Biography,'" *USA Today*, November 29, 2005, http://www.usatoday.com/news/opinion/editorials/2005-11-29-wikipedia-edit_x.htm.

[14] Stacy Schiff, "Know it All: Can Wikipedia Conquer Expertise?" *The New Yorker*, July 31, 2006, http://www.newyorker.com/archive/2006/07/31/060731fa_fact.

[15] J. J. Hermes, "The Wiki Watcher," *Chronicle of Higher Education*, October 19, 2007, A6.

[16] Noam Cohen, "Wikipedia May Restrict Public's Ability to Change Entries," *New York Times*, January 23, 2009, http://bits.blogs.nytimes.com/2009/01/23/wikipedia-may-restrict-publics-ability-to-change-entries/?pagemode=print.

[17] See, for example, P. D. Magnus, "Early Response to False Claims in Wikipedia," *First Monday* 13 no. 9 (September 1, 2009), http://firstmonday.org/htbin/cgiwrap/bin/ojs/index.php/fm/issue/view/269.

[18] Gorman, "Jabberwiki."

[19] Axel Bruns, *Blogs, Wikipedia, Second Life, and Beyond: From Production to Produsage* (New York: Peter Lang, 2008).

[20] Ibid., 25–26.

[21] Ryan McGrady, "Gaming Against the Greater Good," *First Monday* 14, no. 2 (2 Feb. 2009), http://firstmonday.org/htbin/cgiwrap/bin/ojs/index.php/fm/article/view/2215/2091.

[22] Defined as "a general spirit of collegiality and mutual understanding among wiki users." See "Wikipedia:WikiLove," *Wikipedia*, http://en.wikipedia.org/wiki/Wikipedia:Wikilove.

[23] McGrady.

[24] See, for example, Pierre Bourdieu, *The State Nobility* (Stanford, CA: Stanford University Press, 1998), 117–118.

[25] Danah Boyd, "Academia and Wikipedia," *Many2Many* January 4, 2005, http://many.corante.com/archives/2005/01/04/academia_and_wikipedia.php.

[26] John Willinsky, "What Open Access Research Can Do for Wikipedia," *First Monday* 12, no. 3 (Mar. 5 2007), http://firstmonday.org/htbin/cgiwrap/bin/ojs/index.php/fm/article/view/1624/1539.

[27] Paul Duguid, "Limits of Self-organization: Peer Production and 'Laws of Quality.'" *First Monday* 11, no. 10 (October 2, 2006), http://firstmonday.org/htbin/cgiwrap/bin/ojs/index.php/fm/article/view/1405/1323.

[28] S. Elizabeth Bird, *The Audience in Everyday Life* (New York: Routledge, 2003).

[29] James Carey, "A Cultural Approach to Communication," *Communication as Culture* (New York: Routledge, 1989), 14–36.

[30] Bird, 48.

[31] "List of Wikipedias," *Wikimedia*, http://meta.wikimedia.org/wiki/List_of_Wikipedias.

[32] Norbert Wiener, *The Human Use of Human Beings: Cybernetics and Society* (New York: Da Capo, 1988), 34–35.

[33] Ibid., 35.

[34] Ibid., 187.

[35] Ibid., 25.

[36] Ibid., 192.

[37] Ibid., 193.

[38] Ibid., 93.

[39] Ibid., 93.

[40] Ibid., 110.

[41] Ibid., 108.

42 Ibid., 111.
43 Bruns, 131–132.
44 Shirky, 119.
45 Bruns, 114.
46 Ibid., 28.
47 See, for example, Charles S. Peirce, "How to Make Our Ideas Clear," *Popular Science Monthly* 12 (January 1878): 286–302, http://www.peirce.org/writings/p119.html.
48 Charles S. Peirce, "The Fixation of Belief," *Popular Science Monthly* 12 (November 1877): 1–15, http://www.peirce.org/writings/p107.html.
49 Henry Jenkins, "Astroturf, Humbugs, and Lonely Girls," *Confessions of an Aca-Fan*, September 13, 2006, http://www.henryjenkins.org/2006/09/astroturf_lonely_girls_and_cul.html.
50 Joseph Margolis, "Peirce's Fallibilism," *Transactions of the Charles S. Peirce Society* 34 (1998): 535–569.
51 Joseph Margolis, "Rethinking Peirce's Fallibilism," *Transactions of the Charles S. Peirce Society* 43 (2007): 246.
52 McGrady.
53 "Troll (Internet)," *Wikipedia*, http://en.wikipedia.org/wiki/Troll_%28 Internet%29.
54 See, for example, discussions of edit reverts and retaliation in "Criticism of Wikipedia," especially the subsection on Social Stratification.
55 "Wikipedia: General sanctions/Climate change probation," *Wikipedia*, http://en.wikipedia.org/wiki/Wikipedia:General_sanctions/Climate_change_probation.
56 Bruns, 120.
57 Matthew Fuller and Andrew Goffey, "Toward an Evil Media Studies," in *The Spam Book*. ed. Jussi Parikka and Tony D. Sampson (Cresskill, NJ: Hampton, 2009), 141–159.
58 Mattathias Schwartz, "The Trolls Among Us," *The New York Times*, August 3, 2008.

Jam

Chapter 10

Contingent Operations: Transduction, Reticular Aesthetics, and the EKMRZ Trilogy

Michael Dieter
University of Melbourne

There is no diagram that does not also include, besides the points it connects up, certain relatively free or unbounded points, points of creativity, change and resistance, and it is perhaps with these that we ought to begin in order to understand the whole picture.

Gilles Deleuze[1]

UBERMORGEN.COM's work is unique not because of what we do but because how, when and where we do it. The Computer and The Network create our art and combine every aspect of it.

Hans Bernhard[2]

The EKMRZ Trilogy (2005–2008) is a series of artworks by UBERMORGEN. COM (Hans Berhard and Lizvlx), in partial collaboration with Paolo Cirio and Alessandro Ludovico, produced by "targeting" several major corporations specializing in online services: Google, Amazon, and eBay. While each project differs thematically, they are all formed by a common technique based on channeling flows of networked data gleaned from the materialities of the actual sites and remolding them into alternative configurations. For instance, *Google Will Eat Itself* (2005) was created by collecting revenue through Google's AdSense software and investing in shares of the company by an automated program; *Amazon Noir* (2006) harnessed vulnerabilities in the online retailer's "Search Inside This Book" software to assemble full versions of texts and host them on peer-to-peer filesharing networks; and finally, *The Sound of eBay* (2008) generated electronic music from user-data supplied from the goods trading site. Taken together, the works are notable as investigations of informational distribution and the extraction of value from Internet-enabled services. Moreover, they mark an important transition in Internet art from an earlier phase of seemingly unfettered exploration; as Florian Cramer observes, "it is the net.art of an Internet that is no longer an open field of experimentation, but a corporate space."[3]

One possible reading of the trilogy can be developed through an emphasis on aberrant or tactical resistance to commercial presence on digital networks ("EKMRZ" being short for e-commerce). However, to the extent these projects are concerned with the economic dimensions of network services, their political content is far from self-evident. On the contrary, *Google Will Eat Itself*, *Amazon Noir*, and *The Sound of eBay* should be understood foremost as an exploration of already existing socio-technical controversies regarding commercial presence on digital networks.[4] These issues arise through processes of info-valorization, whether in the case of intellectual property, the filtering and aggregation of content, or the machinations of user profiling. That the projects have been created via digital entities or machinic processes (operating on the very terrain where exchange is differentiated), and that the outcomes are framed by satirical modes of narration by the artists, conveys something of the indeterminacy or immensurability of these events. *The EKMRZ Trilogy*, in this respect, could be termed as what Matthew Fuller and Andrew Goffey describe as "evil media studies"—an exploratory approach to the trickery, deception, and manipulation performed by object-orientated agencies. Evil is used in a semi-ironic sense, mainly because this mode of working suggests dimensions that are badly understood from the perspective of "autonomous rationality" and "the ideal of knowledge." More specifically, their proposal calls for the recognition of nonrepresentational forces and "the strategies of the object, for what things do in themselves without bothering to pass through the subjective demand for meaning."[5] If the work of Bernhard and Lizvlx can be defined by this category of evil, it would be as pragmatic studies of the necessary contingency and agonism of networked media. The projects, accordingly, are co-emergent with the "problematic" qualities of digital networks as socio-technical ensembles. Through the philosophy of Gilbert Simondon, this chapter argues for a shift toward thinking about communication networks in an ontogenetic register and accounting for the more-than-human agencies in these contexts. Such a reassessment suggests that "tactical media" projects should not be understood exclusively as "jamming" media, but as exploring ways of taking account for the already existing tensions held together and perpetuated by network uncertainties.

Software: Emeragency and Agential Mess

As a force of distribution, the Internet is continually subject to controversies and uncertainties concerning flows and permutations of agency. While often directed by discourses cast in terms of either radical autonomy or control, the technological form of networked digital systems is more regularly a case of establishing structures of operation, codified rules, or conditions of possibility—that is, of guiding socio-material processes and relations.[6] Software, as a medium by which communication unfolds and becomes organized, is difficult to theorize as a result of being so event-oriented. There lies a complicated logic of contingency and calculation at its center, an aspect exacerbated by the global scale and intensive complexity of informational networks, where the inability to comprehend an environment that exceeds the limits of individual experience is frequently expressed through anxieties and paranoia. Somewhat unsurprisingly, cautionary accounts and moral panics on the dark side of identity theft, email fraud, illicit pornography, surveillance, hackers, and computer viruses are as commonplace as those narratives advocating user interactivity and empowerment.

When analyzing digital systems, cultural theory often struggles to describe forces that dictate movement and relations between disparate entities composed by code, an aspect heightened by widespread flows of informational networks where differences are expressed through a constant exposure to contingency.[7] Beginning to understand this problematic landscape is an urgent task, especially in relation to the technological and software-based dynamics that organize and propel such tensions. Certainly, any conceptual or critical perspective on Internet distribution requires an acknowledgement of the nonhuman relations that make such exchanges durable, even if contentious.[8] It requires an understanding of the permutations of agency carried along by digital objects. Software studies has proven an insightful response to this challenge as an attempt to provide new methods and vocabularies devised in dialogue with the nomenclature of computer science and the histories of computing.[9] While stressing the materialities or "stuff" of software, this emerging field has additionally emphasized the messy arrangements that accumulate as code intersects, bleeds into, and couples with other realities.[10] The volatility of these dynamics has equally been registered in a recent turn to logistics in media theory, as once durable networks for constructing economic difference— organizing information in space and time ("at a distance"), accelerating, or delaying its delivery—appear contingent, unstable, or consistently irregular.[11] Attributing actions to users, programmers, or the software is an increasingly difficult task when facing these entanglements, especially in the context of the aggregate behaviors through which network value emerges. These kinds of dynamics are the source material of *The EKMRZ Trilogy*. By drawing from these complexities, the projects touch on widespread concerns with free labor, the illegal distribution of intellectual property, and the processes by which user behaviors are surveyed, accumulated, and recompiled. This chapter suggests that the artistic work of UBERMORGEN.COM is congruent with such tensions, rather than being the source of malevolent errors or disturbances. *Google Will Eat Itself, Amazon Noir*, and *The Sound of eBay*, accordingly, offer generative recompositions of the issues at stake by connecting with the material uncertainties of networks.

Whether or not, following Friedrich Kittler, we agree that software exists as such, it is clear that the technological actions facilitated through programmable media coordinate socioeconomic forces and allow for certain agencies to emerge, even if understood as a cascade of affordances involving "local string manipulations" and "signifiers of voltage differences."[12] A frequent response to the question of software as a force of agency is an approach whereby the composition of a digital object is framed as a condensation of social relations.[13] For instance, according to Lawrence Lessig's influential argument, code is not merely an object of governance, but has an overt legislative function itself. Within the context of software, "a law is defined, not through a statue, but through the code that governs the space."[14] These points of symmetry are understood primarily as concretized social values: they are standards that regulate flow. Similarly, Alexander Galloway describes computer protocols as non-institutional "etiquette for autonomous agents," or "conventional rules that govern the set of possible behavior patterns within a heterogeneous system."[15] In his analysis, these homogeneous actions operate as a style of management fostered by contradiction: progressive though reactionary, encouraging diversity by striving for the universal, synonymous with possibility but utterly predetermined, and so on.[16] Protocols are, as a result, constructed for making data exchange durable by allowing for heterogeneous formations in terms of content and context. Galloway's theory outlines how network infrastructures exist as concentrations of power, but also suggests that these standards contribute to different individuations at alternating scales and relations. Uncertainties, therefore, emerge when the protocological as a set of possibilities operates in an assemblage. From the perspective of Wendy Hui Kyong Chun, the constant twinning of freedom and control at the core of these Cold War systems is the source of this political ambiguity: a problematic that is further exacerbated by the tendency in computer science to reduce social issues to questions of technological optimization.[17] In confrontation with these seemingly ubiquitous regulatory structures, there is a need for a framework of power distinguished from computer engineering and communication theory to account for the controversies that permeate through and concatenate these socio-technical realities.

In the influential short essay "Postscript on the Societies of Control," Deleuze famously described the historical passage from modern forms of organized enclosure (the prison, clinic, factory) to the contemporary arrangement of relational apparatuses and open systems as being materially provoked by—but not limited to—the mass deployment of networked digital technologies. In this frequently referenced analysis, the disciplinary mode described by Foucault is spatially extended to informational systems based on code and flexibility. For Deleuze, these cybernetic machines are connected into apparatuses that aim for intrusive monitoring: "in a control-based system nothing's left alone for long."[18] Such a continuous networking of behavior is described as a shift from "molds" to "modulation," where controls become "a self-transmuting molding changing from one moment to the next, or like a sieve whose mesh varies from one point to another."[19] Contemporary crises underpinning civil institutions are interpreted as consistent with the generalization of disciplinary logics across social space, forming an intensive modulation of everyday life, but one now ambiguously associated with complex socio-technical ensembles. The nature of this epistemic shift is significant in terms of political agency and legitimacy: while control implies an arrangement capable of absorbing massive contingency, for Deleuze, a series of instabilities mark its operation. *Noise, viral contamination,* and *piracy* are identified as key points of discontinuity; they appear as divisions or errors ("counter-information") that force change by promoting indeterminacies in a system that would otherwise appear infinitely calculable, programmable, and predictable.

Deleuze's "postscript" on control has proven hugely significant in debates in new media studies by introducing a series of central questions on power (or desire) and networked relations. As a social diagram, however, control should be understood as a partial rather than totalizing map of relations, referring to the augmentation of disciplinary power in specific socio-technical settings. While control functions as a conceptual motif that refers to open-ended territories beyond the architectural locales of enclosure, implying a move toward informational networks, data solicitation, and cybernetic feedback, there remains a contingent quality to its limits. There is a specifically immanent and localized dimension to its actions that might be taken as exemplary of control as a modulating affective materialism. The outcome is a heightened sense of bounded emergencies that are flattened out and absorbed through reconstitution; however, these are never linear gestures of containment. As Tiziana Terranova observes, control operates through multilayered mechanisms of order and organization: "messy local assemblages and compositions, subjective and machinic, characterised by different types of psychic investments, that cannot be the subject of normative, pre-made political judgments, but which need to be thought anew again and again, each time, in specific dynamic compositions."[20] This notion of stabilizing contingencies can also be related back to the governmental regime and the apparatus of security posited by Foucault, particularly insofar as both are aimed at establishing a milieu and guiding the events that traverse it toward a variety of ends.[21] The vitalistic productivity of such network topology accounts for the ambitions of projects aimed at sculpting communication channels through transversal targeting. *The EKMRZ Trilogy*, for that reason, is pitched specifically against the materialities of communication: the projects both resonate with and differentiate the emergence of socio-technical issues through their immanent taking-form.

Engaging with questions of network value, ownership, and control, the work of UBERMORGEN.COM is characterized by finding niches or uncertainties between overlapping milieus of control, governance, and economics. The name of the group—translated either as "the day after tomorrow" or "super-tomorrow"—conveys their interest in exploring potential scenarios that can develop out from open-ended or problematic formations. Tokyo-based curator Yukiko Shikata offers the neologism "emeragencies" to describe the particular kind of activity that is harnessed in these conditions by referring to a specific mode of agency that occurs in the conjunctures of network uncertainties.[22] A key example is the UBERMORGEN.COM project *[Vote]-Auction* (2000) that offered a web site for citizens of the United States to sell their vote in the federal election to anyone with Internet access. For Shikata, this project is noteworthy for "highlighting the variance between the global economy and domestic politics accelerated by the Internet that eventually stirred up controversy through mass media publicity and the intervention of the

FBI."[23] In fact, the controversial aspects of *[Vote]-Auction* culminated in a media storm, with up to 30 interviews per day and a 27-minute primetime feature on CNN with expert lawyers, journalists, and Bernhard participating via phone. For the artists, this coverage offered further opportunities to recompose their work by feeding back legal documents and press coverage into the project. In this way, their approach resembles a variant of the *Gesamtkunstwerk* for informational contexts, emerging from the agential mess of networked systems as a projective intensity of forces. Bernhard and Lizvlx see these projects as "research-based"—"we are not opportunistic, money-driven or success-driven, our central motivation is to gain as much information as possible as fast as possible as chaotic as possible and to redistribute this information via digital channels."[24] By following the eventfulness of these medial actions, UBERMORGEN. COM explore uncertainty through the actualization of differing potentials, including those between networked machines, digital objects, logical structures, and other nonhuman agencies.

SoE: Individuation and Peer-to-Peer Capitalism

Bringing together teletext pornography, generative music, and online user profiling, *The Sound of eBay* is the final installment in the e-commerce "action tryptich" by UBERMORGEN.COM. Essentially, the project works by translating data from commercial eBay profiles into electronic music, resulting in simple dance tracks that recall TB-303 style acid bass lines with heavily processed vocals.[25] I want to start with this final work for a number of thematic reasons: on the face of it, *The Sound of eBay* seems markedly different from the other projects in the trilogy for taking a celebratory and playful, though no less ironic, stance toward its target.[26] The web site features garish pixilated porn in a highly stylized retro aesthetic, while the "story" praises the establishment of the trading forum on the web: "eBay is romantic and seductive, not like the local fleamarkets in Paris, but sexed up a million times bigger and spherically transcended, much more effective and thoroughly commercialized. We love it!"[27] *The Sound of eBay* was additionally created with a different group of collaborators (Paolo Cirio and Alessandro Ludovico were not involved), and its promotion was comparatively low-key and aimed mainly at a media art audience.[28] Nevertheless, in important ways, the project explores sociotechnical issues around data aggregation, software encoded identity-as-assemblage, and the creation of economic value through digital networks.

UBERMORGEN.COM themselves describe it as "the affirmative high-end low-tech contribution to the atomic soundtrack of the new peer-to-peer hyper-catastrophic shock-capitalism."[29] One aspect of their engagement with the trading web site is the promiscuity of data: a tendency for digital information not only to propagate through multiple associations, but also become transformed through various algorithmic processes. This movement complicates any notion of bounded individual identities and can be perceived as underpinning a series of controversies that stem from informational overflows. Here, this section offers a framework for the material actions raised by *The Sound of eBay* by introducing the concept of transduction: a philosophical account of individuation and technological action based on the work of Simondon.

As a generative work, *The Sound of eBay* pushes harvested user behaviors through the logic of a relational database, allowing transactions from online activity to be made available for reconfiguration and the production of electronic music. The project is based on an interface that solicits personal information from the eBay site, simply requesting a username and email address. Once supplied, according to the *Sound of eBay*, an automated program scrapes the web for relevant data—including credit card and banking details, passwords, and non-sensitive information like items bought and sold—which go into producing a unique MP3 delivered to the email address. The software itself was created by programmer Stefan Nussbaumer and is based on an earlier prototype that worked with the South American eBay-clone Mercadolibre.[30] Now re-titled the "eBay Generator," the application was originally written in SC3 SuperCollider and functions by translating the profiled data into a hard-coded or predefined spectrum of scales and harmonics. Rather than aiming for a diverse range of musical output, however, it works on a uniform style laid out in advance with each set of profiled data producing slightly different variations within these prearranged restrictions. As Nussbaumer explains: "although there is an underlying harmonic structure and a rather conventional rhythmic structure, not every generated song will be a 'hit'—it's a bubbling sea of eBay-userdata."[31] Through this schema, the project can be understood as working from a preset grid within which tracks are assembled in a manner emblematic of the database as a cultural form.[32] The eBay-Generator, moreover, relies on hash sums to reconfigure the data already retrieved into a history or "remembrance" built up over time. This effectively allows for the relational processing of user-profiles into unique musical compositions (entering the same username can, for instance, bring about different

MP3s), causing exchange-based formations that mimic the peer-to-peer transactions of eBay. Beyond these internal mechanics, however, what remains unique about *The Sound of eBay* is the distributed production of the MP3s, the strange capacity for alternative expressiveness to emerge out of networked flows of user information. As Grischinka Teufl explains in her idiosyncratic take on the work, the discovery of "this undetected entropolis" allows for "a suspicious synaesthesia within the neatly structured columns of profit-orientated socio-technological networks."[33]

From a theoretical perspective, the reconfiguration of user-data in *The Sound of eBay* recalls a central observation of the Deleuzian control society thesis regarding the changing status of disciplinary-based compositions of the social in the wake of computational encoding and management. This is memorably described as a transition from the individual to "dividuals," and from masses to "samples, data, markets, or 'banks.'"[34] The term dividual relates to the sub-division of the individuated and amassed subjects of disciplinary societies into aggregate and hybrid formations: a modulation. To the extent that *The Sound of eBay* offers a playful engagement with dynamics of control—modulating commercial user-data into sonic compositions—the question of how this translation of materials operates as a force of differentiation is significant. What becomes of the "username" subjected to these processes: to what does it now refer? Can the project be legitimately described as "counter-information?" Could it be re-titled "the *noise* of eBay?"

Here, I want to consider the concept of transduction from Simondon as one possible framework for considering the socio-technological actions at stake. His thought offers a detailed conceptual vocabulary for the question of individuation as "the becoming of being," a concern with how things come into existence and proceed temporally as projective entities. Developed in part through a philosophical account of the functioning of technical artifacts—combustion engines, electron tubes, hydraulic turbines, among other industrial and scientific standard technologies—Simondon describes how individuality itself emerges by concatenating or drawing together a field of distributed potentials in order to instantiate particular operative processes.[35] His project, moreover, is shaped by an engagement with communication theory and cybernetics, but differs in critical respects through the emphasis placed on the "excluded middle" of *ontogenesis* as that which exceeds any representational logic of coded messages or unit-measures ("bits").[36] Simondon's variant of processual and distributed materialism can, therefore, assist by linking together the differentiations achieved by

The EKMRZ Trilogy as a serial artwork, while also reframing important issues for the control paradigm in terms of the problematic hybrid constitution of digital networks.

Transduction refers to individuation ("the becoming of being") as an operation that draws together disparate things into composites of relations that propagate a structure throughout a domain. Conceived by Simondon under a general science of operations (described as *allagmatics*), this notion can be usefully applied to the work of socio-technical assemblages and, in particular, the power structures in which digital networks are enmeshed as uncertain arrangements of potentials. The process of transduction is, accordingly, described as being an expansive materiality, a movement of continuous iteration or variable steps of contagion in a heterogeneous context:

> By transduction we mean an operation—physical, biological, mental, social—by which an activity propagates itself from one element to the next, within a given domain, and founds this propagation on a structuration of the domain that is realized from place to place: each area of the constituted structure serves as the principle and the model for the next area, as a primer for its constitution, to the extent that the modification expands progressively at the same time as the structuring operation.[37]

These innovative contagions, importantly, work by establishing dimensions of communication between pre-individual states of difference or incompatibilities. As Adrian MacKenzie foregrounds in a commentary on Simondon, this activity arises from a particular type of disjunctive relation between an entity and something external to itself: "in making this relation, technical action changes not only the ensemble, but also the form of life of its agent. Abstraction comes into being and begins to subsume or reconfigure existing relations between the inside and outside."[38] Here, reciprocal interactions between two states or dimensions actualize disparate potentials through *metastability*: a kind of directional equilibrium that proliferates, unfolds, and drives individuation. In doing so, newly discovered structures are elaborated by inverting this problem of intensive difference as a positive material expression. Crucially, this particular variant on technological action is founded on dissymmetry. The reality of pre-individual disparation is never fully surpassed or resolved, but becomes integrated as an ontogenetic force: it becomes a metastable process through which a surrounding milieu unfolds.

In this philosophical rereading of communication theory and cybernetics, Simondon's account on transduction can be extended to describe the operation of networked informational technologies throughout control societies as a whole, particularly in terms of illuminating their operative status as projective systems that thrive on difference and tension. This standpoint is directly opposed to the idea that a digital network can be delineated into a series of nodes or edges, since trans-ductive operations relate to the ongoing concatenating movement of socio-technological ensembles (including the capacity to interweave imagination, perception, memories, and related inter-human attach-ments). In this respect, there is something uncertain, even *problematic*, about digital networks in their function. The concept of the pre-individual as an overflow held in tension complicates any presumption of a stable essence or substance enlisted toward instrumental ends. While commer-cial or political value can be extracted by becoming further enmeshed in exterior material structures, there remains a key dimension of instability that enables their very operation as socio-technical things, as MacKenzie emphasizes: "they provisionally and intermittently generate something thoroughly contingent yet important because it cannot be reduced to existing forms of subjectivity, subjectification or socialization."[39] *The Sound of eBay*, I want to argue, effectively illustrates how this bridging of initially unrelated domains works in the context of data aggregation, algorithmic processes, and digital networks. The problem of identity within the folds of software is precisely an issue perpetuated within the work. Just as Simondon describes transductions as "amplifying reticular structures"—the coming together of potentialities to crystallize forms and structures—the deployment of technological action in *The EKMRZ Trilogy* can be described as a mode of *reticular aesthetics*: a connective ges-ture with the material instabilities that perpetuate and spread uncertain issues.[40] Certainly, problems are not surpassed or resolutely solved by these artworks, but are actively maintained as the key characteristic dynamic of the project itself. While *The Sound of eBay* works with this mode through the translation of user-data, other pieces that make up the trilogy explicate further complexities regarding the production of economic value from contingent operations.

Amazon Noir: Intellectual Property and Digital Objects

In 2006, UBERMORGEN.COM, Alessandro Ludovico, and Paolo Cirio instigated the theft of some 3,000 digital copies of books from the online retailer Amazon.com by targeting vulnerabilities in the "Search Inside the Book" feature from the company's web site. Titled *Amazon Noir*, the project instigated a specially designed software program that bombarded the Search Inside!™ interface with multiple requests over several weeks between July and October, assembling full versions of texts and distributing them automatically across peer-to-peer networks (P2P). This installment in the trilogy explores the tension between intellectual property regimes and the materialities of digital systems.

The rendering of piracy as a tactic of resistance or a technique capable of leveling out the uneven economic field of global capitalism has become a catch-cry for political activism. In their analysis of multitude, for instance, Antonio Negri and Michael Hardt describe the contradictions of post-Fordist production as conjuring forth a tendency for labor to become common. That is, because flexible productivity depends on communication and cognitive skills directed toward the cultivation of an ideal entrepreneurial subject, greater possibilities are engendered for self-organized forms of living that significantly challenge its operation. For Hardt and Negri, intellectual property regimes exemplify such a spiraling paradoxical logic, since "the infinite reproducibility central to these immaterial forms of property directly undermines any such construction of scarcity."[41] The implications of the filesharing program Napster, accordingly, are read as not merely directed toward theft, but in relation to the private character of the property itself; a kind of social piracy is perpetuated that is viewed as radically recomposing social resources and relations. Ravi Sundaram, a co-founder of the Sarai new media initiative in Delhi, has meanwhile drawn attention to the existence of "pirate modernities" capable of being actualized when individuals or local groups gain illegitimate access to distributive media technologies. These are worlds of innovation and non-legality, of electronic survival strategies that partake in milieus of dispersal and escape simple forms of classification—"piracy destabilizes contemporary media property and, working through world markets and local bazaars, both disrupts and enables creativity, and evades issues of the classic commons while simultaneously radicalizing access to subaltern groups in the Third World."[42] Magnus Eriksson and Rasmus Fleische, meanwhile—associated with the notorious Piratbyrn—have promoted the bleeding away of Hollywood profits through fully deployed peer-to-peer networks, with the intention of pushing filesharing dynamics to an extreme in order to radicalize the potential for social change.[43] Such theories are complemented by appropriation art, a movement broadly conceived in terms of antagonistically liberating knowledge from the confines of intellectual property: "those who pirate and hijack owned material, attempting to free information, art, film, and music—the rhetoric of our cultural life—from what they see as the prison of private ownership."[44] These "unruly" escape attempts are pursued through a variety of engagements, from experimental performances with legislative infrastructures (i.e. Kembrew McLeod's patenting of the phrase "freedom of expression") to musical remix projects, such as the work of Negativland, John Oswald, RTMark, Illegal Art, and the Evolution Control Committee. *Amazon Noir* works in concert with these frameworks—both theoretical and practical—but is distinguished by working with the materialities of digital distribution.[45]

The system used to harvest the content from "Search Inside the Book" is described as "robot-perversion-technology," based on a network of four servers around the globe, each with a specific function: one located in the United States that retrieved (or "sucked") the books from the site, one in Russia that injected the assembled documents onto P2P networks, and two in Europe that coordinated the action via intelligent automated programs. According to the artists (referring to themselves as "villains"), the main goal was to steal all 150,000 books from Search Inside!™ then use the same technology to steal books from the "Google Print Service" (the exploit was limited only by the amount of technological resources financially available, but there are apparent plans to improve the technique by reinvesting the money received through the settlement with Amazon.com not to publicize the hack). In terms of informational culture, this system resembles a machinic process directed at redistributing copyright content. A schematic entitled "The Diagram" on the project web site and displayed in gallery contexts visualizes key processes that define digital piracy as an emergent phenomenon within an open-ended and responsive milieu.[46] That is, the static image foregrounds something of the activity of copying being a technological action that complicates any analysis focusing purely on copyright as content. In this respect, intellectual property rights are revealed as being entangled within information architectures as communication management and cultural recombination—dissipated and enforced by a measured interplay between openness and obstruction, resonance and emergence.[47]

This later point is demonstrated in recent scholarly treatments of filesharing networks as media ecologies. Kate Crawford, for instance, describes the movement of P2P as processual or adaptive, comparable to technological action, marked by key transitions from partially decentralized architectures such as Napster, to the fully distributed systems of Gnutella and seeded swarm-based networks like BitTorrent.[48] Each of these technologies can be understood as a response to various legislative incursions, producing radically dissimilar socio-technological dynamics and emergent trends for how agency is modulated by informational exchanges. Indeed, even these aberrant formations are characterized by modes of commodification that continually spillover and feedback on themselves, repositioning markets and commodities in the process, from MP3s to iPods, and from P2P to broadband subscription rates. However, one key limitation of this approach is apparent when dealing with the sheer scale of activity involved, where mass participation elicits certain degrees of obscurity and relative safety in numbers. This represents an obvious problem for analysis, as dynamics can easily be identified in a conceptual sense, without any understanding of the specific contexts of usage, political impacts, and economic effects for participants in their everyday consumptive habits. Large-scale distributed ensembles can be interpreted as problematic in their technological constitution. They are sites of expansive overflow that provoke an equivalent individuation of thought, as the Recording Industry Association of America observes on their educational web site: "because of the nature of the theft, the damage is not always easy to calculate but not hard to envision."[49] The politics of the filesharing debate, in this sense, depends on the command of imaginaries; that is, being able to conceptualize an overarching structural consistency to a persistent and adaptive ecology. As a mode of material engagement, *Amazon Noir* dramatizes these ambiguities by framing technological action through the fictional sensibilities of narrative genre.[50]

The extensive use of imagery and iconography from "noir" can be understood as an explicit reference to the increasing criminalization of copyright violation by digital technologies. However, the term also refers to the indistinct effects produced through this engagement: who are the "bad guys" or the "good guys?" Are positions like "good" and "evil" (something like freedom or tyranny) so easily identified and distinguished? In this respect, *Amazon Noir* is congruent with the agonism between the object-orientated agencies and "moral" demands of representational frameworks posited in Fuller and Goffey's evil media studies.[51] If anything, a weird "innocent" character to the project is conveyed through noir. As Paolo Cirio explains: "it's a representation of the actual ambiguity about copyright issues, where every case seems to lack a moral or ethical basis."[52] While user communications made available on the site clearly identify culprits (describing the project as jeopardizing arts funding, as both irresponsible and arrogant), the self-description of the artists as "failures" further highlights the uncertainty regarding the project's qualities as a force of long-term social renewal:

> Lizvlx from Ubermorgen.com had daily shootouts with the global mass-media, Cirio continuously pushed the boundaries of copyright (books are just pixels on a screen or just ink on paper), Ludovico and Bernhard resisted kickback-bribes from powerful Amazon.com until they finally gave in and sold the technology for an undisclosed sum to Amazon. Betrayal, blasphemy and pessimism finally split the gang of bad guys.[53]

Here, the adaptive and flexible qualities of informatic commodities and computational systems of distribution are knowingly posited as critical limits; in a certain sense, the project fails instrumentally in order to succeed conceptually. This might be interpreted as guaranteeing authenticity by grounding the aesthetic qualities of the work. Through this process, however, *Amazon Noir* illustrates how forces confined as being exterior to control (virality, piracy, non-communication) regularly operate as points of distinction to generate change and innovation. This draws into focus the significance of a Simondon-inspired perspective on socio-technical ensembles as always predetermined by and driven through disparation as difference. Indeed, this was an important characteristic that Deleuze would emphasize in an early review of Simondon, stressing how "a metastable system thus implies a fundamental *difference*, like a state of dissymmetry. It is nonetheless a system insofar as the difference therein is like *potential energy*, like a *difference of potential* distributed within certain limits."[54] Just as hackers are legitimately hired to challenge the durability of network exchanges, malfunctions and errors are relied upon as potential sources of future information. Indeed, the notion of demonstrating "autonomy" by revealing the shortcomings of software is entirely consistent with the logic of control-based systems as a modulating organizational diagram. These so-called "circuit breakers" become points of bifurcation that open up new systems and encompass an abstract machine or tendency governing contemporary capitalism.[55]

GWEI: Clickonomics and the Common

Google Will Eat Itself engages with the mechanisms through which profit is extracted from informational networks, focusing in particular on the complex dynamics by which user-generated data is collected, stored, and incorporated into Google digital services, resembling a recursive form of technologically enabled labor. This disjuncture of partial incorporation and exclusion can be broadly seen as a broader issue from which a series of problems arise, from privacy concerns and the visibilities enabled by the storage of vast quantities of user data to the non-transparency of infrastructural access to search engine software.[56] These tensions have only increased under the paradigm of cloud computing and in the seemingly endless diversification of services offered by the expanding corporation—from browsers to digital mapping services and mobile phones. *Google Will Eat Itself* is a work that operates through an engagement with the emergent and intertwined dimensions of value and content generation harnessed by the online corporation. Specifically, this celebrated project engages with the reticulation of user actions in the machinations of Google for the management of informational flows and aggregation of attention.

The project essentially works by generating money through AdSense placed on a network of "hidden" sites. The AdSense application allows for individuals to sign up for video, text, and image-based advertisements to be posted to their site. These generate revenue on a "per-click" or "per-impression" basis, sent as micro-payments to the web site owner while Google retains a percentage as profit. The novelty of *Google Will Eat Itself* is based on the accumulation of income through these services to purchase shares in the company, which are consequently handed over to the common ownership of GTTP Ltd. ("Google To The People Public Company")—"we buy Google via their own advertisement! Google eats itself—but in the end 'we' own it!"[57] In this respect, the project delves into the dynamic and co-emergent qualities of the informational economy carefully managed by the company for profit. These particular performative dimensions of the project make it distinctive from related artistic engagements with Google, which have tended to focus on the characteristics of the PageRank algorithm as a structural mechanism; works such as Cory Arcangel's *Dooogle* site, which alters all queries to only return pages relevant for the words "Doogie Houser", or Constant Dullaart's more recent *Disease Disco* which returns searches by color for the term "disease" while playing the number one Billboard hit song in the background when the piece was created ("Boom Boom Boom" by the Black Eye Peas).[58] These projects tend to satirize Google search as an imposition on the diversity of content online, a kind of predetermined aperture that imposes a partial ordering of informational environments.

Significantly, *Google Will Eat Itself* operates on a different level by investigating the interlinked dimensions of users, economics, and software, between the complexities of value generation and the material infrastructures that constitute digital cultures. The project, in other words, reveals how informational flows are directed by metastabilities harnessed in particular directions. Such dynamics are integral to the operations of the corporation, but arise through the interwoven actions of human and nonhuman agencies beyond the commercial territories of Google.

Beyond a critique of Google based on panoptic surveillance or the "imperial nature of its monopoly," Matteo Pasquinelli has provided a useful theoretical model that draws from post-autonomia thought, focusing in particular on cognitive labor power as a force of ontogenetic potentiality.[59] With reference to Paolo Virno, Antonio Negri, and Carlo Vercellone, his approach is distinctive for reversing the prevalent critique of Google as a top-down corporate "Big Brother" to draw attention instead to how the kinds of network value accumulated by the company co-emerge relationally with the practices of users. In his conceptual sketch, Pasquinelli argues that the technical mechanisms central to the PageRank software create wealth in a parasitic manner by harnessing the operations of the "common intellect" online. The work of web users— from the uploading of material to linking and conducting searches—is retroactively fed back into maintaining the efficiency of Google software-based services as a concatenation of forces pertaining to the organization, legitimacy, and relevancy of information. This was, of course, the key innovation for the original project by Larry Page and Serge Brin for large-scale searches that make use of the "additional information" maintained by hypertext structures already present on the web.[60] As opposed to the assumption of networks as radically horizontal, implying a democratic equality of status and privilege, PageRank draws from the emergent dissymmetry of hyperlinks as a linear or mono-directional flow of establishing hierarchical orders. For Pasquinelli, arguing for a case of parasitic surplus, Google extracts value as a tertiary node that reroutes and absorbs the collective work of socio-technical ensembles in a specific direction. Accordingly, the PageRank algorithm is interpreted as having "installed itself precisely on this movement that shapes the collective sphere of knowledge and the Internet in molecular vorticles of value."[61]

In relation to the mechanisms of enterprise deployed by Google to engage with the generation of value, the reference to Virno is instructive as a theoretical link to the role of individuation and the common. The pre-individual of Simondon is interpreted in *A Grammar of the Multitude* as referring to potential latent structures that constitute the general intellect. This is perceived as referring to factors that precede valorization and the controlling measures imposed by capital, such as the vitalistic human capacity to produce new things by linguistic statements, "generic biological endowments," and cooperative cognitive innovation.[62] This perspective follows a well-known premise of post-autonomia thought that emphasizes the anteriority of labor from processes of valorization. However, the notion of the preindividual as a processual and uncertain dimension of socio-technical action requires thinking through "the unequal" as the ground of progressive change. To this extent, *Google Will Eat Itself* demonstrates how mechanisms of accumulation and the foundation of the common co-emerge through an equivocal terrain of networked action, including the problematic political status of non-human agencies. By working on the disparation of large-scale digital systems, and on the emergence of measurement itself, *Google Will Eat Itself* provides a demonstration of "the invention of a common that is not given in advance and which emerges on an ontological background of inequality."[63] Here, to consider the pre-individual as the source of the common implies the necessity to confront the contingent and uncertain aspects of more-than-human ensembles.

Conclusion

The EKMRZ Trilogy relays controversies that surround the extraction of value from digital networks within informational economies. While these can be framed through subversive or aberrant tactics geared toward a critical intervention, this chapter has suggested that the three projects—*Google Will Eat Itself, Amazon Noir,* and *The Sound of eBay*—can be more accurately interpreted as exploring socio-technical ensembles as inherently problematic systems. The mode of technological action found in these projects by UBERMORGEN.COM is described as reticular aesthetics: a connective engagement with the materials that determine the site of controversy itself. This open-ended quality is manifested in the processual dimension of the works; drawing on the tensions that drive informational infrastructures to demonstrate a reconstitution of outcomes. As a consequence, the ambiguities of the trilogy emerge not just from the contrary articulation of proprietary modes of accumulation and the materialities of digital networks, but also by thinking of a consistent program for political agency simultaneously with control. This complication is apparent, for instance, in Alexander Galloway and Eugene Thacker's study of the machines that are synonymous with the operation of Deleuzian control societies—in terms of protocol and bioinformatics—where tactical media are posited as modes of contestation against the tyranny of code, exploiting flaws in protocological systems in order to sculpt such technologies into something new, "something in closer agreement with the real wants and desires of its users."[64] While pushing a system into a state of hypertrophy to reform architectures might be one technique to imagine mastering the internal incompatibilities of software toward progressive ends, it still leaves unexamined the desire for reformation itself as nurtured and produced by the ongoing instabilities of these systems. As Chun notes, there is no difference between resistance and capture in these regimes, since "control and freedom are not opposites, but different sides of the same coin: just as discipline served as a grid on which liberty was established, control is the matrix that enables freedom as openness."[65] The desire for freedom as openness is, therefore, entirely consistent with the uncertainties of digital networked infrastructures as engines of transduction. Moreover, to a significant extent, they *already* effectively function as hypertrophic systems. While political or economic outcomes may be developed out of their problematic status, there can be no guaranteed emancipatory conclusion for an individual, especially given a consideration of the assemblages through which outcomes are expressed.

If *The EKMRZ Trilogy* has any political currency in this context, it lies in a capacity to recognize how the informational networks move to arrange and organize collective desires, memories, and imaginative visions, rather than just antagonisms and counterintuitive economics. The projects effectively demonstrate how the technical architectures of software cannot be fully understood without considering the ongoing *social drama* that permeates through them. As a final point, it is worth emphasizing how these dimensions are heightened through the obscure performative dimensions of the projects: the hidden nature of the AdSense web sites in *Google Will Eat Itself,* in the fact that *Amazon Noir* was publicized only after the settlement with the company, or the manner in which *The Sound of eBay* recomposes "highly sensitive, inaccessible and unreachable data" as MP3s. While providing diagrams and documentation, there remains an aspect of obfuscation carried through by the abstract operation of digital agents in these contexts, even a sense in which these "invisible events" might not have even occurred. To the extent that we believe

UBERMORGEN.COM, that such actions are even possible, is a gauge by which the operations of control interconnect with a desire or longing for change in confrontation with the operations of network societies—these are the resonant forces through specific ensembles that envelope inter-human relations described by Simondon as transindividual. *The EKMRZ Trilogy*, with all its underlying ethical issues, presents us with a challenge to rethink this uncertain terrain by considering our own political attachments with digital networks. It provides a unique opportunity to conceive of a future that begins with limits and limitations as immanently central, even foundational, to our inherent interconnection with socio-technical things.

Notes

[1] Gilles Deleuze, *Foucault*, trans. Seán Hand (Minneapolis: University of Minnesota Press, 1986), 37.

[2] UBERMORGEN.COM, "We-Make-Money-not-Art Interview with Régine Debatty," http://www.we-make-money-not-art.com/archives/2006/10/-uber-morgen-wha.php.

[3] Florian Cramer, "Entering the Machine and Leaving It Again: Poetics of Software in Contemporary Art" (2006), http://gwei.org/pages/press/press/Florian_Cramer/fullversion.html.

[4] Michel Callon, Pierre Lascoumes and Yannick Barthe, *Acting in an Uncertain World: An Essay of Technical Democracy*, trans. Graham Burchell (Cambridge, MA: MIT Press, 2009).

[5] Matthew Fuller and Andrew Goffey, "Towards an Evil Media Studies," in *The Spam Book: On Viruses, Porn and Other Anomalies from the Dark Side of Digital Culture*, ed. Jussi Parikka and Tony Sampson (New Jersey: Hampton Press, 2009), 143–144.

[6] Adrian MacKenzie, *Cutting Code: Software and Sociality* (New York: Peter Lang, 2006), 1–19.

[7] Tiziana Terranova, "Of Sense and Sensibility: Immaterial Labour in Open Systems," in *DATA Browser 03: Curating Immateriality*, ed. Joasia Krysa (New York: Autonomedia, 2006), 27–37.

[8] Bruno Latour, *Reassembling the Social: An Introduction to Actor-Network-Theory* (Oxford: Oxford University Press, 2005).

[9] Lev Manovich, *The Language of New Media*, (Cambridge MA: MIT Press, 2001), 48; Lev Manovich, *Software Takes Command* (2008), 2–29, http://softwarestudies.com/softbook/manovich_softbook_11_20_2008.pdf.

[10] Matthew Fuller ed. *Software Studies: A Lexicon* (Cambridge, MA: MIT Press, 2008).

[11] Sean Cubitt, "Distribution and Media Flows," *Cultural Politics* 1, no. 2 (2005): 194.

12 Friedrich Kittler, "There is No Software," *CTheory* (1995), http://www.ctheory. net/articles.aspx?id=74; Matthew Kirschenbaum, *Mechanisms: New Media and the Forensic Imagination* (Cambridge, MA: MIT Press, 2008).

13 Adrian MacKenzie, "The Strange Meshing of Impersonal and Personal Forces in Technological Action," *Culture, Theory and Critique* 47, no. 2 (2006): 198.

14 Lawrence Lessig, *Code and Other Laws of Cyberspace* (New York: Basic Books, 1999), 20.

15 Alexander R. Galloway, *Protocol: How Control Exists After Decentralization* (Cambridge, MA: MIT Press, 2004), 7.

16 Ibid., 243–244.

17 Wendy Hui-Kyong Chun, *Control and Freedom: Power and Paranoia in the Age of Fiber Optics* (Cambridge, MA: MIT Press, 2006), 1–30.

18 Gilles Deleuze, *Negotiations 1972–1990*, trans. Martin Joughin (New York: Columbia University Press, 1992), 175.

19 Ibid., 179.

20 Tiziana Terranova, "Of Sense and Sensibility," 34.

21 Michel Foucault, *Security, Territory, Population: Lectures at the Collège de France, 1977–78*, trans. Graham Burchell (New York: Palgrave Macmillan, 2007); Tiziana Terranova, "Another Life: The Nature of Political Economy in Foucault's Genealogy of Biopolitics," *Theory, Culture & Society* 26, no. 6 (2009).

22 Yukiko Shikata, "UBERMORGEN.COM: To the Philosophy of the Future," in *UBERMORGEN.COM: Media Hacking vs. Conceptual Art*, ed. Alessandro Ludovico (Christoph Merian Verlag, 2009), 174.

23 Shikata, 175.

24 UBERMORGEN.COM, "We-Make-Money-Not-Art Interview with Régine Debatty."

25 For a sample of 1001 songs, http://www.cronicaelectronica.org/?p=043.

26 As Ed Halter emphasizes, writing for Rhizome.org, "the parasitic *Sound of eBay* has a relatively benign relationship to its host organism." Ed Halter, "Bidding and the Beat," *Rhizome.org* (2008), http://rhizome.org/editorial/407.

27 UBERMORGEN.COM, "Story: The Sound of eBay," http://www.sound-of-ebay.com/500.html.

28 UBERMORGEN.COM, "Manifesto," http://www.ubermorgen.com/manifesto/.

29 UBERMORGEN.COM, "Story: The Sound of eBay."

30 UBERMORGEN.COM, "The Sound of Mercadolibre.com," http://www.sound-of-mercadolibre.com/.

31 Stefan Nussbaumer, "The Generator," http://www.sound-of-ebay.com/pdfs/SoE_Technology.pdf .

32 Manovich, *The Language of New Media*, 212–243.

33 Grischinka Teufl, "The Sound of eBay," http://www.sound-of-ebay.com/pdfs/SOE_UBERMORGEN.COM_grischinka_teufl.pdf.

34 Deleuze, *Negotiations*, 180.

35 Gilbert Simondon, "Technical Individualization," trans. Karen Ocana, in *Interact or Die!*, ed. Joke Brouwer and Arjen Mulder (V2_Publishing/NAi Publishers, 2007); Gilbert Simondon, *On the Mode of Existence of Technical Objects*, trans. Ninian Mellamphy (London: University of Western Ontario, 1980), available

online from http://accursedshare.blogspot.com/2007/11/gilbert-simondon-on-mode-of-existence.html.

36 Alberto Toscano, *The Theatre of Production: Philosophy and Individuation Between Kant and Deleuze* (Basingstoke: Palgrave Macmillan, 2006), 142–147.

37 Gilbert Simondon, "The Position on the Problem of Ontogenesis," trans. Gregory Flanders, *Parrhesia* 7 (2009): 11.

38 MacKenzie, "The Strange Meshing of Impersonal and Personal Forces in Technological Action": 203.

39 Ibid., 201.

40 Simondon, "The Position on the Problem of Ontogenesis", 11.

41 Michael Hardt and Antonio Negri, *Multitude: War and Democracy in the Age of Empire* (New York: Penguin, 2004), 180.

42 Ravi Sundaram, "Revisiting the Pirate Kingdom," *Third Text* 23, no.3 (2009): 338; see also Ravi Sundaram, *Pirate Modernity: Delhi's Media Urbanism* (London: Routledge, 2009).

43 Magnus Eriksson and Raymond Fleischer, "Copies and Contexts in the Age of Cultural Abundance," posted to the nettime mailing list, 2007, http://www.mail-archive.com/nettime-l@bbs.thing.net/msg04203.html.

44 Christine Harold, *OurSpace: Resisting Corporate Control of Culture* (Minneapolis: University of Minnesota Press, 2007), 114.

45 Other projects dealing with digital distribution and copyright have been curated by *The Kingdom of Piracy* exhibitions, see http://kop.fact.co.uk/.

46 UBERMORGEN.COM, Alessandro Ludovico and Paolo Cirio, "The Diagram," http://www.amazon-noir.com/diagram_large.html.

47 Tiziana Terranova, *Network Culture: Politics for the Information Age* (London: Pluto Press, 2004), 6–38.

48 Kate Crawford, "Adaptation: Tracking the Ecologies of Music and Peer-to-Peer Networks," *Media International Australia* 114 (2005): 30–39.

49 RIAA, "Piracy Online and On the Street' (2009), http://www.riaa.com/physicalpiracy.php?content_selector=piracy_details_online.

50 Inke Arns, "Storytellers for the Information Age: On the Role of Narrative in UBERMORGEN.COM's work," in *UBERMORGEN.COM*, ed. Dominico Quaranta (Brescia: FPEditions, 2009), 78–89.

51 Fuller and Goffey, 143–144.

52 UBERMORGEN.COM, Alessandro Ludovico and Paolo Cirio, "Amazon Noir Interview with Franz Thalmair for Cont3xt/furtherfield," http://www.amazon-noir.com/TEXT/CONTEXT_INTERVIEW/interview_amazon_noir.pdf.

53 UBERMORGEN.COM, Alessandro Ludovico and Paolo Cirio, "Amazon Noir: Home," http://www.amazon-noir.com/index0000.html.

54 Gilles Deleuze, "On Gilbert Simondon," in *Desert Islands and Other Texts: 1953–1974*, ed. David Lapoujade (New York: Semiotext(e), 2004), 87.

55 Jussi Parrika, "Contagion and Repetition: On the Viral Logic of Network Culture," *Ephemera: Theory and Politics in Organization* 7, no. 2 (2007): 300.

56 Anna Munster, 'Welcome to Google Earth', in *Critical Digital Studies*, ed. Arthur Kroker and Marilouise (Toronto: University of Toronto Press, 2009), 400.

57 UBERMORGEN.COM, Alessandro Ludovico and Paolo Cirio, "Google Will Eat Itself," http://gwei.org/index.php.

58 D*ooogle*, http://www.dooogle.com/; *Disease Disco*, http://constantdullaart.com/site/html/new/diseasedisco.html.

59 Matteo Pasquinelli, "Google's PageRank Algorithm: A Diagram of Cognitive Capitalism and the Rentier of the Common Intellect," 2009, http://matteopasquinelli.com/docs/Pasquineli_PageRank.pdf.

60 Sergey Brin and Lawrence Page, "The Anatomy of a Large-Scale Hypertextual Web Search Engine," *Computer Networks and ISDN Systems* 30 (1998): 107–117.

61 Pasquinelli.

62 Paolo Virno, *A Grammar of the Multitude*, trans. Isabella Bertoletti, James Cascaito, and Andrea Casson (New York: Semiotext(e), 2004), 76–80.

63 Alberto Toscano, "Gilbert Simondon," in *Deleuze's Philosophical Lineage*, ed. Graham Jones and Jon Roffe (Edinburgh: Edinburgh University Press, 2009), 393.

64 Alexander R. Galloway and Eugene Thacker, *The Exploit: A Theory of Networks* (Minneapolis: University of Minnesota Press, 2007), 98.

65 Chun, 71.

Chapter 11

Queer/Error: Gay Media Systems and Processes of Abjection

Chad Parkhill
University of Queensland

Jessica Rodgers
Queensland University of Technology

To link, as we have done in our title, the notions of queerness and error is a *risky* strategy—an adjective to be read with all of its resonances within safer sex and HIV-prevention literature intact. In doing so, we risk misapprehension: are we saying that queer bodies are errors in the otherwise perfectly functioning system of heterosexual social reproduction? And wouldn't such an assertion, working as it does against decades of gay liberal struggle to paint homosexuality as a morally-neutral matter, count as politically retrograde? After all, the modern gay movement owes its existence to the struggle, inaugurated by Karl Heinrich Ulrichs in the late nineteenth century, to define homosexuality as a character trait possessed by a certain class of persons inclined, through a "riddle of nature," to love members of their own sex.[1] According to this argument, homosexuality must be understood as permissible owing to its naturalness, if not its normality, and its ethical neutrality as a series of private acts between consenting adults. Furthermore, the concatenation of the terms "queer" and "error" has had a long and rich history of homophobic uses. The term "pervert" comes from a theological context, in which it functions as an antonym to "convert," that is, "one that is turned from good to evil,"[2] while to err in the theological context is to sin, or to go astray morally: to be perverted, therefore, is to be in error. This connection persists despite massive conceptual changes in how we understand sexual perversion:[3] in 1970 Joseph Epstein claimed that homosexuals "are an affront to our rationality, living evidence of our despair of ever finding a sensible, an explainable design to the world."[4] In Epstein's formulation, homosexuality is an error in nature that not only defies explanation, but renders nature itself inexplicable; unsurprisingly, he confesses that if he "had the power to do so, [he] would wish homosexuality off the face of this earth."[5]

Despite this history of homophobic concatenation of queerness and error, queer theory itself constructs queerness as something of an error. For David Halperin, "[q]ueer is by definition *whatever* is at odds with the normal, the legitimate, the dominant. *There is nothing in particular to which it necessarily refers.*"[6] In this case, the adjective "queer" has no proper object, and it wanders errantly from object to object: "it could include some [heterosexual] married couples without children, for example, or even (who knows?) some married couples *with* children—with, perhaps, *very naughty* children."[7] Although other queer theorists, prominently Eve Kosofsky Sedgwick, would disagree with Halperin's formulation because it ignores the homosexual specificity of the term "queer"—for Sedgwick, "given the historical and contemporary force of the prohibitions against *every* same-sex sexual expression, for anyone to disavow those meanings, or to displace them from the term's definitional center, would be to dematerialize any possibility of queerness itself"[8]—they nonetheless embrace its potential to signify that which works outside of, works against, or subverts from within systems and logics of the sexual. As Sedgwick notes, "[t]he word 'queer' itself means *across*—it comes from the Indo-European root *-twerkw*, which also yields the German *quer* (transverse), Latin *torquere* (to twist), English *athwart.*"[9] Cutting across, twisting, at odds with the normal, against systems and systematicity: is it any wonder those systems themselves construe the queer as an error that must be gotten rid of, often with fatal consequences? Or a bug in the genetic code of Human 1.0 that, thanks to advances in biotechnology, might be removed when we download the next firmware update?[10]

It is depressing enough when mainstream heterosexual culture understands queerness as such and therefore abjects it;[11] perhaps more disheartening is the repetition of this same gesture in a nominally queer-friendly system, such as that of gay community media. In this chapter we will examine how systems of exclusion in one such gay community medium, the *Sydney Star Observer* (hereafter *SSO*), construct a gay male body—and a gay male *social* body—based on the abjection of what remains "queer" to that system: namely, nonwhite bodies, female bodies, and, importantly for us, the HIV-positive (hereafter HIV+) body.[12] Queer bodies are excluded from the discourses that construct and reinforce both the ideal gay male body and the notions of homosexual essence required for that body to be meaningful. We then argue that abject queerness returns in the *SSO*'s discourses of public health through the conspicuous *absence* of the visibly HIV+ body, since this absence paradoxically conjures up a trace of that which the system tries to expel. We conclude by arguing that because the "queer error" is integral to the *SSO*, gay community media could better serve their constituent communities with a politics of queer inclusion rather than exclusion.[13]

Gay Media Systems: the *Sydney Star Observer*

To argue that gay media systems abject certain forms of queerness is not to argue against the validity or necessity of these systems or the media they produce. Historically, gay and lesbian publications have played an important role in the development of the queer community from the early days of twentieth-century queer activism. U.S. homophile group the Mattachine Society launched *One* in 1953. *One* featured mostly fiction but also research findings and personal essays.[14] Similarly, early lesbian print media in Britain played an organizational and supportive role in community and identity development.[15]

On June 27, 1969, repeated police raids of the Stonewall Inn, a gay bar in Greenwich Village, New York, came to a head when those at the bar fought back. From the resistance to this raid, days of rioting and protests followed; gay men, lesbians, and drag queens fought against police harassment and made their voices heard. The Stonewall Riots are considered by many to signify the birth of the modern, more radical gay liberation movement. However, as Julie Prince argues, "Stonewall, while significant, occurred after decades of struggle in the bars and after the groundbreaking work of early organizing groups. The role that Stonewall played was that of a catalyst."[16] Rodger Streitmatter argues that *Ladder*, *Drum*, and *Homosexual Citizen*, launched in the 1960s, were the first militant gay and lesbian publications, and suggests that these differed from the earlier publications in a number of ways. These include featuring homoerotic and lesbian images, being amusing and informing, and publishing the first news articles written from a queer perspective.[17] These set the tone for future publications of that era, suggesting that gay and lesbian community media formed a significant part of queer community development, identity, and activism in the Stonewall period. Throughout the 1970s and 1980s gay and lesbian community media played an integral role in queer politics and formulating community.[18] A number of historians note the mutual growth of the gay market and gay media.[19] As queer media grew, it attracted larger audiences, and this drew interest from advertisers beyond the queer community. Gay men were seen as a profitable audience because they were thought to have higher-than-average and more readily disposable incomes, and because they were considered to be trendsetters. These connections between the history of gay and lesbian media and the broader gay liberation movement of the late twentieth century indicate the nexus between audiences, community, and consumers.

When thinking about the historical development of gay media, we might be tempted to think that a stable, preformed lesbian or gay subject, with an underlying sense of identity, encounters a text and recognizes themselves in it. However, queer theoretical approaches to gay and lesbian media systems speak about the construction of gay and lesbian identities in line with culturally governed discourses. Thus Rob Cover suggests that the recognition a subject experiences while reading a text is "always a *re-cognition* (or re-thinking) of that subject's sexuality 'in accord' with the significations of the hetero/homo binary."[20] This theory of "re-cognition" is developed from Judith Butler's theory of gender performativity, in which gendered identities, and human subjectivity in general, only come into being through "a stylized repetition of acts."[21] Thus lesbian/gay subjects are reconstituted as such in and through in the process of reading lesbian/gay community media. During this process, readers come to see themselves as identifiable consumers of the text.[22] In reading they are constructed as part of an audience and community.

Cover defines community as "a cultural (discursive) formation which actively produces the identities it claims to have gathered."[23] We can view the gay market as a similar formation. The process of constructing the gay market and the audience-consumer is never complete but rather an ongoing mutually constituted discursive process. *Contra* the Frankfurt School's "hypodermic" model of ideology transmission, audiences do not unthinkingly consume texts.[24] Their preexisting subjectivities contribute to the meanings they make from texts.[25] In reading, audience members encounter the advertising and representations which suggest that consuming particular products cohere with particular identities, creating means of "socially correct participation" in the market.[26] Audiences consolidate their sense of community and identity by participating in particular consumer behaviors and being addressed by advertisers. Alexandra Chasin notes this, arguing that "in the 1990s, market mechanisms became perhaps the most accessible and the most effective mechanisms for many gay people in the process of individual identity formation and entrance into identity-group affiliation."[27] Although some activists raise concerns about assimilation and the loss of politics, for some market engagement is an important part of community engagement and consolidation, and identity formation. Additionally, community engagement can be conceived as political engagement.[28]

Australia's first queer publication was *Camp Ink*, launched in Sydney in 1971. The *SSO* was founded in 1979 and still plays a prominent role in the Sydney queer community today. Former *SSO* editor Marcus O'Donnell classifies its development between 1979 and 1985 as representative of the "clone culture" of the period—a culture that emphasized a subversive mimesis of working-class masculinities, often through an abjection of effeminate men, in opposition to stereotypes that linked homosexuality with effeminacy—and the publication worked toward developing gay community and supporting local gay business.[29] In the past 25 years it has embodied a professionalization of queer journalism and captured the symbiosis that is gay community and market development.[30] In that period many Australian queer community publications with a variety of aims, such as more artistic and literary or politically oriented publications, have come and gone.[31] Other states such as Queensland have weekly or monthly publications similar to the *SSO*. There is also the lesbian-specific gloss magazine *Lesbians on the Loose* and the newer *Cherrie*, for the "sex-positive queer chick." During the 1990s the market grew along with advertising opportunities in the form of more media productions. Today glossy magazines, "bar rags," and an array of web sites stemming from these publications provide the queer community with news, health information, social opportunities, and business contacts and serve as a discursive space for the negotiation of community, identity, audience, and citizenry engagement.[32] In this system, gay media mediates between the marketplace and a community of consumers and sustains dialogic, ongoing relationships with both. While such media have laudable goals and are important institutions in the queer communities they help constitute, they are nonetheless run on principles that maximize efficiency (such as carefully crafted advertisements that appeal directly to gay, lesbian, or bisexual readers) and that, as such, must abject certain kinds of erroneous figures. Such figures, including the HIV+ body, thus become peculiarly productive sites for "jamming" the system and renegotiating the terms of community, marketplace, and the media that helps to constitute them.

Making the Gay Male Body

The entity that publishes the *SSO*, Gay and Lesbian Community Publishing Limited, explicitly positions itself as part of such a system: one of its stated aims is to "serve the Australian gay, lesbian, transgender, bisexual and queer communities" through the *SSO*, which aims to be "the pre-eminent provider of access to news and information through various media."[33] However, as our research indicates, the *SSO*'s understanding of who constitutes this community emphasizes certain correct bodies and abjects others. Although we have identified six main discursive frameworks in the *SSO*—those of community, health, hedonism, equality, affluence, and beauty—our focus in this section will be on beauty. Given that, as Foucault argues, "[s]ilence itself . . . is less the absolute limit of discourse . . . than an element that functions alongside the things said,"[34] we will also examine the silences or absences in the *SSO*'s representation of beauty. In the case of the *SSO*'s discourse on beauty, this absence is the invisible bodies of HIV+ men and women. Although the *SSO* contains a great deal of textual discussion about HIV/AIDS, the visibly HIV+ body is a conspicuous absence from the *SSO*'s gay media system. As an unrepresentable excess that must be abjected in order for the *SSO* to run smoothly, the HIV+ body is properly *queer* to this system.

As gay men in Australia have a statistically higher rate of HIV/AIDS infection than their straight counterparts,[35] and as HIV/AIDS is frequently configured in Western cultures as a "gay disease,"[36] it is not surprising that HIV/AIDS and the battle against it underpins the discourse of community produced by the *SSO* (as well as the more obvious discourse of health). This grounding is present in advertisements, public-service announcements, and stories. Safer sex advertisements, while promoting the obvious idea of safer sex and health maintenance, challenge readers to be active in the fight against HIV/AIDS and to be aware of the risks of unsafe sex. Advertisements encouraging sexual health testing also fulfill these roles and provide information on where to get tested. For example, an advertisement for Holdsworth House Medical Practice in the May 25, 2006 edition of the *SSO* offers "[d]octors who check STIs." However, such advertisements and related stories primarily target gay males. The Holdsworth House advertisement prominently features the Mars symbol (♂), indicating that its intended audience is limited to gay men. Several issues from 2006 feature a well-known series of full-page AIDS Council of New South Wales (hereafter ACON) public service

announcements, each of which pictures an erect condom outline pointing towards a set of buttocks-shaped objects (a pair of disco balls, for
example) and the text "up ya bum." Given both the announcement's
language (the blokey colloquialism implies a male audience) and its
placement in an explicitly queer newspaper, we can rule out the possibility that it targets heterosexual women who engage in penetrative anal
sex. The predominantly gay male audience of these advertisements
and announcements thus limits the scope of who can and cannot engage
in the fight against HIV/AIDS and, therefore, who can and cannot
participate in the *SSO*'s community.

The presence of advertisements encouraging safer sex practices and
regular STI testing for PLWHA (people living with HIV/AIDS) could
suggest that the *SSO* deals with a variety of identities within its discourses
about HIV/AIDS. However, again, these advertisements continue to
target the gay, white male. The ACON treatment updates campaign,
designed to inform PLWHA of new medications and treatment options,[37]
features the Bugle Boys: buff white males, photographed for the April 20,
2006 issue at the campaign launch in a pose of mutual erotic absorption.
(While the Bugle Boys appeared "in the flesh" at the campaign launch,
the advertisements themselves deployed cartoon renditions of the Bugle
Boys.) Stories encouraging PLWHA to engage in safer sex practices,
and reporting the dangers of PLWHA not doing so, complement these
advertisements. Again, these stories also appeared to target mostly gay
men. An example is the story "Sex Strategies Pose Risks," from the April 13,
2006 issue, which reports that "more men appear to be choosing casual
sex partners with the same HIV status."[38] While encouraging safer sex
practices for PLWHA and others, the story's gendered language suggests
that, for the *SSO*, gay males have a larger stake in HIV/AIDS information
and prevention strategies than other parties. Although gay men have
been disproportionately hit by the HIV/AIDS epidemic, particularly in
the early stages of its history (it was initially dubbed GRID, for gay-related
immune deficiency), such gendered understandings of who needs to be
addressed by HIV/AIDS information have the unfortunate side-effect of
contributing to "the medical myth of 'lesbian invulnerability' to STIs."[39]
Undoubtedly the *SSO* and its constituent community are aware of, and
fight against, HIV/AIDS, and in doing so emphasize notions of responsibility, care, and concern for people living with HIV/AIDS. However, the
SSO limits who gets to be aware of and fight against AIDS to gay men,
which creates limited opportunities for other PLWHA or their partners
and family to participate in the *SSO*'s community of responsibility, care,

and concern. These data intimate that the *SSO*'s ideal "queer" citizen is, in fact, the gay male.

Examining the body images presented in the *SSO*'s advertising and feature photography confirms this intimation. The *SSO* citizen revealed in our analysis must possess a number of very specific qualities: *he* must be highly sexed, party often, regard the drag queen highly, be affluent, and appreciate the arts and "high" culture in order to participate in the *SSO*'s community. Importantly, the *SSO* citizen must possess what Johann Joachim Winckelmann has termed "Athenian beauty."[40] Our research shows that white males make up over half of the bodies pictured within the *SSO* in 2006—a powerful imbalance when one considers that all women, transgender people, and intersex people, as well as all nonwhite males, make up the rest of the bodies in the *SSO*, which claims to "serve the Australian gay, lesbian, transgender, bisexual and queer communities."[41] In addition, the kind of white male body displayed in the *SSO* is rigorously policed—overwhelmingly, the bodies on display were able, muscled, toned, smooth (depilated), and in proportion. These bodies bear a striking resemblance to ancient Greek *kouroi*, as well as later Roman marble recreations of Hellenistic statuary, and the male bodies on display in neoclassical portraiture. Indeed, these classical allusions are played with in several advertisements, most prominently for Hairstop Permanent Hair Removal in the June 22, 2006 issue. The advertisement displays a naked, well-muscled, and depilated white male standing atop a pile of freshly removed body hair. In his upheld right palm sits a miniaturized, well-dressed woman with a halo and angel wings—a visual pun on the advertisement's headline, "Every FAIRY needs a Godmother!" The body's shape, proportion, and musculature is almost exactly that of the later Hellenistic *kouroi* and their Roman replicas, on which Winckelmann's model of Athenian beauty is based. In a self-reflexive turn, the advertisement seems aware of its classical heritage through the figure's posture, which closely resembles Myron of Elutherae's *Diskobolos*. Of course, the hairlessness of the figure finds an echo in the smoothness of the marble used for both Hellenistic *kouroi* and their later Roman replicas.

The *SSO*'s insistence on Athenian beauty creates an exclusionary community in several ways. Since Athenian beauty is predicated on the proportion and symmetry of the ancient Greek male, non-white men are highly unlikely to possess the stature, musculature, or facial features to fit its criteria. Similarly, transgender or intersex men are also unlikely to possess these features. Furthermore, since the hair removal treatments that the figure advertises are relatively expensive—starting at AU$50 for small facial areas—a properly Athenian figure, with no unsightly hair to get in the way of those proportioned, symmetrical muscles, is only available to middle- or upper-class men. Indeed, the costs of maintaining the Athenian body are more than just hair removal treatments—such a body also requires a rigorous gym routine, a specialized diet, and the available leisure time to devote to such pursuits.

The Absent Presence of the HIV+ Body

Intriguingly enough, despite the fact that the health discourse of the *SSO* predominantly focuses on HIV/AIDS, and that the *SSO*'s ideal body is that of the gay white male, the *SSO* presented no images of visibly HIV+ or AIDS-afflicted bodies in 2006. Where bodies were textually identified as HIV+, they were presented as attractive and completely healthy-looking. This is in marked contrast to media representations of HIV+ bodies in the early stages of the epidemic, which focused on the corporeal effects of the disease such as Kaposi's Sarcoma (a highly visible type of skin cancer that manifests as purple blotches) and AIDS-related wasting. Although there are several highly pragmatic reasons for this marked shift in representations—including the need to combat homophobic media constructions of the HIV+ body, increases in life expectancy for HIV+ individuals, and the manner in which drug companies market antiretroviral therapies—the removal of the visibly HIV+ body from the pages of the *SSO* indicates that the HIV+ body itself is queer to—and therefore erroneous for—the *SSO* in particular and gay media systems in general.

There is extensive work on the media representations of HIV/AIDS. Most of these focus on mainstream media constructions of AIDS and people with AIDS.[42] These articles argue that the media represented gay men, sex workers, and intravenous drug users with AIDS as deviant and immoral. Those who were infected with HIV via blood transfusion or heterosexual contact were represented as "innocent" victims. Jan Zita Grover, for example, considers images of PLWHA in medical journals, mainstream and queer media, and art photography, all of which fall into a binary of "innocent" or "guilty" bodies. Grover states that the media used images of the "moribund AIDS victim, who was also (magically) a demon of sexuality, actively transmitting his condition to the 'general population' even as he lay dying offstage."[43] "Innocent" victims, such as white heterosexual men and women, were often pictured with their families.[44] There is less work about HIV/AIDS in queer community print media. Most of these document how the queer community press reported the crisis, including brief discussions about images portraying PLWHA.[45] Community publications featured first-person stories showing that PLWHA had loved ones and later showing them living active lives. Strcitmatter details one early image:

> Dramatic images of men struggling with AIDS appeared in *Christopher Street*. The upscale literary magazine painted its first portrait of the disease in 1981 with a first-person article by a man stricken with Kaposi's Sarcoma. Three years later, Phillip Lanzaratta returned to the pages of the magazine to update readers of his life. Lanzaratta appeared on the cover wrapped in the arms of John Lunning, his lover of nineteen years; the strength and defiance expressed on Lunning's face, ushered readers into his lover's inspirational portrait of the resilient human spirit.[46]

Graham Willett briefly recounts how the Australian queer community press responded at the beginning of the crisis and later.[47] Much like U.S. publications, Australian queer community publications featured no-nonsense safer sex information as it became available. These publications also offered the latest research and treatment information, and garnered support for community activism and fund-raising. Community health publications played a similar role. Some research in the area of health communication considers how queer publications construct a sense of identity and community, which is drawn on to communicate safer sex information.[48] Community media research also examines media produced by and for PWHA, such as zines that provide alternative representations and perspectives to popular queer community media.[49] These contributions discuss how such publications project notions of PLWHA as part of a loving community (as opposed to isolated and shunned), as living with a manageable condition, not a death sentence, and as survivors who can take control of their bodies and health and be labeled victims.[50] Scholarship detailing AIDS activism and its use of media activism contributes to the collection of work on representations of PLWHA.[51] AIDS activists such as AIDS Coalition to Unleash Power (ACT-UP) used confronting tactics to draw attention of the media and its audiences to the urgency of the situation. Some research considers the representation of PLWHA in advertising directed at PLWHA.[52] As we have already noted, advertisements for HIV treatment products in both the *SSO* and else-where feature healthy, active bodies. Some feel that these advertisements glamorize HIV, create unrealistic expectations, and downplay the delete-rious side effects of antiretroviral medication.[53] From this summary a gap arises in scholarship that analyzes representations and constructions of HIV/AIDS bodies in queer community media. Further, there is no Australian work on the topic and little scholarship which considers HIV+ bodies. Some of the discussion on advertising considers this. Rodney H. Jones notes that the majority of studies into media representations of AIDS "focus on discourse directed towards the uninfected (the prover-bial 'general population')."[54] He also states that emerging community discourses have received less critical attention. This is where our work can be located, in the gaps between community representations of PLWHA, the images used by these publications and the dearth of current research in this area, and research which documents Australian publications.[55]

Bodies identified as HIV+ in the *SSO* are by and large seen in the con-text of public health messages to the effect that "anyone can have HIV/ AIDS." Historically, the necessity for these safer sex campaigns emerged from the depiction of PLWHA discussed above, where mainstream photojournalists would selectively present certain forms of graphic AIDS-related illnesses in order to construct the alternately pitiable and horrifying body of the "AIDS victim." As such, in the early stages of the epidemic it was necessary to promote the message that one could not "serosort" potential partners simply by examining them for the well-known stigmata of the disease (such as Kaposi's Sarcoma, oral thrush infections, or AIDS-related wasting). Such campaigns feature aggres-sively *healthy* and attractive bodies in order to demonstrate that HIV is an "invisible illness"—a message that neatly dovetails into later representa-tions of HIV+ bodies as active and healthy thanks to advances in HIV treatment such as HAART (highly active antiretroviral treatment). One such campaign is the aforementioned ACON treatment update campaign featuring the Bugle Boys. In ACON's print advertisements, these Bugle Boys are cartoon figures with large biceps and tasteful naval tattoos, cheerfully spreading the good news about updated treatment regimes with fewer side effects in a catty, jocose manner (one headline reads "Still partying like it's 1999?"). The implicit promise of this advertise-ment is that, by updating their treatment regimes, PLWHA can "pass" as healthy HIV-negative (hereafter HIV–) bodies. Other advertisements featuring bodies marked as PLWHA offer images of equally healthy bodies without such celebratory overtones. A sombre advertisement in the June 22, 2006 issue of the *SSO* features two neat, bearded young men drinking coffee. One is relatively animated and speaking towards the other, who looks glumly into the middle distance. "Think it's easy for someone to say they're HIV? [*sic*] Think again," the advertisement's body reads. Here the absence of corporeal signifiers of HIV infection is not celebrated (as in the ACON treatment update campaign), but rather depicted as introducing ethical complications regarding disclosure.

Paradoxically, many explicitly HIV– bodies appear in HIV awareness advertisements. An advertisement in the June 22, 2006 issue, as part of a campaign searching for HIV– volunteers to participate in a HIV vaccine trial, features images of a smiling young Asian-Australian man, a smiling young Anglo-Australian man, a young Anglo-Australian woman, and a middle-aged gay male couple. A disclaimer reads, "The people depicted are models and are used for illustrative purposes only," thus further muddying the reader's ability to distinguish HIV+ and HIV– bodies visually. Is the disclaimer intended to ensure that the models depicted aren't mistakenly understood to be HIV+ themselves? Or does it function as an admission that the advertisers and the modeling agency cannot guarantee that all of the bodies pictured are, in fact, HIV–? In either case, the disease is literally *unrepresentable*.

Tellingly, many of the safer sex campaigns in the *SSO* avoid these quandaries simply by bypassing the depiction of bodies of any sort. A preponderance of the advertisements feature no human bodies, either photographically reproduced or in stylized or cartoon form. In their place the advertisements rely heavily on text and on tasteful design to attract the reader's attention. *In extremis*, these advertisements eschew graphics entirely, as does an advertisement looking for HIV+ men to take part in a long-running study that appears in the May 25, 2006 edition. In this advertisement, stark black text on a white background renders the HIV+ body as radically unrepresentable. Tellingly, the white page space, which has been evacuated in order to avoid the ethical quandaries that revolve around how PLWHA and HIV+ bodies ought to be represented, is more suited to a memorial or funeral notice than an advertisement. In its attempts to abject unsuitable media representations of HIV+ bodies as "AIDS victims," both the *SSO* and its advertisers fail to negotiate the representation of HIV+ bodies, opting instead to follow the path of least resistance—and maximum efficiency—by avoiding the question altogether. While it is clear that a return to the ghostly specter of the "AIDS victim" would not be politically desirable, to abject the HIV+ body from the pages of the *SSO*—or any other gay community medium—would be to waste the potential of that figure to jam the otherwise-smooth functioning of gay community media systems and, in so doing, renegotiate what counts as "community," and at what cost.

Conclusion: On the Queerness of HIV/AIDS

How, then, might we reconsider the entanglement of the terms "queerness," "error," and "HIV/AIDS," as outlined in our introduction? If gay media systems such as the *SSO* function in some way like a body—insofar as they recognize themselves as a "self" and contain mechanisms that abject undesirable "others" in order to live as such—then we must ask what autoimmune diseases such as HIV/AIDS might do to these "bodies." How is HIV/AIDS queer, and how is it queer to gay media systems such as the *SSO?* If the HIV+ body is abjected from these systems, how might these systems more productively engage with the HIV+ body, leaving room for queerness and error?

The potential of HIV/AIDS to queer previously monolithic understandings of what constitutes "self" and "other" can be seen in Donna Haraway's essay "The Biopolitics of Postmodern Bodies: Constitutions of Self in Immune System Discourse," which examines the impact of post-AIDS research into the immune system. A common understanding of the immune system construes it as that which differentiates between the cells and organisms that collectively constitute the "self" and invasive "other" cells, and takes action to destroy "other" cells before they replicate and threaten the integrity of the "self." This perception of the immune system construes it as a barrier, that which stands between "self" and "other" and abjects the "other" from the "self" in order for the "self" to live. AIDS is thus construed as a disease that weakens, and eventually destroys, the immune system's ability to distinguish between "self" and "other," allowing invasive microbial and viral agents the ability to replicate freely inside the self. Such an understanding of the immune system consequently privileges the "self" at the expense of the "other," as Haraway notes:

> When is a self enough of a self that its boundaries become central to entire institutionalized discourses in medicine, war, and business? Immunity and invulnerability are intersecting concepts, a matter of consequence in a nuclear culture unable to accommodate the experience of death and finitude within available liberal discourse on the

collective and personal individual . . . The perfection of the fully defended, "victorious" self is a chilling fantasy, linking phagocytotic amoeba and moon-voyaging man cannibalizing the earth in an evolutionary teleology of post-apocalypse extra-terrestrialism. It is a chilling fantasy, whether located in the abstract spaces of national discourse, or in the equally abstract spaces of our interior bodies.[56]

This fully defended, invulnerable "self" can only understand alterity as a threat, and must use its full array of resources—be they technological, such as the biohazard suit, or immunological, such as T cells—to abject and destroy that which threatens its integrity.

However, the human self revealed by recent biological science bears little resemblance to the "self" of this fantasy of complete individuation. As Haraway notes, evolutionary science presents the "self" as an individual—literally, as Julian Huxley defines it, that which cannot function if cut in half. However, "Huxley's definition does not answer *which function* is at issue. Nothing answers that in the abstract, it depends on what is to be done."[57] Certain bodily functions are preserved if the body is massively damaged, others are not; whether the body that has suffered the damages still counts as an individual depends on which functions are deemed to be most salient to human selfhood, as we can see in the contested legal and ethical terrain of brain death and persistent vegetative states. Thus the "self" revealed by evolutionary biology is highly pluralistic: there are many different types of individuals, depending on the functions that are perceived to be most salient; human selves are a strategic assemblage of other kinds of individuals, or in Richard Dawkins's words, "a partially bounded local concentration, a shared knot of replicator power."[58] When the self is conceived as this strategic assemblage that may or may not be an individual, the notion of an "other" that threatens the "integrity" of the self, and is abjected by that self's immune system, becomes deeply problematic. This is further compounded by post-HIV/AIDS research that reveals the immune system to be anything but a passive system of defense waiting for the invading other. Drawing on Niels Jerne's earlier "network theory" of immune function, Leo Buss's work portrays the immune system as a dynamic series of competing cell lines, "each engaged in its own replicative 'ends.' "[59] These cell lines may have different reproductive interests to the organism itself, thus any harmony between the immune system and the organism that it protects is "highly contingent."[60] Advances in immunological science fueled by the HIV/AIDS crisis render previously common sense distinctions between the self, the other, and the mechanisms that abject the other from the self deeply problematic; in this sense, HIV/AIDS does not so much effect a collapse in the body's ability to distinguish between "self" and "other" but reveals that such distinctions are always already contingent, strategic, and highly vulnerable. As Haraway puts it, "[l]ife is a window of vulnerability. It seems a mistake to close it."[61]

What Haraway's work makes clear is that just because the "self" is a strategic assemblage we must not necessarily affirm absolute alterity and openness. The immune system's processes of abjection may be deeply ambiguous, but this does not negate the fact that it must operate in order for human life to exist, a fact that has been rendered painfully obvious through the HIV/AIDS crisis. Abjection *tout court* is not problematic; whether or not the individual ought to be open to alterity depends on its concrete circumstances and which functions of the self are most salient in those circumstances. If we transpose this understanding of the immune system onto gay media systems' processes of abjection, we can understand that the problematic exclusion of the HIV+ body from the *SSO* does not necessitate a return to media images of the skeletal, blotchy "AIDS victim." Such images are, as we have noted previously, deeply homophobic in intent; for gay media systems to open themselves up to the return of such images, even in the name of a policy of nonexclusion, would be near-suicidal. However, the "AIDS victim" is not the only kind of HIV+ body that the *SSO* could portray. Advances in HIV/AIDS treatment regimes, including HAART, have not only extended the lifespans of HIV+ individuals, but, owing to their side-effects, have also created new somatic markers of HIV+ status. One such somatic marker is lipodystrophy, "a relatively poorly understood metabolic disorder that involves an unusual process of fat redistribution (resulting in an accumulation of fat in certain body parts) and fat loss (*lipoatrophy*)."[62] HIV+ individuals on certain treatment regimes can experience the loss of or redistribution of body fat, typically away from the face (leaving them with a characteristic gaunt look) and buttocks, towards the abdomen ("Crix belly," named after Crixivan) and the back of the neck ("buffalo hump"). These somatic changes are entirely limited to HIV+ individuals on certain antiretroviral therapies, and do not occur in the HIV– population. (As Persson notes, "[a]s is the case with many HIV treatment-related effects, it is . . . unclear to what extent these conditions are caused by the drugs, by HIV itself, or by a combination of both.")[63] Furthermore, these somatic changes are not unknown within the queer community that the *SSO* claims to represent. Persson reports that the recognition of lipodystrophy in the body of another is a common method of "serosorting" potential sexual partners. As one of her informants, "Alf," puts it, "[p]ositive gay men, a lot of them will look for other positive gay partners. Now there are all sorts of mechanisms that people use intuitively and otherwise to actually facilitate that. But the way people look is part of it and to some degree lipodystrophy and lipoatrophy figure in that."[64]

Despite the fact that lipodystrophy and lipoatrophy are well-known in Sydney's queer community, and that these disorders significantly alter the bodies of HIV+ individuals, they are strangely *invisible*, both in the pages of the *SSO* and in medical literature in general. Indeed, Persson herself co-wrote an information pamphlet about these disorders, and opted not to include images of the somatic changes that lipodystrophy and lipoatrophy effect upon the HIV+ body; when she presents conference papers about her work about these disorders, she refrains from using images.[65] Given the problematic history of images of HIV+ bodies in mainstream media, and the potential for images of lipodystrophy or lipoatrophy to be misused by non-queer communities intent on rendering the HIV+ body visible so that it can be eradicated, Persson's decision is justified. However, this absence of representation has dramatic effects on those HIV+ individuals who experience lipodystrophy and lipoatrophy: as one informant puts it, "[y]ou don't have any role models and positive people with 'lipo' coming out and saying, 'I have lipo and this is the effect it's had on me' . . . You hear about what it's like to be [HIV] positive and all that stuff. But you don't hear about what it's like to have lipodystrophy . . . There is a silence around it."[66] While the *SSO* has published articles about lipodystrophy,[67] in 2006 there were no issues that visually portrayed bodies with this disorder. If the *SSO* were to include images of lipodystrophy in its pages, these would necessarily have to be approached with tact and a sensitivity to how these images could be misused both within and without Sydney's queer community. Nevertheless, to actively abject HIV+ bodies with lipodystrophy and lipoatrophy in the name of "protecting" the queer community radically underestimates that community's ability to actively renegotiate what can appear to be negative images (such as those presented by mainstream media at the beginning of the HIV/AIDS crisis) and use these images as the basis of community building. As Dennis Altman notes, "there is little doubt that the creation of an international apparatus to respond to the [HIV/AIDS] epidemic has enabled a growing number of young men across the world to develop a shared perception of themselves as part of an international gay world."⁶⁸ While it is clear that gay media systems must abject certain images and discourses in order to survive, the exclusion of visibly HIV+ bodies problematically neglects to deal with issues that the very community the *SSO* helps to constitute faces.

Notes

1. For a brief history of Ulrichs, see Hubert Kennedy, "Karl Heinrich Ulrichs: First Theorist of Homosexuality," in *Science and Homosexualities*, ed. Vernon A. Rosario (New York: Routledge, 1997), 26–45.

2. Arnold I. Davidson, *The Emergence of Sexuality: Historical Epistemology and the Formation of Concepts* (Cambridge, MA and London: Harvard University Press, 2001), 63.

3. For more on this, see Michel Foucault, *The History of Sexuality, Volume One: The Will to Knowledge*, trans. Robert Hurley (London: Penguin, 1998), esp. 17–49; and Davidson, 30–65.

4. Cited in Colm Tóibin, *Love in a Dark Time: Gay Lives from Wilde to Almodovar* (Sydney: Picador, 2001), 8.

5. Cited in Jonathan Ned Katz, *The Invention of Heterosexuality* (New York: Plume, 1996), 3.

6. David M. Halperin, *Saint Foucault: Towards a Gay Hagiography* (Oxford: Oxford University Press, 1995), 62, emphasis in original.

7. Ibid, emphasis in original.

8. Eve Kosofsky Sedgwick, *Tendencies* (London and New York: Routledge, 1993), 8, emphasis in original.

9. Ibid., xii, emphasis in original.

10. Gay men and women have frequently welcomed genetic studies that "confirm" the existence of a "gay gene" (or, more realistically, a series of genetic variations that predispose one to be sexually attracted to the members of one's own sex), as the scientific confirmation that their sexual identity is immutable could potentially lead to increased civil protections from a judiciary that has typically understood homosexuality as a matter of "lifestyle choice." Nonetheless, as Sedgwick notes, "it is the conjecture that a particular trait is genetically or biologically based, *not* that it is 'only cultural,' that seems to trigger an estrus of manipulative fantasy in the technological institutions of the culture . . . a medicalized dream of the prevention of gay bodies seems to be the less visible, far more respectable underside of the AIDS-fueled public dream of their extirpation." Eve Kosofsky Sedgwick, *Epistemology of the Closet* (Berkeley and Los Angeles: University of California Press, 1990), 43.

11. In this chapter we deploy the verb "to abject" in its broadest, common-language sense: etymologically, *ab-iacere*, "to cast away." Such an understanding of abjection is not incompatible with Julia Kristeva's famous elaboration of the concept in her *Powers of Horror: An Essay on Abjection* (New York: Columbia University Press, 1982), but our use of the term should not be read as relying

upon Kristeva's post-Lacanian understanding of ego-development. Our concern here is less with psychic abjection (the process of expelling matter that threatens the coherence of the ego) but with bodily abjection (the immune system's processes of neutralizing and expelling foreign antigens), and systemic abjection (the gay media system's processes of expelling representations of the HIV+ body).

The abjection of the queer by mainstream heterosexual cultures need not necessarily take the crude form of anti-gay violence, although it does frequently enough. Other forms of abjection include the sublimation of this violence in a concerted effort to cure homosexuality as a "diseased" form of sexual expression, as in late nineteenth-century sexology, or in the desire to ensure that no child has "the gay lifestyle" "forced" upon them—a form of abjection that has become increasingly salient in the wake of same-sex marriage debates and the pervasive fear that children with same-sex adoptive parents will be more likely to grow up gay themselves. See Eve Kosofsky Sedgwick, "How to Bring Your Kids Up Gay," *Social Text* 29 (1991): 18–27.

[12] Throughout this chapter we will deploy the terms "HIV" and "AIDS" in the singular for the sake of expediency. However, it should be noted from the outset that "HIV" and "AIDS" name a series of different illnesses with different symptoms, causes, and effects. We can note from the outset that there are, in fact, two species of HIV, named HIV-1 and HIV-2, each of which has distinct geographical reaches (HIV-2, being less virulent, remains mostly restricted to West Africa; HIV-1 occurs globally). Within each of these species exist several different strains, some of which break down the immune system more swiftly than others. AIDS itself is a syndrome indicated by the presence of several opportunistic infections such as Kaposi's Sarcoma or *Pneumocystis jirovecii* pneumonia. As such, as Cindy Patton notes, "AIDS is historically specific, arising (presumably) at a moment when advanced technology could locate a primary causative agent to a set of extremely diverse symptoms." See Cindy Patton, *Inventing AIDS* (London and New York: Routledge, 1991), 27. Furthermore, one of the historical conditions for emergence of AIDS as a known syndrome is the existence of a robust and visible gay subculture that had contested previous discursive constructions of homosexuals as constitutionally diseased, so that a virus decimating its population could be perceived statistically. See Patton, 28. Our discussion of "HIV" and "AIDS" throughout this chapter, therefore, refers to a particular white, Western formation of HIV and AIDS and this interest is marked by our own subject positions as white, Western queers within the academy. It may well be impossible for any author or authors to fully represent HIV and AIDS in its medical, sociological, and personal complexity; as such, our deployment of the terms "HIV" and "AIDS" should not be read as arrogating any epistemic privilege of "authoritatively" representing this complex of diseases within this chapter.

[13] Although we recognize, with Alan McKee, that the marketplace is a legitimate place to construct notions of citizenship, this does not mean that all markets are beyond critique, especially markets whose basis is economically, sexually, and ethnically exclusionary. See McKee, "I Don't Want to Be a Citizen (If It

Means I Have to Watch the ABC)," *Media International Australia Incorporating Culture and Policy* 103 (2002): 14–23.

14 Rodger Streitmatter, "*The Advocate*: Setting the Standard for the Gay Liberation Press," in *The Columbia Reader on Lesbians and Gay Men in Media, Society and Politics*, ed. Larry P. Gross and James D. Woods (New York: Columbia University Press, 1999), 450.

15 Rebecca Jennings, *A Lesbian History of Britain: Love and Sex Between Women Since 1500* (Oxford: Greenwood World Publishers, 2007), 161–162.

16 Julie Prince, "This Queer History: An American Synthesis," *Culture, Society and Praxis* 3, no. 1 (2004): 62.

17 Rodger Streitmatter, "Lesbian and Gay Press: Raising a Militant Voice in the 1960s," *American Journalism* 12, no. 2 (1995): 142–162.

18 See: Streitmatter, "*The Advocate*"; Fred Fejes, "Advertising and the Political Economy of Lesbian/Gay Identity," in *Sex and Money: Feminism and Political Economy in the Media*, ed. Eileen R. Meehan and Ellen Riordan (Minneapolis: University of Minnesota Press, 2002), 196–208; Marcus O'Donnell, "*Star* Wars: Patterns of Change in Community Journalism at the *Sydney Star Observer*," *Australian Studies in Journalism* 13 (2004): 139–163; and Katherine Sender, *Business, Not Politics: The Making of the Gay Market* (New York: Columbia University Press, 2007), 24–63.

19 In addition to Streitmatter, Fejes, and O'Donnell (cited above), see also: Dan Baker, "A History in Ads: The Growth of the Gay and Lesbian Market," in *Homo Economics: Capitalism, Community, and Lesbian and Gay Life*, ed. Amy Gluckman and Betsy Reed (New York: Routledge), 11–21; Alexandra Chasin, "Interpenetrations: A Cultural Study of the Relationship between the Gay/Lesbian Niche Market and the Gay/Lesbian Political Movement," *Cultural Critique* 44 (2000): 145–168; and Katherine Sender, "Gay Readers, Consumers and a Dominant Gay Habitus: 25 Years of the *Advocate* Magazine," *Journal of Communication* 51, no.1 (2001): 77–99.

20 Rob Cover, "Re-sourcing Queer Subjectivities: Sexual Identity and Lesbian/Gay Print Media," *Media International Australia incorporating Culture and Policy* 103 (2002): 115, emphasis in original.

21 Judith Butler, *Gender Trouble: Feminism and the Subversion of Identity* (London and New York: Routledge, 2007), 191, emphasis removed. See *Gender Trouble* 175–193 for a more complete description of gender performance; Angela McRobbie's *The Uses of Cultural Studies* (London, Thousand Oaks, and New Delhi: SAGE, 2005), 68–90 clears up some of the more tendentious misreadings of Butler's theory as voluntaristic.

22 Cover, 113.

23 Ibid.

24 Janice A. Radway, *Reading the Romance: Women, Patriarchy and Popular Literature* (Chapel Hill: University of North Carolina Press, 1984); Angela McRobbie, *Feminism and Youth Culture: From "Jackie" to "Just Seventeen"* (Boston, MA: Unwin Hyman, 1991); Alan McKee, "Introduction," in *Beautiful Things in Popular Culture*, ed. Alan McKee (Malden, MA: Blackwell, 2007), 4.

25 Alexander Doty, *Making Things Perfectly Queer: Interpreting Mass Culture* (Minneapolis: University of Minnesota Press, 1993); Herman Gray, *Watching*

Race: Television and the Struggle for "Blackness," (Minneapolis: University of Minnesota Press, 1995). The *locus classicus* for this argument is Roland Barthes, "The Death of the Author," in *Image–Music–Text,* ed. and trans. Stephen Heath, (New York: Hill and Wang, 1988), 142–148.

26 Richard Ohmann, cited in Sender, "Gay Readers, Consumers, and a Dominant Gay Habitus," 76.

27 Chasin, 151.

28 McKee, "I Don't Want to Be a Citizen"; Leila J. Rupp, *A Desired Past: A Short History of Same-Sex Love in America* (Chicago: University of Chicago Press, 1999).

29 O'Donnell, 144–149.

30 Martyn Goddard, "The Whole Truth: Limits on Gay and Lesbian Journalism," in *Gay and Lesbian Perspectives III: Essays in Australian Culture,* ed. Garry Wotherspoon (Sydney: University of Sydney, 1996), 1–16.

31 Anne Scahill, "Queer(ed) Media," in *Queer City: Gay and Lesbian Politics in Sydney,* ed. Craig Johnston and Paul van Reyk (Annandale: Pluto Press), 183.

32 Cover, "Re-Sourcing Queer Subjectivities"; McKee, "I Don't Want to Be a Citizen," 17; O'Donnell; Rob Cover, "Engaging Sexualities: Lesbian/Gay Print Journalism, Community Belonging, Social Space and Physical Place," *Pacific Journalism Review* 11, no. 1 (2005): 113–132; Alan McKee, *The Public Sphere: An Introduction* (Cambridge: Cambridge University Press, 2005), 140–171; Amit Kama, "Israeli Gay Men's Consumption of Lesbigay Media: 'I'm Not Alone . . . in This Business,' " in *Media/Queered: Visibility and its Discontents,* ed. Kevin G. Barnhurst (New York: Peter Lang, 2007), 125–142; Shirleene Robinson, "Queensland's Queer Press," *Queensland Review* 14, no. 2 (2007): 59–78.

33 Gay and Lesbian Community Publishing Limited, "About Our Company," http://www.starobserver.com.au/about.

34 Foucault, 27.

35 Marian Pitts, Anthony Smith, Anne Mitchell, and Sunil Patel, "Private Lives: A Report on the Health and Wellbeing of GLBTI Australians" (La Trobe University, Victoria: Gay and Lesbian Health Victoria, The Australian Research Centre in Sex, Health & Society, 2006), 11.

36 See Lee Edelman, *Homographesis: Essays in Gay Literary and Cultural Theory* (London and New York: Routledge, 1994), 79–92.

37 See John Burfitt, "The New Word on HIV Meds," *Sydney Star Observer,* April 20, 2006, 5.

38 Ian Gould, "Sex Strategies Pose Risks," *Sydney Star Observer,* April 13, 2006, 5.

39 Pitts, Smith, Mitchell, and Patel, "Private Lives," 43.

40 Cited in George L. Mosse, *The Image of Man: The Creation of Modern Masculinity.* Oxford: Oxford University Press, 1998), 26.

41 Gay and Lesbian Community Publishing Limited, "About Our Company."

42 See Leo Bersani, "Is the Rectum a Grave?," *October* 43 (1987): 197–222; Simon Watney, "The Spectacle of AIDS," *October* 43 (1987): 71–86; John Tulloch, "Australian Television and the Representation of AIDS," *Australian Journal of Communication* 16 (1989): 101–124; Allison Fraiberg, "Of AIDS, Cyborgs, and Other Indiscretions: Resurfacing the Body in the Postmodern," *Postmodern Culture* 1, no. 3 (1991); Deborah Lupton, "Apocalypse to Banality: Changes to

Metaphors about AIDS in the Australian Press," *Australian Journal of Communication* 18, no. 2 (1991): 66–74; Dorothy Nelkin, "AIDS and the News Media," *The Milbank Quarterly* 69, no. 2 (1991): 293–307; Deborah Lupton, Simon Chapman, and W. L. Wong, "Back to Complacency: AIDS in the Australian Press, March to September 1990," *Health Education Research: Theory and Practice* 8, no. 2 (1993): 5–17; Valerie Sacks, "Women and AIDS: An Analysis of Media Misrepresentations," *Social Science and Medicine* 42 no. 1 (1996): 59–73; Simon Watney, *Policing Desire: Pornography, AIDS, and the Media* (London: Cassell, 1997); Brian Ray Fruth, *Media Reception, Sexual Identity, and Public Space* (Ph.D. dissertation, University of Texas, Austin, 2007).

43 Jan Zita Grover, "Visible Lesions: Images of the PWA," in *Out in Culture: Gay, Lesbian, and Queer Essays on Popular Culture*, ed. Corey K. Creekmur and Alexander Doty (Durham: Duke University Press, 1995), 362.

44 Ibid., 364.

45 See Dennis Altman, *Power and Community: Organisation and Cultural Responses to AIDS* (London: Taylor and Francis, 1994), 21; Grover, "Visible Lesions"; Rodger Streitmatter, *Unspeakable: The Rise of the Gay and Lesbian Press in America* (London: Faber and Faber, 1995), 243–267, 273–275, 286–294, 318–320, 330; Graham Willett, *Living Out Loud: A History of Gay and Lesbian Activism in Australia* (St. Leonards: Allen and Unwin, 2000), 170, 177–178.

46 Streitmatter, *Unspeakable*, 264.

47 Willett, 170, 177–178.

48 Thomas A. Morton and Julie M. Duck, "Social Identity and Media Dependency in the Gay Community: The Prediction of Safe Sex Attitudes," *Communication Research* 27, no. 4 (2000): 438–460; Michael Hurley, *Then and Now: Gay Men and HIV*, Monograph series no. 46 (Melbourne: Australian Research Centre in Sex, Health, and Society, La Trobe University, 2003).

49 Steven Dunne, "Inter/Erupt! Queer Zine Scene?" *Media International Australia Incorporating Culture and Policy* 78 (1995): 53–68; James Gillett, "The Challenges of Institutionalization for AIDS Media Activism," *Media, Culture and Society* 25 (2003): 607–624.

50 Ibid., 617–618.

51 David Roman, *Acts of Intervention: Performance, Gay Culture, and AIDS* (Bloomington: Indiana University Press, 1998); Douglas Crimp, "AIDS: Cultural Analysis/Cultural Activism," in *Theory in Contemporary Art Since 1985*, ed. Zoya Kocur and Simon Leung (Malden: Blackwell, 2005), 141–149.

52 Streitmatter, *Unspeakable*, 319–320; Rodney H. Jones, "Marketing the Damaged Self: The Construction of Identity in Advertisements Directed Towards People with HIV/AIDS," *Journal of Sociolinguistics* 1, no. 3 (1997): 393–418; Blaine J. Branchik "Queer Ads: Gay Male Imagery in American Advertising," *Consumption Markets and Culture* 10, no. 2 (2007): 147–158; Sender, *Business, Not Politics*, 36–39, 47–49, 158–162.

53 See, for example, Richard Goldstein, "Climb Every Mountain: The Art of Selling HIV Drugs," *Poz* 40 (1998): 64–67; Sean Slavin, "Positive Aspirations: HIV, Representation and Stigma," *Positive Living* (February 1, 2003), http://napwa.org.au/pl/2003/02/positive-aspirations-hiv-representation-and-stigma.

54 Jones, 394.

55 Perhaps this gap also signifies errors within other systems—an exclusion
    of the consideration of HIV+ bodies in scholarly literature about queer com-
    munity media. As such, our own chapter can be read as an attempt to
    productively jam systems of knowledge-production by reintroducing the
    abjected, and highly troubling, figure of the HIV+ body into a specialized
    branch of academic discourse.
56 Donna Haraway, "The Biopolitics of Postmodern Bodies: Constitutions of Self
    in Immune System Discourse," in *Simians, Cyborgs, and Women: The Reinvention
    of Nature* (New York: Routledge, 1991), 224.
57 Ibid., 216, emphasis in original.
58 Cited in ibid.
59 Ibid., 220.
60 Ibid.
61 Ibid., 224.
62 Asha Persson, "Facing HIV: Body Shape Change and the (in)Visibility of
    Illness," *Medical Anthropology* 24 (2005): 243, emphasis in original.
63 Ibid., 259 n. 4.
64 Ibid., 250.
65 Ibid., 254.
66 Ibid., 253.
67 See ibid., 248–249.
68 Dennis Altman, "On Global Queering," *Australian Humanities Review* 2 (July–
    August 1996), http://www.australianhumanitiesreview.org/archive/Issue-July-
    1996/altman.html.

Chapter 12

Error-Contagion: Network Hypnosis and Collective Culpability*

Tony D. Sampson

University of East London

This chapter introduces the concept of error-contagion and applies it to two examples of contemporary social collectivity. It argues accordingly that so-called *botnet* cybercrime and *financial contagion* are not errors generally defined as a "deviation from truth or accuracy,"[1] but are rare, non-periodic events, inseparable from the psychological, biological, and technological permeability of the social collectives in which they self-propagate. The chapter begins by exploring two alternative sociological approaches, which may help to explain how social collectives become prone to both criminal and financial error-contagions. My aim is initially to draw attention to how both examples can be framed, on the one hand, by Émile Durkheim's theory of *anomie*, and on the other, Gabriel Tarde's notion of *imitation-suggestibility*. Ultimately though, my intention is to fast-forward these two old thinkers of the collective to the new age of networks. From there the discussion assesses how these distinct approaches might help to establish culpability for error-contagions, which seem to increasingly piggyback relations between end-users of a network and the institutions of the political economy.

Defining Error-Contagion: Anomie or Imitation?

Durkheim's theory of anomie is a *necessary evil* a corrupted society needs to suffer in order to regulate itself and become whole again. Paradoxically

* An earlier version of this chapter was presented as a paper at the ATACD Changing Culture: Cultures of Change Conference, Barcelona, Spain, 10–12 December, 2009. Thanks go to those who commented on it and to University of East London for funding the visit.

perhaps, without anomie there is no social need to express moral or legal rules, which are always worked out via the averaging of the opinion of the collective consciousness.[2] Crime and financial crisis can, as such, be defined as indispensable anomalies that function to correct instabilities and return equilibrium to the collective. In other words, the emergence of deviations from the norm results in social actors formally coming together to reject what is contrary to the common good. In the case of acts of criminality, Durkheim argues that by offending the sentiments of collective consciousness, the crime provokes punishment. "We must not say," Durkheim states, "that an action shocks the *conscience collective* because it is criminal, but rather that it is criminal because it shocks the *conscience collective.*" He continues: "[w]e do not condemn it because it is a crime, but it is a crime because we condemn it."[3]

In times of economic disasters, when individuals go bust and become unemployed, or in periods of boom in which power and wealth grow out of proportion, Durkheim contends that the affected anomic actors need to recommence their "moral education." The outcome of which is that they either learn greater self-control, and thus fit back into to the equilibrium of the collective consciousness, or drift further towards a state of anomie, ultimately committing the act of anomic suicide.[4]

Tarde was apparently inspired to make imitation-suggestibility central to his late nineteenth-century microsociology following his time as a lawyer observing how crime waves spread through social environments.[5] The "social man," Tarde argues in *The Laws of Imitation* (1890), is like a somnambulist, imitating the thoughts and actions of others, mostly unawares. In Tarde's epidemiological model of the social, comprising of periodic repetitious and habitual unconscious associations, imitation-suggestibility is an ever-present factor. Imitation radiates through the porous relations established between the self/other divide, triggering the occasional rare, non-periodic monstrosity.

The spread of the invention of novel crime relates to both the contagious density of social environments and an imitative tendency for influential criminals to spread their innovations, from person-to-person, like a fashion that unconsciously usurps older forms of criminal performance. Later in 1902, Tarde drew on similar ideas when he published his contribution to political economy-*Economic Psychology*-in which he argued that value is not the only yardstick by which to measure the rise and fall of stock markets. For Tarde, biological and psychological desires have an imitative momentum of their own, which can equally affect value. Taking into consideration the role of a range of new technologies, including networks, that can amplify and potentially steer imitative tendencies, Nigel Thrift recently argued that the "reach and complexity [of imitative momentum] has expanded inordinately since Tarde's time, allowing them undreamt of generative powers that precisely follow a logic of 'mimetic rationality.' "[6] Thrift's updating a Tardean "propensity" to imitate, not only re-approaches political economy, pointing to the importance of biological and psychological social factors affecting value, but also, I contend, lends itself just as well to the herding practices evident in botnet crime.

Consequently, on the one hand, through Durkheim, error-contagions are defined as the anomic regulation of social stability and collective consciousness. On the other hand, following Tarde, error-contagions are regarded as contagious momentum, cascading through the unconscious associations of networked collectives. So as to further explore anomic and imitative understandings of error-contagions, the chapter initially focuses on a specific example of a botnet called *Storm*, before moving on to look at the role share sliding practices have played in the recent *Credit Crunch* financial crisis.

The Storm Worm

Although much of the hyperbole surrounding networks tends to portray them as the perfect medium for sharing information, they are also the ideal medium for digital contagions.[7] Indeed, network architectures support emergent forms of so-called *collective intelligence*, but they also cultivate a virus-friendly environment that readily takes advantage of *collective gullibility*. Botnets (robot networks) are the latest innovation of this kind. As a derivative of the computer virus and worm phenomenon, beginning in the mid-1980s, the botnet starts by using social engineering techniques to spread a Trojan-like virus to a networked PC, with the intention of tricking the end-user into unwittingly turning their machine into a *zombie*, which then involuntarily infects other machines, and so on. Eventually large networks of zombie PCs (millions in some cases) are controlled, at-a-distance, by a *bot herder* in order to attack targeted web sites with floods of erroneous data, distribute billions of spam mail, and potentially steal financial data from online *bank accounts*.

The virus analyst Vitaly Kamluk claims that botnets have been around for at least 10 years.[8] In the late 1990s hackers developed automated "backdoor" programs that could infect and enable remote control of one or many computers. Initially these early bot programs were limited to the direct control of individual hackers, who would send commands to each machine. Over time however, empirical tinkering has resulted in a more decentralized and indirect exploitation of Internet Relay Chat, Instant Messaging, World Wide Web, and Transport-Layer Protocols. Echoing to some extent the benevolent viral ecologies proposed by Fred Cohen in the late 1980s,[9] botnets tap into the networks they infect in order to generate enormous amounts of computing power, allowing the anonymous control of vast collectives of infected machines.

Kamluk identifies two major developments in botnet technology that exploit network vulnerabilities. Both begin by infecting machines using viral techniques, but the first uses what is termed as a command and control center(C&C) method, which waits for an infected machine to connect to a central server and then registers it in a database. All of the addresses of the infected zombie machines are then made visible to the C&C, allowing the bot herder to send commands from a central point. There are various commands that can be sent to a botnet. Commands can update the executable file (.exe) that infected the PC in the first place and trigger a Distributed Denial of Service (DDoS) attack. Botnet commands can control the spreading of millions of spam messages and adware programs on the network and also make a machine a proxy server, with the aim of concealing the control center from network security programs. However, since the release of *Storm Worm* in 2007, a second type of peer-to-peer (P2P) botnet has materialized. Instead of the centralized database, each bot has a list of neighbors that it can pass commands on to. In other words, in place of the control center, commands can be transferred from bot to bot. Although, according to Kamluk, most botnets begin by directing a bot to a central server and then switching to p2p mode (a mixed topology approach), increasing decentralization of control makes it almost impossible to trace the source of a botnet.

Share Sliding: Market Equilibrium and the Bursting of Bubbles

A similar level of hyperbole surrounds the advent of the networked economy. "Special" empowerments are afforded to consumers connected to value chains via, for example, price comparison web sites and ready access to stock market trading.[10] However, the catastrophic spreading of financial contagion through networks of traders increasingly dependent upon automated and algorithmic transactions exposes an imperfect medium for value exchange. Indeed, as I will argue below, networks become highly promiscuous vectors for risk-driven increases in market value, which on rare, non-periodic occasions develop into speculative bubbles that build and build until they inevitably burst. Pointing to one potential contributing factor to the current financial crisis, the software theorist Matthew Fuller describes a process in which the algorithmic abstraction of money "develops its own mutant systems of self-referentiality . . . disturbing, inventing or wrecking what it is putatively 'equivalent' to."[11] One recent news report claims that 40 percent of the trading decisions in the recent financial crash were automated,[12] while other industry insiders suggest it is considerably higher.[13] The escalating use of algo-trading, ICTs speeding up transactions and the incorporation of market sentiment into electronic feeds intensifies error-contagion, pushing it towards an epidemic threshold point.

However, the virality of financial contagion not only contains the self-referentiality of automated value, but also interweaves the hormonally charged atmosphere of the market mood into collective decision-making processes. In times of economic boom erroneous trader decisions can chaotically inflate markets to a point of self-destruction. Recent events in the world market have, as such, been linked to the abuse of financial instruments designed to spread profitable risk, but propagate perilous contagions triggered by greed and hesitation instead.[14] Moreover, the spurious flows of financial contamination, rumor, and testosterone fuelled trading practices do not merely influence exchange values, but impact upon the emotional psychologies and physical well-being of the individuals whose livelihoods and desires are interwoven with the network economy. The error here is not solely about the looping logic of a mostly automated value system, but questions a political economy model wherein it is assumed that rational self-interest of economic collectives, unaffected by the "noise" of the social environment, guarantees that the market always knows best.[15] Economic crisis brings to light a considerable glitch in this decision-making process. *Homo Economicus* and the electronic marketplace are, evidently, erroneous administrators. The rise and fall of value is steered as much by the collective psychologies and biologies of risk, greed, hazard, anxiety, and panic as it is by rational, cognitive self-interest.

Significantly, error-contagions in financial markets are not simply reducible to the contagion in itself, or indeed the volatile bubbles that develop prior to the dramatic collapse of markets. The error relates instead to, as some economic researchers argue, the puzzling question of why, if nearly all professional traders agree that there is a bubble present in the economy, is the anomaly actively encouraged to persist until it reaches a point where it overspills into crisis?[16] In a research project carried out at Cornell University designed to test out how this scenario might come about, human participants were asked to act as "smart traders" competing in the market against a robot model of the bullish irrationality of "sentiment traders." The smart traders, as expected, begin by "fronting running" the sentiment traders, buying up stocks before prices begin to rise. Nonetheless, the human actors decide to delay taking advantage of price differentials, because they tend to believe they can make more profit from the arbitrage of price errors than they will selling in periods of market equilibrium. As the resulting bubble starts to build, the decision-making processes of the smart traders tends to veer towards the chaotic, no one knowing whether or not to sell or hold off until prices hit their ceiling. Subsequently, delay after delay continues to inflate the bubble, leading to aggressive short-selling only when the bubble finally begins to burst.

The chaotic accidentalness of these events would seem to lie in the collective irrationalities of the smart trader's desire for riches. However, such illogical behavior is not beyond the exploitation of those who wish to turn the desire for riches into further surpluses of value. Reports referring to the near collapse of *Halifax Bank of Scotland* (HBOS) in 2008 demonstrate (see below) how errors, particularly those fed through automated systems, can be purposefully steered by rumor mongers, forcing rapid slides in share prices. The ensuing financial contagions took just a couple of hours or so to spread through the global network.

8am:	Before share trading began rumors were swirling that Lloyds TSB faced a funding crisis. But this talk—which originated in Asia—quickly turned to HBOS, which had been strong in initial share trading.
8.30am:	HBOS shares start to slide.
8.43am:	The London Stock Exchange's computers automatically stopped trading in the shares because of the huge discrepancies in the prices being fed into the system made the exchange question whether mistakes were being made. When trading resumed, the dramatic fall took place.

9.01am: HBOS moved to deny any problems. "This is complete and
            utter nonsense," a spokesman said. "HBOS is one of the
            strongest financial institutions in the world. We are one of
            the most respected institutions in the capital and whole-
            sale markets. We continue to access the wholesale markets
            whenever we think it is appropriate to do so."
10.18am: The Bank of England denies rumors of a banking crisis.[17]

Collective Culpability

Questions concerning culpability for error-contagion inevitably arise
from both of the above examples. Who should, for instance, assume
responsibility for two potentially catastrophic error-contagions driven,
to some extent it seems, by a collective propensity to involuntarily and
contagiously spread error to a point where an epidemic threshold is
breached? However, as I will argue below, the answer is not approached
consistently in the recent responses of the economic institutions involved.
On the one hand, the collective culpability for involuntarily becoming
infected by a botnet is more often than not inculpated to the end-user of
an online banking system. A recent surreptitious change to the banking
code in the UK, for example, has "quietly" shifted financial responsibility
to those end-users not considered to be adequately protecting themselves
from zombie programs by using "up-to-date" and "reputable" security
software.[18] On the other hand though, the indiscretions of bankers
engaged in so-called casino practices become bizarrely absolved from
the events that trigger monstrous financial contagions. A return of the
billion dollar bonus culture and resistance to regulation would surely
suggest that many professional traders do not see themselves as
accountable for contagions ignited in part by the erroneous collective
decision-making processes of desiring traders: a risky collective propensity toward a crisis that affects the psychological and biological well-being
of all those connected to its flow.

Two Sociologies of the Collective

To clearly distinguish between Tarde and Durkheim, it is perhaps
necessary to return to a much older spat between these two forefathers

of collective sociology. My discussion is, as such, partly framed by a "momentous debate" between Tarde and Durkheim at the *Ecole des Hautes Etudes Sociales* in 1903.[19] Tarde and Durkheim disagreed on a wide range of issues and have become, in recent years, regarded as the polar opposites of their discipline. As one conference blurb recently put it, "Durkheim has been thinned over the years to the point of becoming a straw man," while Tarde, "[o]nce dismissed as a naive precursor to Durkheimian sociology . . . is now increasingly brought forward as the misrecognised forerunner of a post-Durkheimian era."[20] Yet, although clearly slipping under the radar of much of what constitutes northern European sociology, Tarde was not altogether eclipsed by the Durkheim paradigm, but rather bubbled-up in early translations in American microsociology, in Gilles Deleuze's occasional footnotes,[21] and in the work of Maurizio Lazzarato.[22] More recently, Bruno Latour has claimed Tarde as a thinker of networks, and as such, makes him a "forefather" of actor network theory (ANT).[23] Moreover, a host of northern Europeans have recently rediscovered Tarde's approach to political economy: notably Nigel Thrift, Lisa Blackman, and Christen Borch.[24]

Focusing on the question posed in this chapter, I want to fast-forward Durkheim and Tarde as distinct thinkers of contemporary social collectives and assess how they might contrastingly approach these examples of error-contagion. As I have already set out above, there is a fundamental distinction to make between Durkheim's theory of anomie and Tarde's notion of imitative associations. On the one hand, Durkheim's theory corresponds with a notion of the network as a conscious collective "soul" (*l'ame collective*) that influences the consciousness of the individual.[25] On the other hand, Tarde's concern with imitation makes no real distinction between social categories (nature or culture), but instead describes social relations as a monadology of micro-level associations passing through mostly unconscious collectivity. The question, it would seem, is how Durkheim and Tarde would, from these differing perspectives, grasp the examples of erroneous criminality and economic crises, and how would they, ultimately, apportion culpability? I begin with Durkheim.

Collective Intelligence or Error?

If Tarde is a precursor to Deleuze's assemblage theory, as well as the forefather of ANT,[26] then Durkheim's notion of "dynamic density" arguably makes him the forefather of an altogether different theory of social

complexity and collective emergence. By way of his influence on Talcott
Parsons' functionalism, Durkheim has subsequently been claimed by
some authors as an early pioneer of systems theory and cybernetic
approaches to the social, including notions of swarm, collective, and distributed intelligence.[27] As Durkheim contends below, together we are
indeed smart!

> Society is not at all the illogical or a-logical, incoherent and fantastic
> being which it has too often been considered. Quite on the contrary,
> the collective consciousness is the highest form of the psychic life,
> since it is the consciousness of the consciousnesses. Being placed out-
> side of and above individual and local contingencies, it sees things
> only in their permanent and essential aspects, which it crystallizes into
> communicable ideas. At the same time that it sees from above, it sees
> farther; at every moment of time, it embraces all known reality; that is
> why it alone can furnish the mind with the moulds which are applica-
> ble to the totality of things and which make it possible to think of
> them. It does not create these moulds artificially; it finds them within
> itself ; it does nothing but become conscious of them.[28]

As follows, Durkheim was concerned with how an "increase in the volume and dynamic density of societies . . . making social life more intense
and widening the horizons of thought and action of each individual,
profoundly modifies the basic conditions of collective life."[29] The organic
glue that brings the social together is in this light a homeostatic process
founded upon social facts and the downward influence of collective
intelligence.

It can be argued that Durkheim would regard the network that is
conscious as the network that controls the individual. Thought of as
a crime against the collective mind, the botnet becomes, as such, an
offending neurosis, an irregular anomie, like a brain cancer growing
inside of organic solidarity.[30] This cancer would necessitate a new framework of moral and legal rules. Perhaps the admixture of morality and
legality applied in the recent changes to the UK banking code discussed
above reflect to some extent a corresponding interpretation of anomie
in which the health of the majority is seen as more important than the
rights of the minority. By encouraging end-users to keep up-to-date,
and purchase "reputable" security solutions, the individual is certainly
compelled to capitulate to the influence of the collective consciousness.
The banking institutions' implicit support of the billion dollar income

revenues of reputable network security businesses aside, the erroneous deviations of cybercrime are best dealt with, it seems, by the application of epidemic averages: the more widely distributed protection is, the less likely a contagion will spread. But botnets surely question the entire moral legitimacy of Durkheim's definition of what constitutes collective consciousness. Indeed, could this evolved technological version of the collective mind be regarded as conscious at all? Is not this decidedly *technological unconsciousness* the flip side of collective intelligence? The making stupid of the smart mob!

Although not considered a political economist, insofar as he understood the economy as just one of many other determining social facts, Durkheim's theory of anomie is never too far away from debates concerning the damaging fallout and culpability for financial crisis. In a recent blog post addressing the consequences from the current economic crisis, Matthew Taylor, the CEO of the Royal Society of Arts, and a former Tony Blair adviser, argues that "the credit crunch will generate a wave of anomie."[31] Taylor, drawing directly from Durkheim, predicts that as in other periods of anomie, rates of murder, suicide, and depression will inevitably go up. To be sure, according to abundant news reports in the popular press, the *Credit Crunch* has been repeatedly compared and contrasted to the cataclysmic events of *The Great Depression* in the 1930s. Along similar lines, Taylor's blog response to the crash is perhaps typical of the contriteness of the New Labour party in the United Kingdom, who were once content with letting the banks pretty much do as they pleased, until—that is, post-crash—they started to worry about the number of disaffected voters who become unemployed and depressed as a partial result of their neo-liberal deregulation policies. Again, perhaps similar to twenty-first-century Durkheimians, New Labour now treats economic shock like a crime against the collective good, justifying the necessity for reregulation. As Taylor puts it, "[i]f the reaction of those retreating to their mansions in Gerrards Cross is simply to shrug their shoulders and murmur 'shit happens' they can hardly then complain when the hoi polloi . . . treat the ruling class as undeserving parasites."[32]

Suicide

What would Durkheim's positivistic social sciences approach make of the comparisons and contrasts being made between the current *Credit Crunch* and *The Great Depression*? Well, to begin with, the relation between anomic suicide and economic crisis is a mixture of social myth and fact. For example, the popular notion that *The Great Depression* sparked a suicide epidemic among investors who made financial losses in Manhattan on the October 29, 1929 is not supported by the figures. According to Brian L. Mishara from the International Association for Suicide Prevention (IASP), although a very small number of high profile, and widely reported, cases may have given that impression, the suicide rate was in fact higher in the summer before the *Wall Street Crash* than it was shortly after.[33] Nevertheless, the subsequent rise in long-term unemployment following the crash, and the social problems associated with it, do more concretely correspond to a rise in the suicide rate.[34] Indeed, in the *Credit Crunch* a similar overhyping of investment banker suicide is again apparent in the popular press in the United Kingdom, but as the recession universally impacts upon job security and the levels of poverty, the link between recession and depression, or more precisely, the use of antidepressants to ward off depression, becomes less ambiguous. It has been recently reported, for example, that 2.1 million more prescriptions of antidepressants were made in 2008 compared with 2007.[35]

Mostly Unconscious Desiring Machines

In contrast to Durkheim, Tarde's microsociology appeals to point-to-point social interactions, or imitative encounters, which are not exclusively controlled by the downward pressure of the macro level of collective consciousness. Tarde does not completely dismiss the idea of social wholes, but argues that the whole is a manifestation of habitual repetitions of social invention and imitation prone to the occasional monstrous contagion. Such a social process might be summarized in terms of the *networkability* or *momentum* of imitative flows, which Tarde described as repetitious collisions with intervallic and co-causal points of hesitation, opposition, and exchange following a capricious passage to social adaptation.[36] Along similar lines, Deleuze and Guattari's homage to Tarde makes a number of useful observations:

> Tarde countered that collective representations presuppose exactly what needs explaining, namely "the similarity of millions of people." That is why Tarde was interested in the world of detail, or of the infinitesimal: the little imitations, oppositions, and inventions constituting an entire realm of subrepresentative matter. The Durkheimians answered that what Tarde did was psychology or interpsychology, not sociology. But

that is true only in appearance, as a first approximation: microimitation does seem to occur between two individuals. But at the same time, and at a deeper level, it has to do not with an individual but with a flow or a wave.[37]

In lieu of Durkheim's attention on consciousness and social category, Tarde's microsociology comprises of mostly unconscious flows of desire, passion, and imitative radiations of muscular and cerebral activities. Indeed, what distinguishes Tardean from Durkheimian sociology is the latter's attempt to render all things psychological, biological, and neurological categorically distinct from the social. In the aforementioned debate between Tarde and Durkheim, the latter supposedly made a particular issue of how the social sciences needed to make its subject matter separate from these other phenomena. As he puts it:

[T]here is between psychology and sociology the same break in continuity as there is between biology and the physical and chemical sciences. Consequently, every time a social phenomenon is directly explained by a psychological phenomenon, we may rest assured that the explanation is false.[38]

However, it is important, I contend, not to mistake Tarde's appeal to psychological phenomena as akin to a Freudian subliminal space of the unconscious. It is instead a description of a semiconscious that prefigures current neurological explanations. Social somnambulism is an almost liminal process of association, likened to hypnosis, in which categorical distinctions between phenomena become inseparable. As Maurizio Lazzarato argues, Tarde understands associations by making "no distinction . . . between Nature and Society."[39]

Tarde's mostly unconscious associations are certainly not the same as Durkheim's collective intelligence, which is both separable from, and determining of, the organic body of the individual. As Tarde argues against Durkheim, there is no "absolute separation, of this abrupt break, between the voluntary and the involuntary . . . between the conscious and the unconscious . . . Do we not pass by insensible degrees from deliberate volition to almost mechanical habit?"[40] For Tarde then, neither the social nor the individual are a given. He focuses instead on the hypnotic agency of *what spreads* from person to person. That is, an imitative-suggestibility that passes through the collective unconscious so that the affects of the other becomes etched into the body and mind of the porous self.[41]

Economies of Inattention

I contend that both cybercrime and economic crisis can be approached through the somnambulistic tendencies of Tarde's "social man." As Thrift well highlights, Tarde's somnambulist is engaged in the distribution of a noncognitive, rather than cognitive intelligence.[42] Supported by current ideas in neuroscience, which claim that as much as 98 percent of thought derives from the unconscious,[43] the dream-like passage of imitation-suggestibility is indeed comparable in many ways to the experiences of a botnet. Its purposeful attempt to steer what Tarde regarded as the accidentalness of the flows of desire by way of an increasingly technological unconscious is a *mass mesmerism gone bad.*[44] Jonathan Crary similarly points to the hypnotic character of media technology and its subsequent role in the production of attentive subjectivities through induced inattention, distraction, and reverie.[45] In order to fully grasp the implications of these Tardean perspectives on mass persuasion, it is necessary to return to the technical detail involved in setting up a botnet, since what this technology is programmed to do is avoid the attention of the infected end-user, and further evade any attempt he or she makes to defend against zombification.

Storm Worm is an exemplary product of what Mathew Fuller and Andrew Goffey, following Crary, have termed the economy of inattention.[46] By using updated techniques of polymorphism, initially developed by virus writing communities during the late 1980s early 1990s, the worm can mutate its code so as to hide its "digital footprint," which might otherwise be picked up by an antivirus program. Unlike the polymorphism of old though, in which mutation was embedded in the program itself, *Storm Worm* generates new signatures from dedicated proxy server bots. Hourly mutations, occurring on the server side rather than directly on the user's machines, help the bot to escape the attention of antivirus database scans and updates. Another mode of defense is activated if security analysts continue to download copies of the worm for analysis. A high frequency of downloads from a particular address will trigger a DDoS on that address. Furthermore, the worm is programmed to make itself inconspicuous by regulating its activity. It has long been recognized by viral analysts that if a virus or worm can remain dormant on a user's system, and therefore survive unnoticed, it will last longer, spread further, and eventually do more damage.

The coding of a virus is not, however, all there is to exploiting inattention. Successful botnets need to paradoxically engage end-user attention in order to distract them from the goal of infection. In other words, so as to surreptitiously infect a machine, an end-user must be steered unconsciously to open up an .exe file. The user subsequently becomes unwittingly an actor in the sense that they play a role in spreading infected code on the network by becoming an intervallic node of exchange in the botnet itself. This means that end-users unknowingly become part of a topology of command. To some extent, the zombie updates its own .exe file, allowing it to flood other nodes with erroneous requests, harvest data (including financial information), and send out millions of spam messages. Moreover, a zombie will transform another end-user's machine into a proxy server—further concealing the precise location of control. Like this, techniques of inattention are dependent upon the distraction of a seemingly attentive subject by offering socially engineered inducements that aim to spread the infection far and wide. *Storm Worm* is again exemplary in this case. As Kamluk discusses, it was initially distributed as an attachment to a spam mail that looked like a safe and trusted .pdf file. Later on the worm was triggered when a user linked to an infected file inserted in an ostensibly harmless spam message posted to blogs in an attempt to expand the botnet.

The techniques of inattention employed in the design of a botnet relocate the eye of the unsuspecting user, but the infectiousness of the spamming botnet is not entirely in the realm of visual attention. Arguably, viruses, worms, Trojans, and spammers operate on a multimodality of the senses in hope of tapping into the flows of desire. They extend beyond the periphery of conscious cognitive visibility, acting as they do on the distractions of the daydreaming net surfer, moments of reverie, ennui, and lapses in concentration due to overload in the workplace. Indeed, it is from within this technologically driven sleepwalk that the botnet reaches beyond the curtain of consciousness to appeal directly to hardwired desires. A moment of the "evil" manipulation of the fragility of the technological unconscious is perfectly captured in Matthew Fuller and Andrew Goffey's account of the inattention economy:

> The end-user has only finite resources for attention. She will slip up sooner or later . . . A keen interest in the many points at which fatigue, overwork, and stress make her inattentive is invaluable. In attention economies, where the premium is placed on capturing the eye, the ear, the imagination, the time of individuals . . . It is in the lapses of vigilant, conscious, rationality that the real gains are made.[47]

As Crary argues, modern distraction isn't a disruption of stable or natural kinds of attention, but a constitutive element of the many attempts to produce (and steer) attentiveness in human subjects.[48] The problem of attention, Crary contends, is hence inseparable from inattention. The two are not polar opposites—they are a continuum of psychological, biological, and social sensations. To be sure, what Crary notes, and what can be applied directly to the botnet example, is the hypnotic nature of this continuum and the indistinct border it forms between focused normative attentiveness and a mesmerizing trance.[49]

A Tardean account of economic crisis similarly draws attention to the mostly unconscious intersections between social, psychological, and biological phenomena and their implication in the propagation and diffusion of desire in the marketplace.[50] A political economy of desire traces the fluctuations in market value to the rise and fall in testosterone and cortisol levels as the bubble begins to build,[51] and likewise tracks the rates of depression that intensify after it bursts. As the Cornell University experiment demonstrates above, it is the spreading of desire and senti- ment that inflates a speculative bubble and the ensuing chaotic results of greed that trigger erroneous cascades of panic selling. The behavior of traders and their algorithmic robots during the invention of speculative bubbles reveals somnambulistic tendencies in the marketplace that further expose the inseparability of biological desires and sentiments from the rational decision-making processes normally associated with the distribution and exchange of value. However, this in no way implies a duality between rationality and irrationality. As Christian Borch argues, instead of . . .

> . . . understanding social processes as a result of free will and socially unrestrained, optimizing decision-making, we should see them as a complex blend and interplay of, on one side, affection, desire and similar features usually associated with the "irrationality" of crowds, and, on the other side, purposeful action. It is this very in-between, the semiconscious state, which [Tarde's] suggestion thesis urges us to analyze.[52]

Economic behavior is constrained by ongoing social relations that in part elide rational, purposive, decision-making processes. Decisions are not therefore embedded in people, but as Borch argues, in the networks that connect them. Economic relations are similarly subject to the same action-at-a-distance of Tardean hypnotics apparent in the botnet. They occur, as such, through fascinations that exceed mere social networks of persons (merely following the market leaders etc.), but point to the magnetizing distractions of the riches, commodities, and prices that compose the network economy.

Financial contagions are unconscious to the extent that they are determined by unfettered desires, but as they cascade downwards from the rich banker to the laborer, the error becomes concretized in unemployment and depression. In a mode distinctly different from Durkheim's downward pressure of the collective consciousness, and the error-correcting force of anomie, the mostly unconscious collective associations that trigger Tardean error-contagions trickle down to become the depressing actualization of poverty, and could therefore be considered in terms counter to Durkheim. It is, accordingly, the consciousness of the network that emerges from the unconscious associations of the collective.

A Collective Sixth Sense? The Defense for the Somnambulist

In order to conclude this chapter I briefly evaluate how anomie and imitation-suggestibility lend themselves to the issue of culpability. Finally, I will set out a defense for the much maligned somnambulist by relating the passing-on of contagious errors and accidents to a collective sixth sense.

First, what of Durkheim's claims that anomie is a necessary evil that rouses the judicious nature of the collective consciousness? In the example of the botnet, the anomic crime becomes the flipside of the collective consciousness. But zombification is a different kind of intelligence to that described in Durkheim's macro-reductionism. It is a mesmerized unconsciousness captured through appeals to desire and unwittingly put to work. Culpability, as the change to the banking code demonstrates, is nevertheless laid at the door of the zombified end-user. Not because of a logical collective sense of injustice, but because the source of the crime (the herder) cannot be found on the network. Thus everybody—except the banks of course—will have to pay! Similarly, the error of the speculative bubble has, as argued above, resulted in calls for more regulation of the market. However, when another bubble inevitably begins to build around another desire-fueled boom time for ICTs, less than ideal housing loans, and the like, will it once again be a failure to regulate or is it that desire has once more overwhelmed the mostly unconscious dealings of economic man?

In contrast, Tarde seems to have profoundly prefigured the role of a factory-like Deleuzian unconscious in the production of what amounts to be a very thin slice of collective consciousness. What made the social open to imitation-suggestibility in the episodes Tarde chronicled may differ considerably from what is becoming an evermore technologically framed and manipulable unconsciousness, but the epidemiological model he presented is nonetheless an apt template on which to chart things anew. In times of boom and bust, for example, it is perhaps useful to be reminded of Tarde's distinction between two kinds of contagious desire. The first are "periodically linked desires." Organic life, Tarde noted, "need[s] to drink or eat," clothe itself to ward off the cold, and so on.[53] These desires become interwoven into the repetitious habits of day-to-day survival. However, when such desires become economically translated into social inventions, they become "special," and as such, take on an imitative life of their own. These are "capricious, non-periodic, desires." They are the much imitated fashions that organic life seems to passionately aspire toward almost unaware of their mesmeric desirability. On occasion, the passion grows to such an extent that it becomes a collective obsession which overspills into social spaces. Tarde is not the only "old" figure to have chronicled such events. Charles Mackay's account of the 1630s European economic crash, for example, illustrates how an earlier error-contagion was triggered by the Dutch obsession with tulips. Tulipomania, as it is called, reportedly involved individuals exchanging their homes for a single flower. Economies, like botnets, produce somnambulists.

There is, it seems, scope for a thoroughgoing analysis of social power relations established between capitalism and somnambulism, and significantly, a sustained probing of the culpability the former has in the production of the latter. Concomitantly, since culpable subjectivities are seemingly produced within an asymmetrical power relation between banks and the end-users of a network, problems concerning error, contagion, injustice, and what might constitute a defense for the somnambulist, come to the fore. What I argue is that these problems draw attention to how error-contagions are possibly linked to an emergent somnambulistic sixth sense. Surprisingly perhaps, a report in an edition of *The New York Times* published in 1879 provides an embryonic notion of a sixth sense on which such a somnambulistic defense may indeed be built. The report recalls a case that came before the Old Bailey in the United Kingdom much earlier in 1686 relating to a "noted somnambulist . . . charged with murder, having shot a guard, as well as his horse . . . The prisoner's propensity for sleep walking was clearly established, and he was acquitted." According to the report, cases of homicidal somnambulism were attributed by experts at the time to a kind of "sixth sense, usually located in the pit of the stomach, and supposed to be active in somnambulists and clairvoyants of mesmerism." It concludes that "[t]he somnambulist is perfectly unconscious of all he does while in that state, and can, therefore, not be held responsible for the occurrence of such accidents as result from his unfortunate propensity."

So what can be said of a present-day sixth sense of the somnambulist collective and how might this extrasensory state become implicated in the tendency toward error and contagion? To begin with, looking at examples of contemporary cases of homicidal somnambulism there are very similar terms employed. That is, the senses of the somnambulist are understood neurologically as functioning in a dream state *stuck* in between consciousness and unconsciousness. The somnambulist is a parasomniac. To put it another way, what the somnambulist senses is neither straightforwardly awake nor asleep. Nonetheless, the transitional state between sleeping and waking does not exclude the self's relation to the other. The social actions materializing from the somnambulist's dream-state are more likely to be prompted by noncognitive influences from the outside than they are by inner cognitive awareness. Indeed, the involuntary actions of the twenty-first-century somnambulistic are perhaps best understood as a Tardean sixth sense *open* to unconscious associations with others, now increasingly experienced at-a-distance via networks. In other words, an illusionary sense of self-determination emerges from affective, visceral collective encounters which stir up collective gut feelings, arousing contagious emotions and prompting mass automatisms, all of which eventually, at a later stage, feedback into cognitive self-reflection.

Significantly, the collective sixth sense is not a transcendent sensibility derived from the effects of paranormal or spiritualist transfers, but instead emerges from affective encounters with events. What contagiously passes through the collective is nonetheless very much like the mesmeric medium of the clairvoyant inasmuch as it is also a space of passage made vulnerable to imitation-suggestibility. It is important to add that although Tarde described the hypnotic force of such encounters in terms of accidental imitative radiations (imitative rays) passing through the crowd mostly unawares, today, as Thrift argues, the capricious and unwitting encounter within electronically mediated flows of influence is evermore automated, measurable, anticipated, and steered from a distance.[54] Affective flows can be tapped into and prompted into action. Like this, botnet zombification and the algorithmic piggybacking of market mood follow a similar trend toward what Thrift describes as the growth of "small cognitive assists," which automatically prompt everyday practices and stimulate new kinds of network intelligence functioning at a "minimal level of consciousnesses."[55] The making available of flows of feelings, obsessions, and desires via networks is leading to new modes of mind reading: "an attempt to produce a world in which semiconscious action can be put up for sale."[56]

Conceivably botnets and algo-trading software are a rudimentary forewarning of a technologically driven clairvoyant mesmerism able to tap into the spontaneous flows of the nearly asleep collective, guiding its intelligence, mood, and physiological movement to new imitative spaces. The somnambulist remains under the illusion that his or her actions are of their own volition, occurring without the influence of a leader-subject. However, imitation-suggestibility is an incorporeal material. It certainly doesn't need a body. The somnambulist is indeed under the sway of a subjectless mode of network hypnosis.

Notes

[1] Definition of error provided by the *Merriam Webster Dictionary*. http://www.merriam-webster.com/dictionary/error.

[2] Émile Durkheim, Preface to the Second Edition of *The Division of Labor in Society*, trans. W. D. Halls (New York: Free Press, 1984), xxxv–lvii.

[3] Mark Kirby, Warren Kidd, Francine Koubel, John Barter, Tanya Hope, Alison Kirton, Nick Madry, Paul Manning, and Karen Triggs, *Sociology in Perspective* (Oxford: Heinemann Educational Publishers, 1997), 601. Cited in Durkheim, Preface to the Second Edition of *The Division of Labor in Society*.

[4] Craig Calhoun and Joseph Gerteis, *Classical Sociological Theory* (New York, London: John Wiley & Sons, 2007), 197. Cited in Durkheim, Preface to the Second Edition of *The Division of Labor in Society*.

[5] Paul Marsden, "Forefathers of Memetics: Gabriel Tarde and the Laws of Imitation," *Journal of Memetics: Evolutionary Models of Information Transmission*, 4 (2000). http://jomemit.cfpm.org/2000/vol4/ marsden_p.html.

[6] Nigel Thrift, "Pass it On: Towards a Political Economy of Propensity." A conference paper delivered at the *Social Science and Innovation Conference* Royal Society of the Arts (RSA), London, February 11, 2009 http://www.aimresearch.org/uploads/File/Presentations/2009/FEB/NIGEL%20THRIFT%20PAPER.pdf.

[7] Sarah Gordon, "Technologically Enabled Crime: Shifting Paradigms for the Year 2000." Computers and Security (Amsterdam: Elsevier Press, 1995). http://www.research.ibm.com/antivirus/SciPapers/Gordon/Crime.html.

[8] Vitaly Kamluk, "The Botnet Business," *Help Net Security* (security news web site), May 28, 2008. http://www.net-security.org/article.php?id=1138.

[9] Tony D. Sampson, "A Virus in Info-Space: The Open Network and Its Enemies," *Media and Culture Journal* 7, no. 3 (2004). http://journal.media-culture.org.au/0406/07_Sampson.php.

[10] Jan Van Dijk, *The Network Society* (London: Sage, 2006), 79–81.

[11] Matthew Fuller, "Executing Software Studies," paper given at *MeCCSA Annual Conference*, National Media Museum, Bradford, January 14–16, 2009.

[12] Sean Dodson, "Was Software Responsible for the Financial Crisis?" *The Guardian*, October 16, 2008. http://www.guardian.co.uk/technology/2008/oct/16/computing-software-financial-crisis.

[13] Matthew Tyler (senior programmer working for *Reuters*), interview by the author, November 16, 2009.

[14] Transcript of a speech to the US Congress by British PM Gordon Brown in Washington, March 4, 2009. http://www.number10.gov.uk/Page18506.

[15] John Stuart Mill, "On the Definition of Political Economy, and on the Method of Investigation Proper to It," *Essays on Some Unsettled Questions of Political Economy*, 2nd edn. (London: Longmans, Green, Reader & Dyer, 1874).

[16] Sanjeev Bhojraj, Robert J. Bloomfield, and William B. Tayler, "Margin Trading, Overpricing, and Synchronization Risk," *Review of Financial Studies*, Forthcoming. http://ssrn.com/abstract=786008.

[17] Jill Treanor, "Authorities Avert Run on HBOS Caused by False Rumours," *The Guardian*, March 20, 2008.

[18] Danny Bradbury, "Banks Slip through Virus Loophole: Is My Money Safe? A Quiet Rule Change Allows British Banks to Refuse to Compensate the Victims of Online fraud if They Do not Have "Up-to-Date" Anti-Virus Protection," *The Guardian*, Thursday June 12, 2008.

[19] A reenactment of which occurred at the *Tarde/Durkheim: Trajectories of the Social Conference*, St Catharine's College, Cambridge, March 14–15, 2008. http://www.crassh.cam.ac.uk/events/2007–8/tardedurkheim.html.

[20] Ibid.

[21] Gilles Deleuze and Félix Guattari, *A Thousand Plateaus* (London and New York: Continuum, 1987), 218–219.

[22] Maurizio Lazzarato, *Puissances de l'Invention: La Psychologie Économique de Gabriel Tarde contre l'Économie Politique* (Paris: Les Empêcheurs de Penser en Rond, 2002).

[23] Bruno Latour, "Gabriel Tarde and the End of the Social," in *The Social in Question: New Bearings in History and the Social Sciences*, ed. Patrick Joyce, (London: Routledge, 2002), 117–132. Archived at: http://www.bruno—latour.fr/articles/article/082.html.

[24] Nigel Thrift, "Pass it On: Towards a Political Economy of Propensity," Lisa Blackman, "Reinventing Psychological Matters: The Importance of the Suggestive Realm of Tarde's Ontology", *Economy and Society* 36, no. 4 (2007): 576

and Christian Borch, "Urban Imitations: Tarde's Sociology Revisited," *Theory, Culture & Society* 22, no. 3 (2005): 83.

[25] Jeffrey C. Alexander and Philip Daniel Smith, *The Cambridge Companion to Durkheim* (Cambridge, New York: Cambridge University Press, 2005), 142.

[26] Latour.

[27] Robert Keith Sawyer, *Social Emergence: Societies as Complex Systems* (Cambridge, N York: Cambridge University Press, 2005), 1–9, 63–124; Elias L. Khalil and Kenneth Ewart Boulding (eds.), *Evolution, Order and Complexity* (London: Routledge Taylor & Francis Ltd, 1996); Jennifer M. Lehmann, *Deconstructing Durkheim: A Post-Post-structuralist Critique* (London, New York: Routledge, 1993), 129 and N. J. Enfield and Stephen C. Levinson (eds.), *Roots of Human Sociality: Culture, Cognition and Interaction* (Oxford, New York: Berg, 2006), 377.

[28] Émile Durkheim, *The Elementary Forms of the Religious Life*, trans. Joseph Ward Swain (London: George Allen & Unwin Ltd, 1915), 444.

[29] Émile Durkheim, *The Rules of the Sociological Method* (New York: The Free Press, 1982 [1884, 1895]), trans. W. D. Halls, Chapter v.

[30] Durkheim, The Division of Labor in Society, 353–354.

[31] Mathew Taylor, "The Credit Crunch will Generate a Wave of Anomie," blog post to *The Daily Telegraph* web site, October 14, 2008. http://blogs.telegraph. co.uk/news/matthewtaylor/5449617/The_credit_crunch_will_ generate_a_wave_of_anomie/.

[32] Ibid.

[33] Brian L. Mishara, "Suicide and Economic Depression: Reflections on Suicide during the Great Depression," *International Association for Suicide Prevention (IASP) New Bulletin*, December, 2008. http://www.iasp.info/pdf/papers/ mishara_suicide_and_the_economic_depression.pdf.

[34] Ibid.

[35] Jamie Doward, "Antidepressant Use Soars as the Recession Bites: Experts Warn on 'quick fix' after a Rise of 2.1m Prescriptions in 2008," *The Observer*, Sunday June 21, 2009.

[36] Gabriel Tarde, *Social Laws: An Outline of Sociology* (Ontario: Batoche Books, 2000).

[37] Deleuze and Guattari, 218–219.

[38] Durkheim, The Rules of the Sociological Method, 129.

[39] Nigel Thrift, *Non-Representation Theory: Space/Politics/Affect* (London, New York: Routledge, 2008), 230. Cited in Maurizio Lazzarato.

[40] Gabriel Tarde, Preface to the second edition *The Laws of Imitation*, trans. E. C. Parsons (New York: Henry Holt and Company, 1903), p. xi.

[41] Thrift, *Non-Representation Theory: Space/Politics/Affect*, 237.

[42] Ibid., 36.

[43] George Lakoff, *The Political Mind: A Cognitive Scientist's Guide to Your Brain and Its Politics* (London, New York: Penguin 2008).

[44] Thrift, *Non-Representation Theory: Space/Politics/Affect*, 243.

[45] Jonathan Crary, *Suspensions of Perception: Attention, Spectacle, and Modern Culture* (London, Cambridge, MA: MIT Press, 2001), 45–46.

⁴⁶ Matthew Fuller and Andrew Goffey, "Toward an Evil Media Studies," in *The Spam Book: On Viruses, Porn, and Other Anomalies from the Dark Side of Digital Culture*, ed. Jussi Parikka and Tony D. Sampson, (Cresskill, NJ: Hampton Press, 2009), 152–153.

⁴⁷ Ibid., 152.

⁴⁸ Crary, 49.

⁴⁹ Ibid., 65.

⁵⁰ Gabriel Tarde, "Economic Psychology," trans. Alberto Toscanao, *Economy and Society* 36, no. 4 (November, 2007): 626.

⁵¹ Thrift, "Pass it On: Towards a Political Economy of Propensity," 14.

⁵² Christian Borch, "Crowds and Economic Life: Bringing an Old Figure Back In," paper presented at the *Centre for the Study of Invention and Social Process*, Goldsmiths College, November 21, 2005, 18–19. Also published in *Economy and Society* 36, no. 4 (November, 2007): 549–573.

⁵³ Tarde, "Economic Psychology," 633.

⁵⁴ Thrift, "Pass it On: Towards a Political Economy of Propensity," 18.

⁵⁵ Thrift, *Non-Representational Theory*, 162–168.

⁵⁶ Thrift, "Pass it On: Towards a Political Economy of Propensity," 5.

Chapter 13

Error 1337*

Stuart Moulthrop
University of Wisconsin-Milwaukee

When people learn to play video games, they are learning a new literacy.
James P. Gee[1]

You have used the word LITERACY in a way I do not understand.
Video Game[2]

Among the many transformations that have come with digital culture, we must include a new understanding of error. Once we undertake work or play with software, the original sense of this term—*errare*, to move without clear direction, departing from truth, norm, or some other analog of unity—gives way to something more complex. In the culture of networked, computational systems, there is never a single path of expression or encounter, but always a large or indefinite domain of possibilities. In traversing the textual spaces of the World Wide Web, or the spatial simulation of a video game, digital subjects must make decisions and take actions in order to progress. They search, navigate, explore, maneuver, and otherwise configure systems that encompass their own understanding and intention, external structures of logic (code and processors), and an emergent set of conditions and relations called *state*.

Espen Aarseth, whose theoretical insights into *cybertext* remain essential after more than a decade, has named this engagement *ergodic*, a word whose roots mean *path work*.[3] From a cybertextual perspective, while one or more paths may be privileged (as in many popular video games), their value depends on the possibility of alternative selection. We enjoy games

---

* As always, I am indebted to Nancy Kaplan for advice, criticism, and timely research.

not because we always win, but because we win only after repeatedly losing. Nor is this design principle limited to video games. As the hypertext theorist Mark Bernstein points out, loops, spirals, matrices, and mazes proliferate in digital poetics, rendering repetition or recursion not a flaw but a crucial constituent of meaning.[4] In the world of digital expression, we are expected to depart from unity. To loop and wander is human, or perhaps incipiently posthuman.

It is tempting to say further that to err, divagate, or otherwise work the paths underlies the *literacy* of cybernetic texts; perhaps this is the most important factor in the new sensibility that James Gee and others have linked to video games and other interactive systems.[5] However, something about this reassertion of an all-too-familiar name—a term which for most academics has distinctly godlike qualities—should give us pause. As Anne Wysocki and Johndan Johnson-Eilola crucially ask: "What are we likely to carry with us when we ask that our relationship with all technologies should be like that we have with the technology of printed words?"[6] Does the invocation of *literacy* implicate the social assumptions that attach to that word in non-cybernetic contexts: for example, expectations about basic levels of competency and performance, as defined by industrial or consumerist regimes? Will any new literacy be as regularizing and normative as the old? If video games imply something like literacy, should there be an authorized canon of Great Games, from which every student is expected to derive a measurable quantity of improving fun? Should wall jumping, combination attacks, and the general theory of mazes be taught in primary schools? Should no child be left at Level 17 in *World of Warcraft*?

This quick descent into absurdity leads to a more serious question: should academics and cultural critics still be thinking about anything like literacy as we have known it? Or to put this somewhat more ambitiously: should we continue to privilege the family name of the letter in thinking about technologies that move radically beyond stable inscription?

These last questions flow from the second, simulated epigraph for this piece, modeled on encounters with early text-adventure games, where suggestions that exceed the system's limited vocabulary earn an all-too-predictable response. To use the word *literacy* in the context of video games, hypertexts, simulations, or other examples of cybertext may indeed be an error; and while error is in some sense a primary signature of cybertext, every error is still worth examining.

We might start by noting the crucial differences between typewriters and Turing machines. Among other things, the *halting states* that afflict the latter are endogenous, produced by the vagaries of processing and the topology of paths, not the imaginative failure of an external operator. Therefore such states are considerably more interesting than in the unplugged condition. Halting states, moments when the continuous operation of the system is suspended, are quite common in encounters with software: think not simply of crashes and lockups, but of search queries that return no results, broken hypertext links, or indeed, any experience of frustration or virtual death in a video game. For Terry Winograd and Fernando Flores, these disruptions of flow provide a phenomenological basis for understanding software itself:

> Following Heidegger, we prefer to talk about "breakdowns." By this we mean the interrupted moment of our habitual, standard, comfortable "being-in-the-world." Breakdowns serve an extremely important cognitive function, revealing to us the nature of our practices and equipment, making them "present-to-hand" to us, perhaps for the first time. In this sense they function in a positive rather than a negative way.[7]

The putatively erroneous application of *literacy* to cybertext could itself have positive value. While it may bring us to a halting state, a condition where our customary vocabulary no longer maps usefully onto conditions of operation, it may also disrupt our habitual ways of thinking and thus permit clearer understanding of our particular circumstances. From this experience may come a way forward. If nothing else, considering *literacy* as a point of breakdown may open the theoretical nomenclature to new candidates. If as Winograd and Flores suggest, the value of breakdowns is largely heuristic, we may find a new concept to set in its place.

Before going further, however, some stipulations are due. Gee is an accomplished theorist of reading, so his understanding of *literacy* is more nuanced than the preceding critique has perhaps allowed. Having made his remark about "new *literacy*," he elaborates:

> Of course, this is not the way the word "literacy" is normally used. Traditionally, people think of literacy as the ability to read and write. Why, then, should we think of literacy more broadly, in regard to video games or anything else, for that matter?[8]

In answer to the rhetorical question, Gee observes that literacy has been substantially redefined by specialists in areas that include linguistics, cognition, cultural theory, and pedagogy. Indeed, a parallel term, *numeracy*, has been seriously advanced by mathematicians.[9] *Literacy* itself can now be applied to far more than "the ability to read and write," and encompasses our relations with a large variety of sign systems, including pop songs, film and video, comics, images in advertising, piercings and body markings, types of dress, transport, and many other forms of material culture, including video games. Gee's "new literacy" represents a further subdivision or expansion of the general field called *multiliteracies*.

Sensibly enough, this movement aligns itself with the general project of semiotics. It further reflects the understanding, increasingly common in the last half of the previous century, that meanings are multiply and flexibly mediated, and that conditions of their mediation cannot be entirely separated from their message content. Gunther Kress, another major exponent of multiple literacies, articulates this complex relationship of message and medium even as he echoes Gee's desire to address media or "modes" beyond writing:

> There is no question of separating form from meaning: the sign is always meaning-as-form and form-as-meaning. The means of dealing with meaning are different; we need to understand how meanings are made as signs in distinct ways and specific modes, as the result of the interest of the maker of the sign, and we have to find ways of understanding and describing the integration of such meanings across modes, into coherent wholes, into texts.[10]

At this point it seems necessary to pause (if not halt) for critical reflection. To suggest error in Kress's eminently sensible doctrine, or Gee's treatment of video games as models for enhanced learning, may seem itself misguided. Indeed, the argument of this chapter represents a step off the main path of current rhetorical studies. We make this move with good reason; but despite this difference in direction, there is no intent here to slight the value of the multiple-literacies approach, as a theory, a direction for empirical research, or a basis for pedagogy. Clearly, those who think most capably about communication must find ways to accommodate diverse ways of making meaning, especially as modes emerge whose underlying difference represents a true break from what has come before. The concept of multiple literacies is one approach, and not without value. However, it has important limits.

Consider in particular the last word of Kress's otherwise compelling statement—that interesting term *text*—and recall Wysocki and Johnson-Eilola's caution about legacies and inheritance, as object-oriented programmers say. *Text* is obviously a property defined in a namespace rather distant from digital communication, though as usages like *cybertext* demonstrate, it is still in play. Down at the roots, *text* is but a small step from texture or textile, a weaving of elements into some discernible pattern or structure. Only history, not some fundamental law, confines its meaning to words or writing.

Still, certain questions persist. Can writing, that traditional way of making texts, be easily identified with other means of manipulating signs? When we set out to *read* a video game, as Gee proposes at least hypothetically toward the end of one of his studies, do we imply some prior act of writing? [11] If so, in what sense can we speak of someone, or some corporate entity, *writing* a video game? True, many games credit one or more persons as writers, but the work of these groups or individuals is similar to screenwriting, where the text in question may be something very different from the overall word-image-action ensemble of the film, or the dynamic and variable experience expected from a game.

There is a profound social difference between what you are reading now, the kind of writing most academic humanists produce—not for nothing called a *monograph*—and the highly compartmentalized, intricately coordinated team efforts that underlie production in the digital realm. Literally speaking, these products belong to different worlds. The writer on a video game project most likely inhabits a cubicle farm or other open-plan work environment. Research faculty often have private offices, the modern equivalent of the monastic cell. This disparity of architecture is far from trivial; it reflects a significant difference of culture.

Like filmmaking, work in new media differs distinctly from literary writing, the sort of production many if not most scholars of *text* are trained to investigate and interpret. Academia is reflective, individualized, and inwardly focused, while game development is intensely and inherently social, strongly attached to the mass market. To a large extent, academics write and publish alone, and even when they collaborate, their work falls within the single technical domain of wordcraft. Game development, by contrast, takes something like a village, and its technical demands are multifarious. Independent production teams can be as small as two or three, but large corporate enterprises employ hundreds, working in a variety of specialties. The muster at a typical game studio includes visual artists of various sorts (2-D and 3-D), animators, motion-capture technicians, voice actors, composers, sound designers, programmers of at least three types (utility, system, and interface specialists), a corps of testers, publicity and marketing people, and high-level managers, often called designers, who coordinate the entire process.

Being roughly equivalent to film directors, game designers are sometimes recognizable by name (e.g., Shigeru Miyamoto, Will Wright, Sid Meier). Still, if we are tempted to think of them as *auteurs*, analogous to literary writers, we should remember Wysocki and Johnson-Eilola's test. What do we carry over in drawing this likeness? Specifically, do we risk confusing what designers and their companies do with the writing of books? Typewriters are not Turing machines, and video games are nothing like books. Games (video or otherwise) have logics of engagement and development which must be actively encountered. Thus Aarseth enjoins those who would write about games to play first, write later:

> Games are both object and process; they can't be read as texts or listened to as music, they must be played. Playing is integral, not coincidental like the appreciative reader or listener. The creative involvement is a necessary ingredient in the uses of games.[12]

Aarseth's advice suggests there may be some wisdom in our decision to stray from the scholarly mainstream. If there is a way of light in multiliteracies, there may be also a more shadowy tendency, a temptation to skimp on the "creative involvement" and stick with what scholars know best, which is writing.

Again, this complaint needs careful qualification. You are, after all, reading one of those scholarly monographs right now, produced not by any digital native, but one of those technological Pilgrims who arrived (like his American ancestor) on a later boat. Further, to clear a particular debt: Gee's study generally satisfies Aarseth's requirement of play, offering extensive, detailed accounts both of observed and personal encounters with games. The same might be said for the efforts of many other scholars who like Aarseth came to new media earlier in life (e.g., Jesper Juul, Mary Flanagan, Katie Salen, Eric Zimmerman, Ian Bogost, Nick Montfort, Noah Wardrip-Fruin, Gonzalo Frasca). Likewise, much good work has come from the multiple-literacy camp, for instance, Cynthia Selfe and Gail Hawisher's ongoing study of "Literate Lives," which uses students' reflective writing as a means to investigate changes in the social construction of signs.[13]

In deference to these colleagues, some adjustments must be made before attempting the notably absurd task of writing away from the letter. If multiliteracies will not entirely serve, perhaps we might reconsider Gee's formulation, "a new literacy," but in a sense not originally intended—a convergence of traditional writing with *code*.

Suppose, while we are hacking other people's expressions, we expand Aarseth's notion of "creative involvement" to include not just play or performance of video games, but the creation of original or derivative structures for interaction. It is after all a longstanding tradition in game design to allow or even encourage user modification of released code. This pattern began with the original *Adventure*, itself a reworking of a cave-crawling simulator, and persists in *Counter Strike*, *Portal*, and a long list of contemporary games built on existing engines. Indeed, game play itself has now largely become derivative work. The genre of massively multiplayer role-playing games depends significantly on so-called player-created content, as do nongame services such as Wikipedia, Flickr, YouTube, and Second Life. While these later examples may not involve programming, they do in most cases implicate some sort of code, from interface scripting to markup languages. Users may not be immediately aware of these code structures, but in an important sense, that does not matter.

Arguably, any publication on the World Wide Web, and increasingly, less formal domains such as RSS feeds, instant messaging systems, and Twitter, affects the cybernetic infrastructure of aggregators and search engines, simply by increasing the field of indexable data, and perhaps also by creating references and explicit links to other bodies of information. Consider the practice of *Google bombing*, in which thousands of participants publish linked pages whose registration by search engines will cause a target page, bearing a particular title, to appear as the first result of certain search queries (e.g., "worse than Satan" for Microsoft, or "miserable failure" for George W. Bush). This intervention, and others like it, demonstrate that in the context of indexed, networked communication, every shared item transforms the system, and thus engenders, or perhaps constitutes, a computable statement. Seen in this way, all writing entered into the web, even the sort covered by the older sort of literacy, participates in a process of configuration. Writing thus becomes *pro-gramming* in the etymological sense of *writing-forward*: a process of arrangement or design that changes the state of its system, and in so doing lays groundwork for further permutations.

Thus even if we resist Gee's assertion that video games embody a new domain of multimodal literacy, we might still argue that they articulate with a major transformation of writing: the opening of alphabetic language to the possibilities inherent in cybernetic processing. If we are looking for a mark of difference to distinguish this development from its origins, we might first attempt the minimum variation. We could begin with a concept very much like literacy, but now forward-leaning, or inherently belated; not simply writing, but writing forward. So we might say that when literacy enters the cybernetic domain, it becomes *later-acy*.

To which certain ungenerous minds, warped by the 1970s, will respond: *Later, dude.*

Aside from the inevitable tendency of portmanteau words to trigger the epiglossal Pun Reflex, there is perhaps a more substantial problem in any reach back to a new forwardness in writing. We might raise the same objections here that we brought against Gee's "new literacy."

What will one have to do in order to pass as *later-ate?* In order to participate in the system of ubiquitous processing, should every competent adult maintain a presence on the Internet? If so, does activity in social networks like Facebook suffice, or, suspending progress at the Millennium, will only hand-coded web sites do? Is "the creative involvement" of game play adequate qualification, or should we use level building to level the playing field? Can anyone be considered *later-ate* without demonstrated mastery of a scripting language? A programming language? Which ones?

Not all these questions are trivial. The issue of communicative competency, in a regime that increasingly fuses play with work, deserves deeper treatment than is possible here. It could be quite important to attempt to define a latter-day literacy that fulfills some if not all the functions of its precursor. Still, the enterprise seems strongly circumscribed, if not entirely pointless. Perhaps *later-acy* is no great improvement on "new literacy"; in which case, we can at least reset the game and continue to play. If *later-acy* is no less erroneous than what it seeks to replace, it too may constitute a productive breakdown, a failing-forward, if not itself the next great thing.

To realize this progress, however, it is necessary to move beyond the focus on competencies and practices *per se*. The idea of a new, belated literacy leaves out of account other, larger dimensions of our cultural situation. As Pierre Lévy explains:

> . . . cyberspace dissolves the pragmatics of communication, which, since the invention of writing, has conjoined the universal and totality. It brings us back to a preliterate situation—but on another level and in another orbit—to the extent that the real-time interconnection and

dynamism of on-line memory once again create a shared context, the same immense living hypertext for the participants in a communication. Regardless of the message, it is connected to other messages, comments, and constantly evolving glosses, to other interested persons, to forums where it can be debated here and now. Any text can become the fragment that passes unnoticed through the moving hypertext that surrounds it, connects it to other texts, and serves as a mediator or medium for reciprocal, interactive, uninterrupted communication.[14]

Lévy's notion of a "universal without totality" throws our prior concern with mere "pragmatics" into sharp relief. The importance of writing-forward lies not simply in its enabling of specific creative involvements—search engine exploits, viral memes, metagames, and a million clever hacks—but rather in their coextensiveness with a dynamic, living medium that poses a substantial challenge to prior forms of cultural encoding.

What might differentiate new literacy from old—or indeed, literacy from what comes next—has less to do with operations (praxis) and more to do with conditions of meaning (episteme). When writing belonged exclusively to book and page, it operated through an attenuation or withholding of context. As the fundamentalist creed goes (East and West alike): *God said it, it's in the Book, and so I believe it.* Never mind that Holy Writ in the aggregate tends to be obscure, parabolic, inconsistent, and deeply marked by historical context. Its truth claims are transcendent, absolute, or "universal," in Lévy's terms. They are meant to produce or embody *meaning.*

Whatever comes later than the old-time literacy must operate in a radically different situation. There may still be a category of the universal, but it can no longer be identified with extra-human truth or divine Providence. Nor will it have the same investment in products of the printing press or other graven images. Lévy continues:

> The ongoing process of global interconnection will indeed realize a form of the universal, but the situation is not the same as that for static writing. In this case the universal is no longer articulated around a semantic closure brought about by decontextualization. Quite the contrary. This universal does not totalize through meaning; it unites us through contact and general interaction.[15]

If the notion of a universal-without-totality marks a primary difference between the old regime of writing and a new scheme of "general interaction," then perhaps the successor term for *literacy* should indicate something more than mere belatedness. Instead of either *literacy* or *later-acy*, we might seek a term that properly registers both the dissolution of the old pragmatics, and the contextually abundant sign spaces of the Net.

With that mixture of trepidation and glee that seems to mark forays into cyberspace, we therefore write the name of that which comes after writing.

L33T

Much needs to be said about this name, but first some attention is due to the act of naming itself. Anyone who crosses the divide between glyph and cybernetic code quickly learns a few things about nomenclature. On the one hand, in systems that do not tolerate ambiguity, the names of objects and procedures demand precision. At the same time programmers, constrained only by words already defined in the relevant namespace, are generally free to use any designation they want when creating new structures. Cybernetic names are thus heavily fraught, or invested with power, and at the same time quite arbitrary. Perhaps this curious double nature accounts for another common property of cybernames. They tend to be deeply ironic. Take for example the programming language Python, named for the British comedy troupe, perhaps in celebration of their unique irreverence for formal logic.[16] Or consider the variables *foo* and *bar*, traditionally used by teachers of programming. The names derive from the classic military acronym *FUBAR* (Fucked Up Beyond All Recognition). As Winograd and Flores might say, software is born under the sign of breakdown.

It is in this sense of self-ironizing power that we invoke the numinous term L33T. The name has been in existence for about three decades, and is most likely a product of the personal-computing revolution. It has no recorded inventor or specific point of origin, though it is thought to have first appeared in electronic bulletin-board systems of the pre-web era.[17] Pronounced *leet*, and sardonically short for *elite*, the word can also be written entirely in numbers (1337), though the hybrid form seems more appropriate here. Strictly speaking, L33T is primarily known today as an alternative, pseudo-cryptographic spelling system, a linguistic fetish invented by the second generation of recreational computer users, picked up by their younger siblings in the online gaming community, and now a fairly common feature of digital writing.

None of which is overwhelmingly important.

At this point in the old, decontextualizing game of pre-L33T lit, your author would be obliged to make further scholarly observations about the name, noting its association with feral computer programming (hacking), with the adolescent impulse to evade censorship (as in the L33T term *pr0n* as a replacement for *porn*), and its celebration of dyslexia (*teh* for *the*) and sheer linguistic perversity (*suxxOr* for *sucks*). One might also mention its presence at the margins of polite literature (Neal Stephenson's character Da5id in *Snow Crash*), and the attention it has drawn from serious commentators on new media (Bruce Sterling, Eric Raymond, Stephen Johnson). As hinted above, something of greater interest could probably be said about the ironic resonance and history of the word, which may have begun as a more or less sincere marker of expertise, but now belongs to the playground one-upmanship of game spaces (see usages like *n00b*, *pwned*, and *F4!1*).

If Lévy's remarks about the cyberspatial difference are correct, though, this old sort of scholarly discourse has diminishing importance for any attempt to think beyond *literacy*. There are plentiful examples of L33T available on the World Wide Web, awaiting your contact and general interaction. This chapter is not the boss of your context.

http://www.urbandictionary.com/define.php?term=leet
http://ryanross.net/leet/
http://www.google.com/intl/xx-hacker/
http://www.usenix.org/event/leet08/
http://icanhascheezburger.com/

As a matter of fact, the current context of L33T has only limited bearing on this discussion. It is a point of departure, not a specification. Though there is something generally identifiable called L33T, it is not invoked here as agent or agency, but something more like signature or symptom: the complex not the name I call, to update Milton.

Mixing letters with numbers or further deforming the grammar of English and other languages does not constitute sufficient pattern for new procedures; though it may provide necessary groundwork. These practices will probably persist as touchstones of what Alan Liu calls "intraculture," and intraculture does appear to have some importance for the immediate future. [18] Nothing new lasts, however. For all we know, the fusion of letter and number might eventually fade away, or turn into something quite different. Still, the project of L33T as imagined here—that is, the name for what comes after literacy; the deeper link between writing and code—will certainly continue. As indicated earlier, the focus of this chapter is less on "pragmatics" than episteme, or as Liu would have it, "ethos."

L33T is thus something of an alias or approximation, a *nomme des nommes*, as someone nearly says in *Ratner's Star*. [19] (In the L33T dialect LOLspeak, *n0m n0m n0m* is an onomatopoeic representation of mindless consumption—a fact that may not be entirely irrelevant.) Such assumed names are never intended to fit perfectly, and their factitiousness is very much the point. In fact, arbitrariness and approximation are precisely what make L33T an appropriate counter-term to *literacy*. As a self-ironizing name for a discipline whose mysteries are abundantly accessible (universal without totality, amen), L33T seems the right sort of designation for the situation in which we find ourselves.

Whatever that situation may be. Liu, perhaps the most astute critic of cybernetic language and culture to date, writes of the current moment in terms of deep uncertainty:

> The contemporary globe, perhaps, is not so much a preexisting object as a standing wavefront of simulation generated by knowledge work as an *idea* of globalism—named, for example, "new world order," "global market," or "World Wide Web." [20]

As Liu notes, names are both arbitrary and powerful. Indeed, standing waves are themselves paradoxical, manifesting what may be the physical equivalent of irony. That is, they are structurally stable until whatever energy input maintains them is cut off—which is to say, until their systems encounter a halting state, or breakdown. At that point, they succumb to uniformity, randomness, or the condition of medium without message.

Limited equally by time, space, and understanding, we cannot offer a complete account of the cultural formation that succeeds literacy, the concept that would allow us to define what happens when people engage in knowledge work or content-creating play, when they navigate hypertexts, level up in video games, or mash up web applications. The best we can do is speculate on a possible interruption of Liu's standing wave, or momentary system of arbitrary names. Passing beyond both original and belated literacy, we reach a point of breakdown, to which we assign the name L33T. From this point, we must involve ourselves creatively in the game's next move; such is, after all, the general condition of players.

However, as Winograd and Flores point out, the moment of breakdown affords fresh understanding of our "practices and equipment": an ability to ask questions about our situation that might not otherwise have occurred.[21] In the present instance, one question seems especially salient. It is not the question Wysocki and Johnson-Eilola pose—what do we carry over when we look back to literacy—but rather its corollary: *if we take up a new name, what can we leave behind?*

Back in the Reign of Error that dominated U.S. political life in the first decade of the new century, there were those who would answer this question in the simple negative, for whom no aspect of long-established tradition should ever be left behind. In a much-discussed report called *Reading at Risk*, the National Endowment for the Arts declared that by mid-century there would be no market for literary writing (curiously defined only as realist fiction and poetry). The result of this decline in readership (or bookselling) would be disastrous, because, the writers averred: "Print culture affords irreplaceable forms of focused attention and contemplation that make complex communications and insights possible."[22] Drawing on correlations from survey data, the authors further argued that this loss of focus would lead to diminished civic engagement, erosion of the social fabric, and the general decline of the Republic.

More recent evidence likens these findings to their administration's claims about Saddam Hussein's arsenal. Reading, it turns out, is decidedly not at risk, at least if one is willing to define the key term more liberally. Reading is in fact *rampant*. A report on the information intake of U.S. consumers finds that consumption of words has increased substantially between 1980 and 2008:

> In the past, information consumption was overwhelmingly passive, with telephone being the only interactive medium. Thanks to computers, a full third of words and more than half of bytes are now received interactively. Reading, which was in decline due to the growth of television,

tripled from 1980 to 2008, because it is the overwhelmingly preferred way to receive words on the Internet.[23]

Once again we confront the category error with which this chapter began: we have used the term *reading* in a way the former government's literacy boosters would not understand. After all, their report targets "literary reading" in its subtitle, while the more recent study focuses on information "received interactively"—certainly embracing Aarseth's "creative involvement," and implicitly fusing reading with writing. The Cassandras of the old NEA would have none of this. They were interested in dissemination and authoritative production, in novels and books of poetry offered for sale at bookstores. The authors of the more recent study are concerned with bytes in transit (actually zettabytes, each of which is equal to a billion terabytes, and 3.6 of which arrived at the average American household last year). Here what seemed a category error may turn into a problem of scale. Prose and poetry are kilobyte-sized; the zettabyte band is the domain of moving images and simulations, primarily video and video games. A world measured in zettabytes differs markedly from the old, alphabetic regime; though obviously, it is capable of containing and transforming it. However, the scale change introduces complexity. Obviously, the reckoning that shows a threefold increase in reading over the last three decades takes in a lot more than novels and poetry, including email, web sites of all descriptions, search queries, text messages, tweets, and perhaps even works of traditional literature contained in electronic books.

Under the new paradigm of multiple pathways, or ergodics, it is of course possible for both the old NEA and the recent consumption study to be right, each in its own domain. Reading—even the sort of sustained, intensive reading Mr. Bush's appointees undertook to mourn—seems to be arguably alive. At the same time, though, "print culture" continues to decline. The future of any business tied primarily to the press, from publishing houses to daily newspapers, looks bleak. Despite these portents of creative destruction, the United States is not about to become a nation of illiterates; but we cannot assume that Twitter can sustain the Great Tradition of English fiction. (A stronger case might be made for poetry.) Much seems open to debate, including whether the affordances of print culture are truly "irreplaceable." To better serve this inquiry, it is probably advisable to disentangle questions about literacy and its posterity from concerns about particular economic and social institutions.

To some extent, though, these entanglements are unavoidable. As we move into the new century, we are not obligated to take along the complete legacy of letters. As object-oriented programmers know, inheritance need not be absolute. The newly instantiated L33T may indeed override certain methods from the superclass of literacy. That possibility raises questions likely to prove troubling, particularly for academics. If reading and writing are rampant, to what discipline can or should they be submitted? Or to put the question most pointedly: in a regime of contact and interaction, web-enabled and ubiquitously networked, is there any reason for L33T to be taught in classrooms?

As we learn everyday in these times of economic blight, the Reign of Error may have ended with an election, but the March of Folly goes on. Liu's standing wave of globalized culture-as-simulation continues to flicker with uncertainty. We are in transition to a new political economy of the sign, and in that process old assumptions and allegiances will continually suffer breakdown, becoming acutely present to attention. Like video gamers, we must learn to thrive in error.

This is not to say our errors will have no consequences. Perhaps the old NEA's worries about the demise of "print culture"—largely meaning agents, publishers, editors, booksellers, reviewers, and other intermediaries—may yet have some resonance for deans, professors, and graduate students. Could the difference between literacy and its successor lie in a transition to self-motivated and self-assembled expertise? As Gee points out, children raised with access to video games and other complex amusements often engage in self-organized, self-directed learning to support their recreation, and these activities may reflect sound educational values.[24] Might formal education thus become—or may it be already—an increasingly arbitrary accompaniment to independent knowledge work?

These questions, centering on the likely relationship between literacy and its successor, seem especially fertile. To revert to the point with which we began—the deconstruction of error in a context of radical plurality—can we even imagine what comes after literacy as a unified, coherent discipline? If we cannot *read* a video game, how can we read a much more nebulous set of cultural practices called L33T? What if The Thing After Literacy remains a perpetually evolving target, an evanescent standing wave, always eluding the embrace of scholarship, and indeed, of typographic language?

If L33T signifies anything—in reference if not meaning, in meaning if not name—it might stand both for continuity and difference: convergence of letter and number, reassertion of expertise and elitism under the sign of an always adolescent unseriousness. It seems to speak to knowledge work, or learning, but in a new and possibly problematic relationship to schools, corporations, governments, and other superstructures that attempt to mediate mediation. In the end it may only be possible to grasp the matter imperfectly, in a transparent exercise in error, doomed to stumble from one halting state to the next.

The game begins.

Notes

1 James P. Gee, *What Video Games Have to Teach Us About Learning and Literacy* (New York: Palgrave Macmillan, 2003), 16.
2 After the error handler response to unexpected input in early text-based computer games.
3 Espen Aarseth, *Cybertext: Perspectives on Ergodic Literature* (Baltimore: Johns Hopkins University Press, 1997), 1.
4 Mark Bernstein, "Patterns of Hypertext." *Proceedings of the ACM Conference on Hypertext and Hypermedia* (New York: Association for Computing Machinery, 1998), http://www.eastgate.com/patterns/Print.html.
5 See: Gee, *What Video Games Have to Teach Us About Learning and Literacy*; Ian Bogost, *Persuasive Games: The Expressive Power of Videogames* (Cambridge, MA: MIT Press, 2007); and Jesper Juul, *Half Real* (Cambridge, MA: MIT Press, 2005).
6 Anne Wysocki and Johndan Johnson-Eilola. "Blinded by the Letter: Why Are We Using Literacy as a Metaphor for Everything Else?" in *Passions, Pedagogies, and 21st Century Technologies*, ed. G. Hawisher and C. Selfe (Salt Lake City: University of Utah Press/NCTE, 1999), 349.
7 Terry Winograd and Fernando Flores, *Understanding Computers and Cognition: A New Foundation for Design* (New York: Addison Wesley, 1986), 77–78.
8 Gee, *What Video Games Have to Teach Us About Learning and Literacy*, 13.
9 John A. Paulos, *Innumeracy: Mathematical Illiteracy and Its Consequences* (New York: Hill and Wang, 2001).
10 Gunther Kress, *Literacy in the New Media Age* (London: Routledge, 2003).
11 Gee, *What Video Games Have to Teach Us About Learning and Literacy*, 204.
12 Espen Aarseth, "Game Studies Year One," *Game Studies* 1:1 (2001), http://www.gamestudies.org/0101/editorial.html.
13 Cynthia Selfe and Gail E. Hawisher, *Literate Lives in the Information Age: Narrative of Literacy from the United States* (New York: Routledge, 2004).
14 Pierre Lévy, *Cyberculture*, trans. Robert Bononno (Minneapolis: University Minnesota Press, 2001), 98–99.
15 Ibid., 99.
16 Mark Lutz and David Ascher, *Learning Python* (New York: O'Reilly Books, 1999), xv.
17 "Leet," *Wikipedia*, http://en.wikipedia.org/wiki/Leet.

[18] Alan Liu, *The Laws of Cool: Knowledge Work and the Culture of Information* (Chicago, IL: University of Chicago Press, 2004), 77.

[19] Don DeLillo, *Ratner's Star* (New York: Vintage, 1989).

[20] Liu, 287.

[21] Winograd and Flores, 78.

[22] *Reading at Risk: A Survey of Literary Reading in America* (Washington: National Endowment for the Arts, 2004), http://www.arts.gov/pub/ReadingAtRisk.pdf, vii.

[23] Roger E. Bohn and James E. Short, "How Much Information? 2009 Report on American Consumers." San Diego: Global Information Industry Center, 2009, http://hmi.ucsd.edu/pdf/HMI_2009_ConsumerReport_Dec9_2009.pdf, 7.

[24] James P. Gee, *Good Video Games + Good Learning: Collected Essays on Video Games, Learning, and Literacy* (New York: Peter Lang, 2007), 2.